Coloniality, Ontology, and the Question of the Posthuman

This book brings together emerging insights from across the humanities and social sciences to highlight how postcolonial studies are being transformed by increasingly influential and radical approaches to nature, matter, subjectivity, human agency, and politics. These include decolonial studies, political ontology, political ecology, indigeneity, and posthumanisms. The book examines how postcolonial perspectives demand of posthumanisms and their often ontological discourses that they reflexively situate their own challenges within the many long histories of decolonised practice. Just as postcolonial research needs to critically engage with radical transitions suggested by the ontological turn and its related posthumanist developments, so too do posthumanisms need to decolonise their conceptual and analytic lenses. The chapters' interdisciplinary analyses are developed through global, critical, and empirical cases that include city spaces and urbanisms in the Global North and South; food politics and colonial land use; cultural and cosmic representation in film, theatre, and poetry; nation building; the Anthropocene; materiality; the void; pluriversality; and, indigenous worldviews. Theoretically and conceptually rich, the book proposes new trajectories through which postcolonial and posthuman scholarships can learn from one another and so critically advance.

Mark Jackson is Senior Lecturer in Postcolonial Geographies at the University of Bristol, UK.

Routledge Research in New Postcolonialisms

Series Editor:
Mark Jackson,
Senior Lecturer in Postcolonial Geographies,
School of Geographical Sciences,
University of Bristol, UK

This series provides a forum for innovative, critical research into the changing contexts, emerging potentials, and contemporary challenges ongoing within post-colonial studies. Postcolonial studies across the social sciences and humanities are in a period of transition and innovation. From environmental and ecological politics, to the development of new theoretical and methodological frameworks in posthumanisms, ontology, and relational ethics, to decolonising efforts against expanding imperialisms, enclosures, and global violences against people and place, postcolonial studies are never more relevant and, at the same time, challenged. This series draws into focus emerging transdisciplinary conversations that engage key debates about how new postcolonial landscapes and new empirical and conceptual terrains are changing the legacies, scope, and responsibilities of decolonising critique.

Postcolonialism, Indigeneity and Struggles for Food Sovereignty
Alternative Food Networks in the Subaltern World
Edited by Marisa Wilson

Coloniality, Ontology, and the Question of the Posthuman
Edited by Mark Jackson

Coloniality, Ontology, and the Question of the Posthuman

Edited by Mark Jackson

Routledge
Taylor & Francis Group
LONDON AND NEW YORK

First published 2018 by Routledge

2 Park Square, Milton Park, Abingdon, Oxfordshire OX14 4RN
52 Vanderbilt Avenue, New York, NY 10017

Routledge is an imprint of the Taylor & Francis Group, an informa business

First issued in paperback 2019

British Library Cataloguing-in-Publication Data
A catalogue record for this book is available from the British Library

Library of Congress Cataloging-in-Publication Data
A catalog record for this book has been requested

ISBN: 978-1-138-92090-3 (hbk)
ISBN: 978-0-367-87278-6 (pbk)

Typeset in Times New Roman
by Apex CoVantage, LLC

Contents

Figures and Box

Figures

Box

Contributors

Carlo Bonura is Senior Teaching Fellow in Southeast Asian Politics at the Department of Politics and International Studies, SOAS, University of London, UK. His research lies in the areas of comparative political thought, contemporary Islamic thought in Southeast Asia, Southeast Asian politics, particularly Thailand and Malaysia, and the intersections of Southeast Asian culture and politics.

Mark Jackson is Senior Lecturer in Postcolonial Geographies in the School of Geographical Sciences at the University of Bristol, UK. His research lies in the areas of postcolonial and decolonial geographies, materiality, political ontology, urban geographies, and political ecology. He is currently completing a monograph called *Decolonising Critique: A Prolegomena.*

Hans-Martin Jaeger is Associate Professor of Political Science in the Department of Political Science at Carleton University, Ottawa, Canada. His research interests are in international political theory and sociology, global governance and international organisation, international public spheres and global civil society, and critical international relations theory. Hans-Martin's current research investigates how 'new master concepts' in International Relations theory such as nomos, world society, or governmentality reflect on the political and post-political, and the global and provincial in international relations.

Chitra Jayathilake is Senior Lecturer in the Department of English, and a Faculty Member of the Faculty of Humanities and Social Sciences at the University of Sri Jayewardenepura, Nugegoda, Sri Lanka. Her research interests lie at the intersection of biopolitics, postcolonial literatures, especially postcolonial theatre, English and cultural studies, and psycholinguistics.

Angela Last is Visiting Research Fellow in the Department of Sociology at the University of Warwick, UK. A geographer with a background in art and design, she is interested in materialism(s) and especially relations between representations of matter in politics. Her current work explores the relations between historical and new materialisms through the 'cosmic materialism' of the interwar period.

Naomi Millner is Lecturer in Human Geography in the School of Geographical Sciences at the University of Bristol, UK. Her research interests focus on food and environmental justice, through the lenses of postcolonial studies, decoloniality, political ontology, and political ecology. She is currently researching agro-ecology, food sovereignty, and permaculture in Latin America.

Sophie Moore is a PhD candidate in Cultural Studies at UC Davis, USA. Her research interests are primarily in the political ecology of rural social change, with a particular focus on Haiti's environmental and agrarian politics. Other academic and practice-based interests include critical pedagogy in environmental education, agroecology, and food sovereignty.

Delacey Tedesco is a Sessional Lecturer at the University of British Columbia (Okanagan), Kelowna, Canada in Geography and the Community, Culture, and Globalisation program. A political scientist, her research interests lie in the areas of political theory, indigenous studies, urban governance, race and settlement, and democratic sustainability.

Lisa Tilley is a Research Fellow in Politics and International Studies at the University of Warwick, UK, where she works on the Newton Funded project 'The Everyday Political Economy of Urban Resettlement.' Her research explores material approaches to 'the colonial question', as well as regimes of racial and gendered difference in relation to processes of accumulation and dispossession in economies of extraction. She has published in relation to debates within political economy, political ontology, post/decolonial thought, and decolonial methodology.

Zoe Todd is Assistant Professor in Anthropology at Carleton University in Ottawa, Canada. A social anthropologist, she researches human-animal and human-environmental relations, Indigenous legal orders and (de)colonial praxis in Canada. Her recent work focuses on fish and Indigenous legal orders. She is also interested in the articulation of Indigenous people's history and rights in relation to municipal development in Canada – specifically, how Indigeneity is expressed through architecture, art, urban planning, and storytelling.

Preface

Coloniality, Ontology, and the Question of the Posthuman responds directly to the emerging demand to rethink and extend the theoretical and empirical grounds of postcolonial studies. The demand comes from interrelated conceptual and empirical advances in the humanities and social sciences: political ontology and posthumanism. It also comes too from within coloniality studies itself. Namely, from developments in decolonial and indigenous studies which seek to enhance or push beyond the somewhat moribund and institutionalised radicality of much postcolonial thinking. Further, ecological, environmental, and technological questions – and pronouncements that we are, and are in, the Anthropocene – increasingly challenge the anthropocentric analyses that dominate the traditional attention of the social sciences and humanities. Human-centred orthodoxies in postcolonial analysis, whose focus has been on topics like identity, cultural hybridity, and political heterogeneity, are now being asked to account for how human beings are entangled ontological aspects of wider relational and ecological processes. The criteria for making these relational and material claims about human entanglement challenge constructionist and textual approaches still taken for granted in postcolonial studies. As a result, postcolonial theory, and postcolonial studies more generally, have struggled to respond effectively to new conceptual and empirical demands. Some authors have even argued that postcolonialism has either run its course, or has entered a contradictory period of decline. Despite this, addressing the global effects of ongoing colonial violences, exclusions, and inequalities continues to be more relevant than ever. It is clear we need postcolonial critique, but we need it also in a form more responsive to contemporary empirical and conceptual demands.

The following book emerges, then, in the need to rethink how discourses addressing coloniality are renewing themselves to meet the theoretical and empirical demands of a more-than-human world. We argue for the continued relevance of postcolonial politics and ethics, but within the need for new analytical questions and approaches. Postcolonial research needs to engage critically with the radical transitions suggested by the 'ontological turn' and its related posthumanist developments. For this volume, many of these questions come, primarily, from posthumanism and the postcolonial critiquing one another, but they also come from decolonial studies, indigenous studies, anthropology, politics, bio-semiotics, international relations, and elsewhere.

While postcolonial studies needs to address some of its more tired limitations, it is also fundamentally important that posthumanisms address and overcome the limiting parochialisms of their 'first world' European theory and often related myopic horizons. We argue, therefore, that critique must also decolonise its approaches to questions of the human and nature. It is not enough to appeal to the more-than-human or materiality, as we have been doing for some time now. We need to recognise that other people, other philosophies, other worlds, and other ideas have been making similar claims on wider ecological relations for hundreds, sometimes tens of thousands, of years, and crucially, in critical ways. What the characteristics of these critiques are, however, is an open question. We often impose criticality as reflexive subjectivity, and imposition inherited within a 'self-critical' Enlightenment tradition. It is a rare day, however, when a critical Yolngu or Yarralin concept is taught alongside Kant, Deleuze, Derrida, Latour, Spivak, or Haraway. Why? We need to ask this, and then not shy away from the causes. And then we need to address the colonial continuities in the classroom and of our knowledge production, by inviting others in, or simply by going out to sit and listen – really listen – in their spaces. This is, after all, Spivak's (1988) oft cited but much under-practiced (here too!), truly radical lesson.

When we say 'philosophy' or 'theory', it must mean more than a continental or Amer-European horizon. In doing so, it also must embrace the possibility that language and their referents like 'ontology', 'materiality', 'affect', 'technology', 'spirit', 'mind', etc. may not be – are not – fit for purpose in translating concepts across worlds. This is not a bad thing. It is a good thing. For, if thinking carefully and rigorously is, most importantly, about creating new possibilities, then the more tools, possibilities, practices, and voices, the better. God knows we need to do something different. The European and modern experiments are not working as promised. Perhaps they never have. Perhaps their promises have always been predicated on their opposites as well (violence, inequality, exclusion, curtailed freedoms, prejudice, etc.).

In engaging any of these hopefully transformative, creative efforts, key assumptions about politics, ethics, subjectivity, knowledge production, critique, and the like – foundational notions – will also, inevitably, be rethought. Of course, the entirety of such a project is far, far beyond the remit of one edited volume. This book attempts, however, to continue conversations begun by many others in their various elsewheres, but which are, perhaps, still as a whole in their nascent stages.

To do this analytical and exploratory bridging work, the volume invokes several interdisciplinary and transdisciplinary bodies of thought, some contiguous with, and some anathema to, postcolonialism and posthumanism. These discourses include decolonial studies; indigeneity; political ontology; cosmopolitics; new materialism; pluriversality; post-development studies; geo-poetics; white settler studies; socio-legal studies; film studies; drama; urban studies; Caribbean studies; film studies; international relations; geo-politics; and, geo-aesthetics.

The book is comprised of ten chapters by ten scholars. The introduction and the first chapter expand on the conceptual and discursive themes related to bridging the postcolonial and the posthuman. They remain unapologetically theoretical and conceptual, even speculative. They seek to explore the tensions and multiplicities invoked by asking the kinds of questions posed when these bridges are built, or, at least proposed and planned. Readers more familiar with the debates and implications of posthuman and postcolonial connections may wish to skip the introduction, and press on to the more substantive and empirically focused chapters. There, we hope, they will be rewarded with fascinating insights from diverse worlds.

For those interested in exploring the conceptual and theoretical implications of political ontology, coloniality, and posthumanism for critique, Chapter One surveys, in more analytical depth than offered in the introduction, many of the mobilising arguments and implications of thinking coloniality and the posthuman. It concludes by posing five key questions which emerge from the analysis. These questions seek to open a wider conversation about decolonising critique, about the status of signs and a material ontology for semiosis, about life and non-life distinctions, about the value of ecological systems, and about the role of the contemporary university in addressing such questions. These are difficult and often immense questions that merit their own many volumes. They are posed at the end of Chapter One simply as indicative trajectories for further thought, which, I know, several of the current contributors are presently exploring.

Thereafter, the chapters are organised into loose, overlapping geographies, first the Caribbean and Latin America, then Canada and North America, then South and Southeast Asia, finally ending with a geo-politics of pluriversality and the void. The chapters are organized not just by regional proximity, but also conceptually to work across and with one another.

The book began as an idea for a conference panel at the Royal Geographical Society's Annual Conference in London in 2014. The response to the call for papers was very encouraging, as was the session at the conference. In the spring of the following year, 2015, at the Association of American Geographers Annual Conference in Chicago, such was the reaction to the first call for papers that one session expanded to four panels and a day-long symposium. Several of the participants at each of the conferences expressed an interest in following up their presentations with chapters to a book. Others who wanted to come to the conferences, but for whom life precluded their participation in its various ways (babies, money, etc.) also contributed. One or two were hit up later either because their work caught my eye or because their scholarship and ideas were recommended by others; these recommendations have been justly rewarded. Some were eventually unable to contribute due to life again intervening in its good and bad ways (babies, divorce, illness, needing to finish the PhD, etc.). Eventually, as is the journey across most bumpy, edited-volume roads, the book has come together in its present form.

The book, as most do, outdates itself. As with all intellectual journeys, if we were to do it again, the outcome would be slightly different. The question of the postcolonial would be framed more explicitly within the decolonial. There is a difference between postcolonialism and decoloniality – several important differences – but it is also important to recognise their similarities and consonances. While the decolonial is a more radical effort to challenge discourses of modernity and refuse them, and whereas the postcolonial might be read more as an attempt to bring multiplicity and difference within the modern, it is also important to recognise that both the postcolonial and the decolonial are critiques of our present coloniality. Which is why the word 'coloniality' appears in the title. (Explanations of our use of the three terms are detailed in the Introduction and in Chapter One.)

The book might also have been bigger and more expansive. The question of the coloniality and the posthuman, of course, entails addressing indigeneities and the enormous insurgence in indigenous studies that is being welcomed (or not) into the academy. It could also have branched out into the digital posthuman. Instead, it circulates around ecological and cosmological posthumanisms (i.e. more environmentally focused questions). The result is a narrower focus. It is also one that calls for the arguments to be extended from bio-semiotics to the silicon and electricity based digital. Doing so might raise important, under-recognised questions that need to be raised about violence, fragmentation, and the semiosis of affect. It could also have branched further into the resurgence and insurgence of Black Studies, and so challenge the hegemonies of white humanism that dominate the discourses of Western liberal and political humanisms, and which also question posthumanisms in interesting ways. We do attend to the critiques of Césaires, of Fanon, of Glissant, and to Wynter's over-representation of Man as humanism. Of course, due to space, we necessarily also leave out other important Caribbean, African, North American, South and Latin American, West Asian, and Pacific thinkers of non-white humanisms.

In all, we hope the reader will forgive some of these absences in the recognition that much work needs to be done to think across and transform problematic humanisms, and so also to decolonise the many hegemonies that continue to constrain asking questions, their many material forms, and their spaces of address. For, as this book goes only a very modest and partial way to arguing: matter, forms, spaces, and modes of address (words, ideas, etc.) are far more implicate than is often assumed.

Mark Jackson

Reference

Spivak, G.C. 1988. 'Can the Subaltern Speak?' In: C. Nelson and L. Grossberg, eds., *Marxism and the Interpretation of Culture*. London: Palgrave Macmillan, pp. 271–313.

Acknowledgements

As the editor, I would like to thank, first, my colleagues who have contributed to this volume. You have each been patient and understanding during a long and sometimes frustrating process of bringing it together. The wait and your patience are rewarded by a stimulating volume that reflects each of your styles, expertise, and commitments. Thank you.

Second, I would like to thank the University of Bristol's Institute for Advanced Study for the award of a University Research Fellowship 2015–2016, during which ideas for this volume were researched, parts of this book were written, and, under whose tenure I was able to organise interesting meetings at the RGS-IBG and the AAG Annual meetings in London and Chicago. My thanks to Paul Bates and Wendy Larner for supporting the application and subsequent research leave.

Third, I would like to thank the following interlocutors for their conversation and ideas in thinking through some of what is presented within the volume: Marc Botha, Dipesh Chakrabarty, David Chandler, Maria Fannin, Anja Kanngieser, Angela Last, Claire Blencowe, Francisco Hernández-Adrián, Tariq Jazeel, Naomi Millner, Walter Mignolo, Jeff Popke, Maria Puig de la Bellacasa, Sarah Radcliffe, Tom Roberts, Pepe Romanillos, Olivia Rutazibwa, Robbie Shilliam, Terri-Ann Teo, Noah Therriault, Amanda Thomas, Lisa Tilley, Zoe Todd, Karen Tucker, Rolando Vazquez, and Elisa Wynne-Hughes.

Fourth, many thanks at Routledge to Faye Leerink for her encouragement and interest, and to Priscilla Corbett for her patience.

Finally, but by no means least, many thanks to Chrystal for her support and love.

Introduction

A critical bridging exercise

Mark Jackson

The broad landscape

Much has been written in the past years about the 'turn to materials'. 'Ontological', 'relational', and 'processual' turns, amongst the more prominent in recent critical scholarship, have attempted to account for how materials and their processes are understood to have 'more-than-human' agency. The argument goes, broadly, that reassembling human sociality through the relational agencies of more-than-human material assemblages enables a better, richer, more honest, and dynamic account of human life. Human life, certainly, is much more diverse than the narrow confines Enlightenment humanism might admit. Much of the research and writing in these fertile arguments has been significant and fascinating; indeed, work under these broad, posthumanist, ontological, and new-materialist banners has become almost de rigueur in some circles for addressing the 'material-semiotics' of planetary contemporaneity. Ecological collapse; the Anthropocene; technological profusion; cyborg embodiment; diverse interdependencies of life and non-life; energetics; globalisation; urbanisation; etc.: they all benefit from the many posthumanist constituencies that have emerged within these convolutions. Traditional Enlightenment humanisms have, thus, taken something of a backseat to the focus on the 'more-than-structural' and 'more-than-economic' pluralities that make thinking and critique possible. Historical materialisms, as many posthumanist discussions posit, also need to think about more-than-human geo-histories and cosmopolitical agencies. What Marx might have called 'forces of production', now observed at micro-political levels (materialities, affects, habits, everyday objects, embodied processes, etc.) have become as essential for contemporary social scientific and humanities research as the macro-political analyses of the more classically oriented modes and means of production (capital accumulation, enclosure, property, structuration, etc.). Examining the interactions, moreover, between these now topologically collapsed 'scales' of the 'micro' and the 'macro' derives much significant research and innovative explanation in the fields of politics, geography, anthropology, environmental studies, gender, and literary criticism.

The profligacy of these material and ontological turns developed from earlier work begun during the so-called cultural turn. Therein, an emphasis on the interpretive and qualitative analyses of language and texts focused critical attention on the often too narrow limits of economic, policy, and environmental investigation. Attending to the importance of culture meant focusing on the inevitable pluralities of human sign-making, representation, and reproduction. Difference became crucial to thinking about the cultural grammars and politics of critique. Gender, sexuality, race, and their numerous intersectional causes and effects developed as key differential constituents in opening analytic categories beyond narrow and naturalised economistic categories. With an increased analytic and interpretive emphasis on texts, language, culture, and difference, it became progressively important for contemporary critique to "cultivate [a] sense of multiple worlds and complex interacting traditions, that inevitable combination . . . of belonging and detachment, reception and resistance" (Said, 2004: 76). Cultures of modernity, hence, became understood as enrolled within cultures of colonialism; where you find modernity, there too you will find colonialism in some form, and vice versa. Power, oppression, enclosure, primitive accumulation, and the energies such forces always already concomitantly foster in response – resistance, opposition, struggle – were read in cultural (textual, representational, linguistic, etc.) terms, as well as politically and economically. Cultures of modern coloniality therefore increasingly emphasised productive, creative, and oppositional social forces; they were cultures of confrontation and hybridity, opposition and renewal, death and resurgence, ruin and creative possibility. Postcolonial studies, one of the key analytic discourses within the 'cultural turn', became an important critical constituency in cultivating, via the grammar and logic of difference, the always already plural sensibilities of societies, and their emergent politics and ethics.

Several features, however, characterise the often-divergent genealogies of these two critical trajectories, posthumanism and postcolonialism. Posthumanisms and, more recently, new materialisms have generally drawn their theoretical and conceptual energies from grammars that, almost exclusively, focus on a very narrow spectrum of reflection on human culture: modern, Amer-European thought. Whether it be materiality, the ontology of objects, affects, assemblages, technology, animals, systems, etc., the terms of debate are almost always drawn within Eurocentric conceptual frameworks. Rarely are these terms and their contexts provincialised. Assumptions and generalisations about 'the human' have characteristically, if also myopically, followed suit. This is not to argue that reflections within the Amer-European tradition are wrong. Far from it. It is simply to suggest two things. First, whether about materials, ontology, affects, or about politics, claims generalised from a narrow theoretical and empirical archive are valid, when they are deemed valid, for that archive. More importantly, second, reading materiality, affect, politics, etc. through an institutionally standardised and legitimised archive necessarily precludes addressing cognate themes, concepts, and approaches through different but equally legitimate lenses. One risk is that an overly narrow approach to thinking and critique thus ensues.

Postcolonial sensibilities have, on the other hand, almost exclusively applied themselves to human problems and, more often than not, human problems as defined through modern, Eurocentric political lenses. These have sought to bring different conceptions of language, culture, and politics into the largely Eurocentric discourses and institutions of modernity, processes like: the nation-state and nationalism; community; subjectivity; identity; freedom; history; and, representation. It is not until more recently that postcolonial studies have addressed themselves to more-than-human questions and topics. When it has, though, with the rise of environmentalism and ecologically focused critical politics, it has largely done so through the lens of eco-criticism, and the study of human representations and narratives of the environment, animals, and the like. Pluralising and, perhaps, thereby hybridising modernity's various institutional reaches has been its overarching focus. And for good reason. Politically and epistemologically, there is much to be done to combat the legacies of colonialism that saturate our present and its many injustices. Conceptually, it has not been until more recently, with developments like the material turn, that theoretical grammars have emerged to challenge assumptions about critique, phenomenology, language, and their limits and anthropocentric roots.

In both cases, the Eurocentric focus of each of these discourses, the posthuman and the postcolonial, stands, somewhat, to reason. Each is, further, critically derived and circulated within modern institutions like universities, wherein discipline-specific analyses in the social sciences and humanities are sedimented in institutionalised histories of coloniality, human enlightenment, reason, and human struggle. Universities, as spaces where the legitimacy of knowledge is circumscribed, have always been one of the key legitimating arms of colonialism, (as well, though less frequently, as a site of its critique). Which may go some way to accounting for the third divergence. Broadly conceived, postcolonial and the posthuman discourses, rarely, until recently, speak much to one-another. This too, as I have intimated already, stands to reason. In a world with so much continuing injustice that is not only the legacy of modernity, colonialism, and their segregationist Eurocentric structures and attitudes, but its engine, coloniality is built into the very grammars of institutional critique and knowledge production itself. Asking the postcolonial scholar to broaden the terms of her political debate and relevance is often interpreted as either unwise or of secondary importance. She might reply, not un-sensibly, to the effect that, 'We've got human suffering and inequality to deal with first, never mind needing to address the conceptual niceties and theoretical complexities raised by dogs, mushrooms, or the effect of ice storms on the electrical grid!' On the other hand, asking the posthumanist, new materialist, or affect theorist to broaden the terms of their investigation to include non-Western philosophy and concepts, also means asking uncomfortable questions about their colonial and Eurocentric complicity, about needing to learn other languages and ways of living, and, in many cases, about relinquishing security, time, prestige, influence, claims to expertise, and power, all of which are crucial and coveted, yet scarce, currencies in the academy. Decolonising the academy is a deeply fraught, often very personal, and, although disruptive and dangerous, necessary struggle to be undertaken. But, it comes with intimate risks; often, it is deemed easier to play it safe and not to entertain such risks.

Finally, and perhaps most important to understanding why the postcolonial and posthumanist discourses rarely speak to one another, they are mutually disruptive.

Asking orthodox posthumanisms to recognise, for instance, an aspect of Nuu-chah-nulth indigenous philosophy that has been making similar arguments for generations is disruptive of the concepts, languages, institutions, legitimacies, and traditions taken for granted in the conventional hegemonies of knowledge making.

> conceiving Amerindian thought in terms of concepts changes not only our concepts but our very concept of concepts, pulling the concept . . . into the orbit of myth and its much greater capacity to effect transformations of not only other myths but also other discursive materials.
>
> (Skafish, 2014: 25)

Let's now transpose the disruptive relation by asking the orthodox postcolonial humanist to consider the aesthetic agencies and sensibilities of cyanobacteria. Doing so will mean disrupting the taken for granted concepts, the political and conceptual categories, never mind the empirical texts, with which the aesthetic, and its outcomes, are normally framed (Jackson, 2016). In each case, opening the two broad discourses to the variable forces of difference at play in the other – the human and the non-human – disrupts the stability, internal coherence, and implications of the valuable ethical, political, and epistemological critiques explored by both the postcolonial and the posthuman.

Yet, the more one digs into the implications of doing pluralising analysis, whether it be on human cultural representation and misrepresentation, or on more-than-human constituencies of material possibility and flourishing, the more one comes to appreciate that difference is not a thing or identity, but a relation (Wheeler, 2014: 70). Difference is an ontological condition of emergence, and it is a relation that precedes and constitutes the possibility of designating something as either human or non-human. Further, the contact zones and events difference makes are always materially embodied processes that create and re-create worlds. Differences and their worlds of relation, in other words, are co-implicated within what non-human subjects and forms make possible. Mind, imagination, and its products emerge in material ecologies that, literally, culture them (see also, Iovino, 2012: 61). Representations, ideas, ecologies, and materialities are not that separate after all; they are all facets of embodied, immanent processes within multiple ecological histories. Knowledge, ethics, and politics are social processes whose legibility through history also entails tracing the different praxes of sociality including their embodiments, stories, and symbols beyond the human.

What this book seeks to do

This book seeks to build bridges – relationships – between the broad analytical discourses of postcolonialism[1] and posthumanism. It is the contention of this volume, and its various authors, that these two broad and often internally diverse trajectories of critical thought are not anathema to one another. Rather, they share as many close affinities and potentials for cross-fertilisation as they do capacities for productive critique and challenge. The object of the contributions presented here is, thus, not to determine either discursive path as more appropriate than the other – if, even, they can be said to exist! 'Postcolonial' and 'posthuman' are more like umbrella terms than either is a cohesive or consistent mode of comprehensive critique. Nevertheless, both terminologies are used consistently to describe important, contemporary, critical energies.

Both also address some of the most significant socio-political currents of our time: the postcolonial and the posthuman ask, in different, yet connected, ways about the contemporary, objective crisis in the concept of the 'human'. This crisis has been framed in several ways: as the much-discussed Anthropocene or Capitalocene; as the increasingly dubious ecological and cultural distinctiveness of human beings; as the contradiction of contemporary technological development and instrumental peril; as a structural inequalities in the global distribution of resources; or, as the racist inequities that shape characterisations of who is more human (i.e. white Europeans and their descendants) and so more deserving of energy, empathy, and equity. Each of these debates articulates around political definitions of what and who constitutes the human. Distinctiveness necessarily entails difference and assertions about self and other, mind and thought, living and non-living. Otherness becomes a key grammar within these conceptual and thus practical logics. One might argue, in fact, that alterity and the question of the 'other' – how it is that otherness is conceived and acted upon – is the defining question for our planetary present, whether it be in how we adjudicate the political and ethical details of equality, or in how we define the boundaries between what is recognised as human and what is not.

Our critical energies in this book are marshalled to examining how cultural and political problematics of representation, affect, embodiment, violence, settlement, and racism share conceptual and geographical grammars with the political problematics of what many term 'more-than-human' processes: ecologies, technologies, materialities, and environments. The critical work in this volume ranges, empirically, from Caribbean geo-poetics, to pigs and post-disaster development, to white-settler urbanisms, to agrarian ecology in Central America, to South Asian and South East Asian dramatics of death, including spirits and ghosts, to fish as indigenous legal agents, international relations and the void, grief, Truth and Reconciliation Commissions, city fabrics, poems, and film. Through these numerous geographies, we argue that decolonising how the other is constructed and engaged also entails addressing fundamental conceptual grammars surrounding humanism, critique, materialism, political agency, the limits of epistemological claims, and, the limits of ontological claims too.

As in all good bridging exercises, our aim is to contribute means by which to transform and extend both approaches, both edifices, both banks, in mutually productive ways. Bridges are generally built to connect mutually agreeable shores, as much as they are also built to embolden existing routes. Good bridges, however, should not commensurate; building dialogue is about recognising and committing to difference. Bridges are 'boundary objects' that emerge from, and create, 'spaces between' (Star, 1995: 32) across which we can communicate, see embankments otherwise, and so come to appreciate different intellectual positions for their strengths and weaknesses, while also endeavouring new possibilities.

Relationships of connection and mutual support are needed in the fast-disappearing, critical academy. New routes for thinking are also desperately needed within the wider world, of which the critical academy remains a crucial part. Academies persist, despite their many assaults, as promissory pausing spaces, spaces between, where, though dwindling, questions are still posed for the sake of the question. We suggest that approaches to coloniality can learn from posthumanisms. Posthumanisms can learn from postcolonial and decolonising efforts. It is in the learning, and so in grappling with the difficult questions that arise for each, that we might be able to think, and so act, carefully towards the many pressing demands of our contemporary moment.

Filling in some landscape details

Before getting ahead of ourselves, let's back up briefly to set out some of the basic terminologies and discourses at issue in the volume. At first blush, it may seem odd to try to connect these two diverse bodies of critical thinking with conceptual linkages. They seem to be concerned with quite different things. What do they need of one-another?

Postcoloniality

Postcolonialism is commonly animated by a largely political and, frequently, aesthetic critique of the structuring conditions – the erasures – perpetrated by both modern colonialism and its legacies. The 'post' within the 'postcolonial' rarely, if ever, signals an acceptance that colonialism is finished and now an historical artefact. Rather, the prefix 'post-', much like its use elsewhere (i.e. posthumanism, post-structuralism, post-secularism, post-development, etc.) signals discursive reflexivity. As Mary Pratt writes, 'post- signals primarily a way of thinking about the scope of one's coloniality' (2008: 460). Colonialism continues, both overtly, as in the contemporary and ongoing occupation of unceded land, for instance in numerous cases across the contemporary Americas, Australasia, Western and Central Asia, Africa, and Europe. And, colonialism continues in the everyday structures of globalised capital, industrialisation, the enclosure of private property, the creation of 'natural resources', and the social and cultural governance apparatuses that facilitate such processes. 'Coloniality' refers, then, to the continuities of colonialism in both the thought and the unthought structures of everyday life (Maldonado-Torres, 2007).

Modernity still very much reproduces, and so reaffirms, constitutive legacies of colonial exclusion and division. Racism, unequal access to economic resources and opportunities, epistemic and representational privilege, the force underpinning capital accumulation, instrumental control and industrialised expansion: all are products of modern colonialism. These dynamics shape the atmospheres within which we, as modern subjects, breathe the coloniality of our present (bid. p. 243). 'Post-', to reiterate, simply signals that reflexive attention to the continuity of inherently violent structures and conditions – in thought and in practice – is necessary if we are to address and over-come the social injustices they predicate.

Reflexive critique attentive to coloniality affirms, therefore, the always already numerous imaginaries of human flourishing that must be brought to bear on the question of living well. As such, the postcolonial and its requirements to decolonise (a more active and, perhaps, more radical imperative than simple textual reflexivity) are largely motivated by ethical and political concerns to differentiate social and cultural accounts of individual and collective human possibility. Humans are, its discourses remind us, far more than the horizons hegemonic European modernity heralds for us and our global futures. We, that is, 'humans', always have been. In response, postcoloniality argues, thinking and knowledge making must decolonise, decentre, diversify, and, in many cases, reject the narratives that have come to over-represent and legitimise continued forms of colonisation, erasure, and violence, epistemic and otherwise.

Yet, for the postcolonial, the 'we' around which this need to provincialise conceptions of sociality articulates rarely extends beyond taken for granted human derived grammars of representation: language, identity, class, form, place, history, and aesthetic expression. Indeed, representations (e.g. literature, art, identities, nations, imagined communities, etc.) are the stuff in trade of orthodox, postcolonial critique. 'We' have never been 'human', if the standards for 'human' are framed by Amer-Eurocentric representational and epistemic criteria, their norms, expectations, and sedimented descriptions of lived possibility. This is because dominant epistemic frameworks are couched in the coloniality of exclusionary logics, divisive rationales, and hierarchical hegemonies, and, importantly, the underlying structural and economic violences that make them possible.

The postcolonial has been a discourse largely responsive to humanism, concerned with humans, and which puts people and the representational politics of people first. There are worries, in fact, amongst influential and important voices (e.g. Braun, 2015; Gilroy, 2015; Lazarus, 2011; Mukherjee, 2010) that shifting the political, epistemological, and critical focus away from the very real human atrocities affecting millions (war, forced migration, slavery, racism, segregation, economic exploitation) to more conceptual debates about the boundaries of the human, actually devalues the significance of colonial violences, their continuities, and, importantly, positing means to resist and overcome coloniality's persistence. Political responses to injustices must not, it is argued, further muddy the conceptual and practical waters by extending political agency – if that is even possible – to non-human actors. It is hard enough to extend thinking and action to embrace the imperatives of representational miscegenation, partiality, indeterminacy, difference, diversity, incompleteness, and imperfection. These are the necessary, and only, epistemic conditions for social possibility; it's all there is. Thinking there is more than this, that it is possible to generate either epistemic completeness or ontological inclusion, that is the problem. Knowledge, postcolonial thinking, is always partial and situated. It gains its honesty and integrity by respecting and acting from ineluctable, yet situated, epistemic partiality.

Posthumanism

Posthumanism is also broadly animated by an epistemological and ontological concern to decentre explanations of human possibility via constitutive difference. Unlike, however, the epistemic leanings (and, in some cases, firm limits) of postcolonial critique, it seeks to affirm and explore the plurality of ontological processes that comprise the multiplicities of human social embodiment (see, e.g. Braidotti, 2016; Wolf, 2007). Partiality means plurality, and situatedness inheres the ecological relations that make it possible. For posthumanism, we have never been 'human', if by 'human' we mean something exceptional to the many material ecologies that make up thinking and action. We are also objects, machines, ecologies, systems, assemblages, networks, agencements, animals, hybrids, 'nature-cultures'. What is termed 'human' is simply a dynamic and ever-changing field of relations within the innumerable cosmic relationships that sustain material complexities like thinking, representation, justice, and beauty. We are, if you like, always already more than ourselves. Human thought and agency is comprised of the agencies afforded through, for example, bees (food and thus energy), bacteria (digestion), affects (sensibility and feeling), neurons (creative connection), geohistories (forces of production), silicone technologies (speed and connection), shared ideas (economics and connection), sunlight, and cosmological immanence. Humans emerge in, and as, the relational potentials these assemblages – these diverse material ecologies – make possible.

Posthumanism, therefore, concerns itself with the processes of the many human and ostensibly non-human relationships that render thought and action potential. In their ostensibly more experimental and boundary pushing forms, posthumanisms challenge even constitutive necessity of concepts like self or subjectivity, terms though rigorously debated, but which, for the postcolonial, are often still assumed to be necessary for political agency. Instead, posthumanist approaches seek to redefine assumptions of the human and non-human, along with those of experience and politics, in terms of the numerous materialities that make up the stuff of more-than-human social relations.

Yet, haven't numerous non-Eurocentric modes of thought always already done this (see, e.g. Sundberg, 2014; Todd, 2016)? Weren't rationalist, Enlightenment, colonialist, and modernist narratives precisely those that attempted to disabuse the noneuropean of such irrationalities? It is quite the condescending claim to appeal now to the same critical traditions in the attempt to over-turn previous conceptual limitations.

Opening institutionalised ecologies to decolonising critique

If the postcolonial is primarily focused on people and their representations and posthumanism has its critical eye on the non-human materials and relations that enact ecologies of place and thought, how and why should they be thought together? The attentive reader will have already anticipated possible connections, and their ultimately disruptive implications.

First, how 'the human' is described and defined varies. Indeed, how the distinction between the human and the so-called non-human or, dare one say, 'natural' is far from assumed. There is a geography to what is meant by 'human' and what is meant by 'not-human' or 'other'. When compared to a modern European, an Amazonian Achuar person will have a very different understanding of how the human differs from the non-human (e.g. Curry, 2008; Descola, 1994; 2013a;

2013b). An Ojibwa person may have a very different understanding of the distinction between living and non-living (e.g. Hallowell, 1975), as will a Belyuen Aboriginal woman have a very different way of rationalising, for instance, the cultural economy of water (e.g. Povinelli, 1995). These are but three, amongst innumerable other, examples.

Second, these differential geographies of the human and non-human have been fundamentally shaped by histories of modern colonialism. Politics and development around the world continue to be fraught by the specific histories that have imposed non-autochthonous derivations of human exceptionalism and their associated notions of bounded individualism. Non-Amer-European others have and use, of course, different ideas about what the West terms political agency, democracy, equality, rights, and responsibility. How these are recognised, understood, translated, and negotiated – or more likely, misunderstood and ignored – remains a significant issue for articulating alter-modern, politics of difference. For instance, characterisations of indigenous law and legal orders are fundamentally distinctive from the legacies of European legal apparatuses that largely shape the global hegemons of property, accumulation, and consumption (see e.g. Napolean, 2013, and Todd, this volume). Notions of kinship (Sahlins, 2013) and intergenerational accountability, too, are radically different within, say, the totemic cosmopolitics of Warlpiri Dreaming (Glowczewski, 2015). Again, these are but a few examples within innumerable others.

Third, the distinctions between what counts as culture, and who and what shares in culture, also vary geographically. If exceptionalism is a function of having 'culture', then the boundaries erected around inclusive and exclusive characterisations will be different for different people. What's more, the supposed hierarchies of truth which label modern, scientific forms of understanding better than non-modern ways of life are increasingly also the same ones that imperil ecological sustainability. While modern exceptionalism can promote extremely productive and beneficial outcomes (e.g. antibiotics, air travel, individualist political autonomy), it would be naïve to argue that the non-modern forms of life are without some value or efficacy. More than naïve, it would be simply incorrect. Other ways of life, other languages, other forms of social interaction, prove extremely insightful and resilient, perhaps because they frame human exceptional capacities and ecological relationships in profoundly different ways.

It is here, then, at the boundaries, in the 'betweening spaces', in negotiating the questions difference makes to decentring ideas of the human, and to re-defining stories of humanness, that there are significant discursive overlaps and potentials for concepts to be created, bridges built, between the two critical discourses and their many facets. Examining how we think about and characterise the nature of the commensurabilities and differences between the postcolonial and the posthuman, and the nature difference makes (indeed, the difference 'nature' makes), we argue, reveals much about the ethical, political, and epistemic resemblances and divergences between these bodies of thought, and the capacities for them to work together to address the pressing problems of our time. Most importantly, bridging their shared and divergent capacities speaks to present social needs.

The following volume commits itself to exploring various conceptual tensions and opportunities that emerge, in their combined attention, so as to forge modest proposals for thinking rigorously about our perilous and fragile present. The 'our' of our present is both and neither human and non-human, it is both and neither postcolonial and posthuman.

It is our view that 'the contemporary imperative of our global interhuman and environmental situation' (Wynter, 1995: 8), our planetary moment, is ontologically, and therefore politically and ethically, imbricated. Human crises are ecological crises. Crises of matter are crises of thought, for thought and matter are an implicate order. They are folds of one and the same. The continuing colonial forces that shape current global traumas of war, racism, extremism, forced migration, capitalist and industrialist violences and their corresponding inequalities – including forms of gendered and sexual violence – emerge from conceptual apparatuses that also destroy possibilities for ecological flourishing. Death, is, of course, intimately interwoven with coloniality (Mbembe, 2001, 2003), and definitions of the bounds of life are themselves functions of coloniality's bio-political, necro-cratic, truth regime.

We argue that how we need to construct 'what should constitute fully modern human beings' (Wynter, 1995: 8) is a matter of fundamentally decolonising normative conceptual and political boundaries of the human, of the living, and, even, of the dead. By decolonising, we mean radically questioning accepted epistemological and ontological precepts about the human, but also about the mode and manner of what questioning itself often assumes. Radical decolonising questioning should extend to critique itself (Jackson, 2017; and ch. 1 this volume). We assert that how we define and what we accept as critique often re-enforces limiting perspectives on human thinking, embodiment, and political agency: a self-conscious agent separated (culture) from a mute and lifeless cosmos (nature). Further, decolonising critique does not entail jettisoning the human, but simply thinking and constituting it transversally. At the very least, we argue, we certainly need to rethink what we mean by 'human' in ways distinct from its over-representation as Man[2] (Wynter, 2003; see also, Tilley, ch. 7, this volume). To do so entails opening the postcolonialisms to posthumanisms, and posthumanisms to postcolonialisms. Decolonising work must provide new means to address false, and now imperilling, dichotomies of social and environmental violence, human and non-human separation, and forms of intercultural and social enmity.

We suggest, therefore, that questioning the ways difference is produced and mobilised by critique requires more than simply challenging humanism's attempt to extend its critical remit to other forms of cultural representation. When we do the work of pluralising the politics of representation, we also see that the ways others always already world their sensibilities are not commensurable with, let alone available to, the imaginaries of dominant Euro-american (and globalising) humanist tropes, or their ontologies, or their consequent politics. Fundamental ontological questions are also raised about the human in the work of asking postcolonial questions and decolonising our practices of thought. Indeed, as suggested, the speciesism that protects a politics of human exceptionalism traditionally mobilised in the humanities and social sciences as 'humanism' has a history that is itself the product of Euro-modern colonial forces and ontological separations that extend beyond the living non-human to the geological, the cosmological, and the spiritual.

Not only are fundamental ontological questions raised, but the very categories we take for granted in much of the Amer-European intellectual framework come into question. 'Ontology', including concepts such as 'gender', 'religion', 'spirituality', 'aesthetics', 'epistemology', 'politics', 'nature', 'culture', 'critique', and the like, also need to be questioned, for they too emerge from, and legitimise, the legacies that also enable coloniality. In many cases, these terms are not only incommensurable or untranslatable across geographies of difference, they are 'radically irreducible' (Mignolo, 2010: liii; see also Latour, 1988). For instance, famously, many peoples do not have words for 'nature' or 'culture'. Likewise, how occidental thinking uses a term like 'gender' might not have a correlate in another context or world. Using the terms as though they do, or as though we can approximate meaning, can be very problematic.

The critique we seek to engage here is a more extensive and decolonising one. It is one that needs 'to think/articulate itself outside the terms of the disciplinary discourses of our present epistemological order' (Wynter, 2003: 331). We argue that we need to expand the questions asked about how to 'introduce invention into existence' (Ibid.), in order to think about forms of solidarity and accountability both to those traditionally marginalised from knowledge production, and to see, in their 'different senses of ontic belonging' (Chakrabarty, 2000: 254), grounds for new sensibilities and associations. Such solidarities must not reduce themselves, either to the privilege of a Eurocentrically represented and defined idea of 'Man', or to a negotiation of predefined epistemic and phenomenological limits experienced by a modular (and colonial) human.

In other words, the idea of critique itself needs to be rethought beyond the humanist, phenomenological structures that contain it to a self-conscious negotiation of cognitive, linguistic, or rational limits, limits which are imagined within the conceptual modes that also continue to legitimise colonial forms of oppression (i.e. distinctions like: ethics/epistemology; nature/culture; word/world; secular/religious; matter/mind, etc.). More controversially, perhaps, we explore the implications of the insight that the Eurocentric politics, ethics, and aesthetics of the human are often reinforced by many contemporary claims about the epistemological limits of human subjects, the textual conditions of critique, and the assumed spaces of the ethical, the political, and the aesthetic.

Most contemporary critique owes its reflexivity to a theoretical deliberation oriented around the limits of cognitive experience. Epistemic and ethical responsibility to those limits circumscribes the legitimacy of the sayable. In a modern European philosophical tradition, whose genealogy originates in the pivotal arguments about human cognition and its possibilities, the critical starting point for explaining the knowable in terms of human experience roots itself in an epistemological negotiation of conscious or phenomenal conditions, as in the case, for example, of Descartes and Kant. The conceptual logic is concerned with building an argument about necessary epistemic limits from reflexive first principles, whether these be transcendental conditions of individual human experience (e.g. Hume, Kant, Locke), collective conditions of historical and material possibility (e.g. Hegel, Marx), or the textual, discursive, and representational formations within which material and phenomenal limits are produced and negotiated (e.g. Derrida, Gadamer, Habermas, Foucault). Critical responsibility is typically constructed herein with respect to a cognitive subject (individual or collective) whose grounding, sustaining parameters are envisioned epistemologically. In other words, they are circumscribed by a narrative that privileges the question of the sayable, and thus the knowable, with respect to a particular idea of human cognitive possibility and, crucially, its self-aware limits.

Epistemological possibility thus comes to ground means of defining alterity. Cognitive possibility within the same tradition is also formed by arguments and assumptions about human exceptionalism, human difference from non-human others, and secular human capacities for control and partial determination of its futures. Starting points around experience, critical capacity, and distinctiveness mean that a certain Eurocentric theoretical privileging of a human construct is mobilised in definitions of cognition and meaning making from an epistemological core. This core enunciates what counts as the limits and responsibilities of critique from the centrality of a particular idea of human limitation or fallibility – we cannot experience the world as it is in itself (noumenal), we can only experience it as experience (phenomenal). Questions of difference and alterity emerge, as a consequence, in respect of others' epistemically constituted limits, rather than, say, their kinship capacities for, or practices in, love and care, or their capacities and practices for compassion, support, or sharing, empathy or sympathy, or their sensibilities of pleasure and play, or their participation in decoration and aesthetic sensibility, or in their propensities to invent and produce forms of future-oriented sustenance or flourishing, etc. Why, for instance, are human distinctions articulated in normative terms by reflecting on them as different cognitive forms, rather than in, say, inter-species and cross-species kinship capacities for nurturing or love (e.g. Rose, 2011)? That different life forms also share affective capacities, that they are shaped by the material resonance of worldly things in ways that can be recognised as similar or even shared, is something long recognised by many people and cultures (e.g. Glowczewski, 2016; Muecke, 2004; Povinelli, 2016; Vivieros de Castro, 2015). Perhaps it is precisely from the fact that difference shares in being affected, rather than conflicting in terms of cognitive intentionality, from which we may be able to build new ecologies for critique.

If we are mindful of other peoples' political worlds in our postcolonial analyses, as the above authors, and many others, encourage us to be, we might see that decolonising the epistemic terms of our critique about the human also emerges from destabilising our present discursive orders about the sources of critical responsibility. Perhaps critical sources lie in positive commitments to shared differential sensibilities, sensibilities that affect numerous living (and non-living) agents. These would be less negatively defined characterisations of differential cognitive (im)possibility than they would be recognitions of shared affects and practices that establish ethical embodiments (e.g. Stevenson, 2014; Puig de la Bellacasa, 2017). For example, consider: shared, care-full capacities to seek and be energised by the sun; reciprocities of touch; shared empathies like hunger; being stirred by aural beauty; experiences and memories of loss; practices of nurturing;

mutual dependence; generosity; play and pleasurable motion; pre-conscious interconnections, as in the case of gut micro-biomes; and, many, many other similar processes of shared embodiment.

The representational forms epistemic boundaries take, and the effects these forms have on producing distinctive sociocultural possibilities as human, narrow, rather than open, the terms of debate about the human and our necessary solidarities (human and non-human). Crucially, they also narrow the conditions of interaction in forming what counts as a shared world of difference. Alterity is, as a result, often imagined and defined from a linear, hierarchically structured logic that places an epistemic vision of human phenomenality at the centre of critique, despite the sometimes desire to overcome such conceptual models and their corresponding logics.

If our point of departure, however, takes alternate accounts of human difference, ones, for instance, that do not privilege cognitive exceptionalism, experiential limits, or human cultural separation as the basis for making ostensibly legitimate claims about our delimitation in the world, then, we ask: what is the effect on how we think about alterity and the meaning of critical responsibility? Why does epistemology need to be the central defining rubric through which difference articulates itself as a basis for living well in plurality? What if we placed an ethics of care, or an appreciation of pleasure, or a responsibility to relations of ecological flourishing, as the terms by which difference is defined and negotiated? Which others, which alterities, would care or pleasure or play or aesthesis (sensibility) throw up? What if we begin not with the humanist other, that is, a notion of critique inherited from a humanist subject as epistemically limited, together with its hierarchical logic, but with an account that attempts to recognise the multiplicities, and so the alterities, that always already make up the possibilities of epistemic awareness? Does an account of the other, defined differently than an assumed human phenomenalist and epistemic centre, shift the terms of analysis? What becomes thinkable about the importance of human agency and responsibility when different starting points are invoked than those epistemically delimited as self-consciously human? To ask these questions we need to engage radical decolonising work in both the domains of postcolonial studies and posthumanism so as to draw out their elective affinities.

The reflexive work of thinking carefully about coloniality, and actively seeking to decolonise theoretical and practical registers, requires a form of radical listening to what the poet and philosopher Édouard Glissant terms the 'rhizome of a multiple relationship with the other' (1997: 16). Relations of otherness can take forms that do not define their alterity in terms of epistemological limits to cognition. In fact, many multiplicities of alterity necessarily co-exist within the innumerable relations that make up human ecologies, but which, despite remaining either unknown or unknowable, nevertheless make themselves felt (or not). They co-exist and interact in ways that need not derive from an alterity defined through hierarchical forms rooted in assumptions about the identity of human self-consciousness, or, the way a story of consciousness represents and identifies its own epistemic limits and the life that it lives (Wynter, 1995: 8). Glissant asks: what happens when we begin by 'prompting the knowledge that identity is no longer completely within the root [i.e. epistemology], but also in [the multiplicities of] Relation' (1997: 18)? What happens when we decentre accounts of otherness through multiplicities of relation, and not simply through a story of epistemic limits to cognitive representation? We might learn to see that the ecologies of others are always already encountering one another in numerous materialities that invent anew and spatialise differently.

Consider, for example, simply the proliferation of planetary difference that has evolved in an immanent totality from the early formation of the earth as molten minerals to the many life forms and ideas that co-exist today. Such a non-human geo-history is the result of multiple material relations immanently constituting what is today the possibility of abstract thought. Articulated in such spaces of intra-action are innumerable modes of geographical vitality that emerge from intimate, ontological inter-connections (McKittrick, 2006: 133). Such vitalities, whose geo-histories are seen to pre-figure, subtend, and constitute human forms are recognised, for instance, in the formative engagements Mary Graham, a Kombumerri and Waka-Waka Elder, Aboriginal scholar, and activist in Australia, argues are attributes common amongst indigenous worldviews. Graham explains these attributes in two profoundly simple principles. First, 'You are not alone.' Second, 'The land is the law' (2008). Both principles inhere a sensibility that extends beyond the Euro-modernist defined human as 'self-conscious knower/reasoner' to one imbricated in dispersed material relations that pre- and post-figure the experience of knowing. Senses of relational engagement emerge, as such, from enfolding material relations, rather than only by being imputed or projected from an assumed internal structure, or set of universalisable conditions by an epistemically oriented modular consciousness, which is then used to demarcate human belonging.

Autonomous and autochthonous forms of living, practice, and imagination are constantly emergent from the material interactions that make our, and others', worlds. Our role as critics is not to channel these emergent poetics, these 'real and imagined geographies' (McKittrick, 2006: 143), these ways of geo-graphing – of 'writing the earth' – in forms that surrender to, or render sense by commensurating terms of critical debate about accepted narratives of what is and is not knowable. Our role, instead, is to open and create avenues of thought and practice from emergent material relations, from what 'geontology' (Povinelli, 2016) makes possible for creative potentials, rather than new regimes of governance. Such forms of invention, experimentation, commitment, possibility, and vitality can cultivate flourishing in terms of how worlds and their resonant domains are variously and differentially bound together (Shilliam, 2015: 31). Our responsibility is to the multiplicities of relation that (already) co-constitute how numerous and multivalent differences make themselves felt as flourishing. Of course, how we do this work has implications as much for the accepted terms of what counts as postcolonial, as much as for the posthuman. Hopefully, in the decolonising work presented here, both will be, at least partially, opened and so creating new critical exercises.

Notes

1 A quick word on the terms 'postcolonial' and 'decolonial'. Much could and has been said about the similarities and differences between postcoloniality and decoloniality (e.g. Bhambra, 2014a, 2014b; Mignolo, 2011; Maldonado-Torres, 2011). My intent in this introductory context is simply to note their similarities as joint critical efforts, rather

than their conceptual distinctions. By 'postcolonial' we mean the theoretical, practical, reflexive, and iterative critique of colonialism, in particular, modern, Western colonialism. This form of colonialism has dominated planetary history since the fifteenth century, and continues to imbricate itself through globalised state apparatuses, through the conjoined neo-liberalisation of global and local economies, values, and cultural forms of representation, and in the proliferation of extinction disasters shaping our planetary present. Postcolonial refers, as Pratt (2008: 460) writes, to a useful way of thinking about the scope of our coloniality. 'Decolonial', of course, entails much the same critique, but it also conveys a more active and perhaps radical sensibility. To decolonise is to engage a postcolonial awareness in changing the specific circumstances of how we think and practice. It refers, like the postcolonial, but perhaps also more demonstrably, to actively doing the critical work of changing our coloniality, rather than ameliorating the negative effects of modernity with diversity. This means that the postcolonial can and *should be* decolonised. Its optic should expand beyond European power relations (McClintock, 1995; Spivak, 1999) to apprehending numerous imperial processes as they move and adjust. Decolonising the postcolonial, though, is also simply thinking rigorously about the scope of our coloniality. We recognise the postcolonial theoretical genealogy through South and Western Asian inflected critique, and the differences its traditional focus on cultural texts has with decolonial emphases on Caribbean, Latin, and South American genealogies of critical economy, indigenous thought, and Iberian imperialism. Although the postcolonial and decolonial are not identical, they are certainly fraternal siblings in critical arms. When we invoke the question of the postcolonial, we are also invoking the recognition of our contemporary coloniality and the imperative to decolonise harmful hegemonies.

2 For readers who may be unfamiliar with Sylvia Wynter's thesis regarding the over-representation of the human as 'Man I' and 'Man II', her argument entails a much more complex and historically specific analysis than the simple use of the gendering noun 'Man' may superficially communicate. See, for instance, Wynter, 2003.

References

Bhambra, G.K. 2014a. *Connected Sociologies*. London: Bloomsbury Academic.

Bhambra, G.K. 2014b. 'Postcolonial and Decolonial Dialogues', *Postcolonial Studies*, vol. 17, no. 2, pp. 115–121.

Braidotti, R. 2016. 'Posthuman Critical Theory', In: D. Banerji and M.R. Paranjape, eds., *Critical Posthumanism and Planetary Futures*. New Delhi: Springer India, pp. 13–32.

Braun, B. 2015. 'New Materialisms and Neoliberal Natures', *Antipode*, vol. 47, pp. 1–14.

Chakrabarty, D. 2000. *Provincialising Europe: Postcolonial Thought and Historical Difference*. Princeton: Princeton University Press.

Curry, P. 2008. 'Nature Post-Nature', *New Formations*, vol. 26 (Spring), pp. 51–64.

Descola, P. 1994. *In the Society of Nature: A Native Ecology in Amazonia*. Trans. N. Scott. Cambridge: Cambridge University Press.

Descola, P. 2013a. *Beyond Nature and Culture*. Trans. J. Lloyd. Chicago and London: University of Chicago Press.

Descola, P. 2013b. *The Ecology of Others*. Trans. G. Godbout and B.P. Luley. Chicago: Prickly Paradigm Press.

Gilroy, P. 2015. 'Offshore Humanism', The 2015 *Antipode* RGS-IBG Lecture. 2 September 2015. http://antipodefoundation.org/2015/12/10/paul-gilroy-offshore-humanism/ [Accessed 19.02.2016].

Glissant, É. 1997. *The Poetics of Relation*. Trans. B. Wing. Ann Arbor: University of Michigan Press.

Glowczewski, B. 2015. *Totemic Becomings: Cosmopolitics of the Dreaming*. Helsinki and São Paolo: n-1 Publications.

Glowczewski, B. 2016. *Desert Dreamers*. Minneapolis: Univocal.

Graham, M. 2008. 'Some Thoughts About the Philosophical Underpinnings of Aboriginal Worldviews', *Australian Humanities Review*, 45. www.australianhumanitiesreview.org/archive/Issue-November-2008/graham.html [Accessed 27.07.2016].

Hallowell, A.I. 1975 [1960]. 'Ojibwa Ontology, Behaviour, and World View', In: D. Tedlock and B. Tedlock, eds., *Teachings From the American Earth*. New York: Liveright, pp. 141–179.

Iovino, S. 2012. 'Material Ecocriticism: Matter, Text, and Posthuman Ethics', In: T. Müller and M. Sauter, eds., *Literature, Ecology, Ethics: Recent Trends in European Ecocriticism*. Heidelberg: Winter Verlag, pp. 51–68.

Jackson, M. 2016. 'Aesthetics, Politics, and Attunement: On Some Questions Brought by Alterity and Ontology', *GeoHumanities*, vol. 2, no. 1, pp. 8–23.

Jackson, M. 2017. 'Commentary III: Decolonising Critique?' *Progress in Human Geography*. Online First. doi:10.1177/0309132517691629 [Accessed 03.04.2017].

Latour, B. 1988. *The Pasteurization of France*. Trans. A. Sheridan and J. Law. Cambridge, MA: Harvard University Press.

Lazarus, N. 2011. *The Postcolonial Unconscious*. Cambridge: Cambridge University Press.

Maldonado-Torres, N. 2007. 'On the Coloniality of Being: Contributions to the Development of a Concept', *Cultural Studies*, vol. 21, no. 2–3, pp. 240–270.

Maldonado-Torres, N. 2011. 'Thinking Through the Decolonial Turn: Post-Continental Interventions in Theory, Philosophy, and Critique – An Introduction', *Transmodernity: Journal of Peripheral Cultural Production of the Luso-Hispanic World*, vol. 1, no. 2, pp. 8–15.

Mbembe, A. 2001. *On the Postcolony*. Berkeley and London: University of California Press.

Mbembe, A. 2003. 'Necropolitics', Trans. L. Meintjes, *Public Culture*, vol. 15, no. 1, pp. 11–40.

McClintock, A. 1995. *Imperial Leather: Race, Gender, and Sexuality in the Colonial Context*. New York and London: Routledge.

McKittrick, K. 2006. *Demonic Grounds: Black Women and the Cartographies of Struggle*. Minneapolis: University of Minnesota Press.

Mignolo, W. 2010. 'Introduction: Immigrant Consciousness', In: Trans. M. Lugones and J. Price and R. Kusch, ed., *Indigenous Popular Thinking in América*. Durham and London: Duke University Press, pp. xii–liv.

Mignolo, W. 2011. *The Darker Side of Western Modernity: Global Futures, Decolonial Options*. Durham and London: Duke University Press.

Muecke, S. 2004. *Ancient and Modern: Time, Culture, and Indigenous Philosophy*. Sydney: UNSW Press.

Mukherjee, U. 2010. *Postcolonial Environments: Nature, Culture, and the Contemporary Indian Novel in English*. London: Palgrave MacMillan.

Napoleon, V. 2013. 'Thinking About Indigenous Legal Orders', In: R. Provost and C. Sheppard, eds., *Dialogues on Human Rights and Legal Pluralism*. The Netherlands: Springer, pp. 229–245.

Povinelli, E. 1995. 'Do Rocks Listen? The Cultural Politics of Apprehending Australian Aboriginal Law', *American Anthropologist*, vol. 97, no. 3, pp. 505–518.

Povinelli, E. 2016. *Geontologies: A Requiem to Late Liberalism*. Durham and London: Duke University Press.

Pratt, M.L. 2008. 'In the Neo-Colony: Destiny, Destination, and the Traffic of Meaning', In: M. Moraña, E. Dussel and C.A. Jáuregui, eds., *Coloniality at Large: Latin America and the Postcolonial Debate*. Durham and London: Duke University Press, pp. 459–475.

Puig de la Bellacasa, M. 2017. *Matters of Care: Speculative Ethics in More-than-Human Worlds*. Minneapolis: University of Minnesota Press.

Rose, D.B. 2011. *Wild Dog Dreaming: Love and Extinction*. Charlottesville: University of Virginia Press.
Sahlins, M. 2013. *What Kinship Is – And Is Not*. London and Chicago: University of Chicago Press.
Said, E. 2004. *Humanism and Democratic Criticism*. New York: Columbia University Press.
Shilliam, R. 2015. *The Black Pacific: Anti-Colonial Struggles and Oceanic Connections*. London: Bloomsbury Academic.
Skafish, P. 2014. 'Introduction', In: Trans. P. Skafish and E. Vivieros de Castro's, eds., *Cannibal Metaphysics*. Minneapolis: Univocal, pp. 9–33.
Spindler, W. 2016. 'Briefing Notes: 31 May 2016', UNHCR. www.unhcr.org/news/briefing/2016/5/574d564c4/mediterranean-death-toll-soars-204000-cross-first-5-months-2016.html [Accessed 01.06.2016].
Spivak, G.C. 1999. *A Critique of Postcolonial Reason*. Calcutta: Seagull Books.
Star, L. 1995. 'Introduction', In: S. Leigh Star, ed., *Ecologies of Knowledge: Work and Politics in Science and Technology*. Albany: SUNY Press, pp. 1–38.
Stevenson, L. 2014. *Life Beside Itself: Imagining Care in the Canadian Arctic*. Oakland: University of California Press.
Sundberg, J. 2014. 'Decolonizing Posthumanist Geographies', *Cultural Geographies*, vol. 21, no. 1, pp. 33–47.
Todd, Z. 2016. 'An Indigenous Feminist's Take on the Ontological Turn: "Ontology" Is Just Another Word for Colonialism', *Journal of Historical Sociology*, vol. 29, no. 1 (March), pp. 4–22.
Vivieros de Castro, E. 2015. *The Relative Native: Essays on Indigenous Conceptual Worlds*. Chicago: HAU Books.
Wheeler, W. 2104. 'Natural Play, Natural Metaphor, and Natural Stories: Biosemiotic Realism', In: S. Iovino and S. Opperman, eds., *Material Ecocriticism*. Bloomington and Indianapolis: Indiana University Press, pp. 67–79.
Wolfe, C. 2007. 'Bring the Noise: The Parasite and the Multiple Genealogies of Posthumanism', In: M. Serres, ed., *The Parasite*. Minneapolis and London: University of Minnesota Press, 2007, pp. xi–xxvii.
Wynter, S. 1995. '1492: A New World View', In: V.L. Hyatt and R. Nettleford, eds., *Race, Discourse, and the Origin of the Americas: A New World View*. Washington, DC: Smithsonian Institution Press, pp. 5–57.
Wynter, S. 2003. 'Unsettling the Coloniality of Being/Power/Truth/Freedom: Towards the Human, After Man, Its Overrepresentation – An Argument', *CR: The New Centennial Review*, vol. 3, no. 3, pp. 257–337.

1 For new ecologies of thought

Towards decolonising critique

Mark Jackson

. . . a new creation, still damp from the vast firmament.

—J. Baldwin, *Another Country*

Signs don't come from the mind. Rather it is the other way around. What we call mind, or self, is a product of semiosis.

—E. Kohn, *How Forests Think*

Interdependency is not a contract nor a moral ideal – it is a condition.

—M. Puig de la Bellacasa, *Matters of Care*

The truth about stories is, that's all we are.

—Thomas King, *The Truth About Stories*

With an eye to developing their critical engagement and implications, this chapter picks up from the preceding 'Introduction' and returns to examine, in more detail, some of the key distinctions at play in this volume. The chapter's main concern is to draw out the implications for critique of bridging relationships between thinking about coloniality, about the questions the posthuman or more-than-human bring, and so also about ontology.

Conceptual affinities: Que(e)rying the production of the human

A quick glance at the operative terms – 'postcolonialism' and 'posthumanism' – reveals the beginnings of some simple discursive kinships. As critical discourses, each circulates around the epistemic, ethical, and political question of the human. What does 'human' refer to? Who counts as 'human'? What are the histories of this designation? And who speaks for the human? Postcolonialism focuses, typically, on modern colonial and imperialist exclusions from political and epistemic legitimacy, by Eurocentric discourses, of non-European peoples, their ways thinking, and their social and cultural practices. For decades, it has centred its penetrating critiques on the historical and continuing colonial productions of race, reason, and right in the largely European designation of normative, modern humanism.

Postcolonial studies argue that modernity and coloniality are fundamentally imbricated aspects of one-another. Further, the modern and the colonial articulate themselves through violent productions of what is human, what counts as human, and, fundamentally, who speaks for what counts as human. Postcolonial critique, thus, seeks means for identifying these violences and resisting them within the insight that we have always already been postcolonial. In other words, the conditions for resisting colonialism precede, subtend, and extend beyond the emergence of the colonial itself. It is not as though those deemed, or deeming themselves, postcolonial learn anew the conditions of resistance, but that the possibility for, and forces of resistance and opposition co-exist, emerge with, and respond to, any exercise of power. Constitutive means for living otherwise, in the very nature of being designated 'other' by the colonial thus precede the circumstances of their exclusion. Crucially, from these always already subtending capacities constantly at play in the field of difference, the postcolonial creates discursive spaces for the diversity of human thinking and practice to shape how sense and value may be recognised in ways other than hegemonic or normative European inheritances. In doing so, it shows how thought may be performed differently and anew. A specifically postcolonial politics emerges, consequently, from the ethical need to situate thinking, speaking, and doing in relation to alterity: other ways of life; other cultures; and, other worldly traditions.

What we have come to call 'posthumanism', arguably a newer and more recent development within the humanities and social sciences, is also 'other' oriented. The term, vaguer and perhaps more diffuse, nevertheless denotes an epistemic and ontological attention to the status of the human. Equally locative as the postcolonial, the resonance of who and what is human centres less, perhaps, on the politics, identities, and representations of non-European human exclusion, and more on incorporating into the performance of thought 'the material heterogeneity that "the human" founds itself upon' (Wolfe, 2007: xii). Its object is to expand the relationships of material inclusion to what embodies the human through its precedence, subtending, and extension, including organic, inorganic, technological, and bio-engineered forces. Where, indeed, do the boundaries of the human animal lie? Does the porosity of human fleshy encounter include the technical as much as it does the preceding and subtending living, biotic non-humans who share the ecologies that make up our embodiment? Like the postcolonial, the posthuman asks us to consider the possibility that we have never been 'human', if by 'human' we mean, 'the free-agent, the citizen-builder of the Leviathan, the distressing visage of the human person, the other of a relationship, consciousness, the cogito, the hermeneut, the inner self, the thee and thou of dialogue, presence to oneself, intersubjectivity' (Latour, 1993: 136). We – that is, those of us setting agendas for knowledge production by often wrongly speaking for others – have always already been more-than-human, and so never human in the terms circumscribed by the normative purview of Enlightenment modernity. Hence, if the human is always already other-made, and other facing, where is the locus of human agency, its essence or uniqueness? Is it even possible to identify some thing or some process that renders humanity unique? What does the search for such an essence or locus reveal about a politics of identity and those making the designation: 'This is "human"; that is "not human" '?

Just as the postcolonial asks us to recognise how we face and navigate those alterities produced as other, socially and culturally, to accepted centralities of thinking and living, so posthumanism extends critique to questions of embodiment, and likewise, to questions of epistemic accountability and responsibility, material and ecological relation, and the dynamic co-constitution of the many forms that make up the complexity of social life. Geology, digitalism, energy, animality, and the affective all become strange, intimate bedfellows in the posthumanist's critical lexicon. Just as for the postcolonial, what the spaces of 'the social, 'the political', and 'the real' mean, as well as who and what articulates these spaces, is a product of the many human and non-human assemblages that make up our sensible worlds.

Both discourses, of course, are mobilised under the prefix 'post-'. 'Post' overtly operationalises a reflexive orientation of critical analysis and self-examination. If as suggested in the Introduction, the postcolonial is most usefully conceived as a means to think carefully about the extent and effect of one's coloniality (Pratt, 2008: 460), it is a means to think about how modern knowledge production is imbricated and complicit within the material conditions of colonialism. Posthumanism, likewise, may also usefully be thought of as a reflexive means by which to reflect on how it is that 'we' moderns demarcate what 'we' mean by 'the human'. The prefix 'post-', in each instance, denotes a critical reflexivity towards how we constitute what count as meaningful statements about ourselves. Again, 'we' and 'our' in each of these instances encompasses those doing the speaking and determining, those in the centres of epistemic and political privilege delimiting the ostensibly necessary conditions for the possibility of knowledge and value. Each mode of analysis is, therefore, formed within an ethos of self-criticism, much more than either makes any contention that we are at an historical moment beyond the colonial or the human.

Quite the opposite. Neither the postcolonial nor the posthuman in any simple sense refers to what comes after colonialism or humanism. 'Post-' obliges our ethical responsibility to interrogate how contemporary knowledge production is complicit within ongoing colonial oppression. If it is the case that our anthropogenic present is Anthropocenic, that is, 'the Age of Man', then the problematics of modern colonialism manifest in the stresses that some human beings place on each other and the planet have, perhaps, never been more pressing. We are certainly not in an age after colonialism and imperialism; the capitalist and industrial social relations responsible for Anthropogenic forcing are predicated in coloniality. In planetary terms, we are certainly not beyond, at least, the question of what some humans can and are doing materially to ourselves and others. Coloniality and its version of the human are still amongst our most insistent and problematic demands.

Two interrelated social phenomena bear this out. One could begin with numerous others. As I write, hundreds of refugees from former colonies and current neo-colonies drown in the Mediterranean Sea. Dozens drown daily. Sometimes hundreds. The United Nations High Commission for Refugees conservatively estimates that well over 7,000 people have drowned in 18 months from January 2015 to October of 2016 (UNHCR, 2016: n.p.). In 2016, one in 88 people died attempting the crossing; one in 47 died if the crossing was between Libya and

Italy. In 2015, 1,015,078 made the crossing. In 2016, 355,361 made the crossing. The dead, and the hundreds of thousands who make the crossing successfully, are refugees attempting to escape the violence and immiseration wrought by modern colonial legacies of war, systematic exploitation, and imperialism in regions like Western Asia and South Asia (Syria, Iraq, Afghanistan, Kurdistan, Yemen, etc.); North and West Africa and the Western Sahel (Libya, Algeria, Nigeria, Mali, Senegal, etc.); and, the Eastern Sahel and parts of Sub-Saharan Africa (Eritrea, Somalia, Sudan, South Sudan, Chad, Congo, Zimbabwe, etc.). Tens of thousands more wait to attempt similar perilous crossings each week. Across Europe, hundreds of thousands find themselves currently detained at shorelines or between borders in temporary, makeshift accommodation. Millions more are housed in now familiar humanitarian spaces of exception: 'permanently temporary' camps. All manifest the bare life of colonial violence and inequality. Yet, debates amongst the Amer-European privileged rage over who should be more compassionate.

At the same time, soaring temperatures are recorded in a parched pre-monsoon South Asia. Never have historically documented temperatures reached 51 degrees Celsius in the Sub-continent. Increasing temperatures are driven by atmospheric concentrations of carbon dioxide that have reached planetary saturation levels not seen in four million years. The atmosphere, every 'corner' of the earth, is now filled with more than 400 parts per million of CO2. Rising temperatures are also part of a pattern of annual increases in global mean temperature that has seen record highs topped every year (and every month since 2014) in the warmest recorded sixteen consecutive years since 1998 (NOAA, 'State of the Climate Report' November 2015). The trend reflects an average rate of increase in the global annual temperature of 0.07°C (0.13°F) per decade since 1880, and 0.17°C (0.31°F) per decade since 1970 (Ibid.). Today's causes of consistent and persistent planetary warming are human driven. Energetic demands behind growth and development strategies are entwined within modern colonial and contemporary capitalist structures of accumulation, production, and consumption. The material possibilities of efficient energy enable some of us to conceive ourselves and our worlds differently. Debates rage amongst the Amer-European privileged – at least by those not in denial – over who should be more flexible in attempting to ameliorate the carboniferous causes of planetary warming, and, therefore, who should transform their conceptual imaginaries and responsibilities.

Thus, the very political logics of inequality that secure the flight of millions are also part of the same architectures of thought that secure an increase in all-too-human atmospheric energies. Melting glaciers, disappearing islands, accelerating bio-diversity declines, and acidifying oceans are connected – one might even argue directly connected – to the political and economic rationalities that lead to drowning refugees. The material conditions of violence and want that drive people, desperate, into rickety, overloaded vessels are causal drivers of anthropogenic environmental destruction. If responses to these connections are imbricated within the global causes and effects of colonialism, then so are they also imbricated within a politics of planetary degradation. The ecologies of material interaction that enrich a global few at the expense of the many are also the ecologies of interaction that imperil human and more-than-human planetary futures. Crucially, the political and ethical responses to both desperations cannot divorce the legitimacy of the human from either its colonial causes, or its human and more-than-human effects. Our contemporary, Amer-European idea of 'the human', embedded as it is in histories of modern colonial production, representation, and violence, is also constituted by the wider material relationships, energies, and forces that made possible those histories, and the character of their futures.

Yet, it is not the case that our present, the purported Anthropocene, is human tout court. It is, rather, the age of a certain idea of the human, one enrolled within capitalism, industrialisation, colonialism, fantasies of control, and patriarchy. Composed within an epistemic and ethical estrangement from the material heterogeneities that make possible human-ness, these are separations which modern processes also predicate as necessary to protect a particular notion of privilege. As the critic and writer Sylvia Wynter notes,

> [I]t is only the capitalist mode of production that can produce and reproduce our present biocentric, and therefore *economic*, integrating conception of being human. That conception is *the* imperative. This is why, however much abundance we produce, we cannot solve the problem of poverty and hunger. Since the goal of our mode of production is *not* to produce for human beings in general, it's to provide for the material conditions of the existence for the production and reproduction of our present conception of being human: to secure the well-being, therefore, of those of us, the global middle classes, who have managed to attain its ethno-class criterion.
>
> (Wynter, quoted in Scott, 2000: 160, emphases in original)

Wynter terms the securitisation of the well-being of a particular conception of the human 'the overrepresentation of Man': '[T]he Western bourgeois conception of the human, Man, . . . over-represents itself as if it were the human itself' (2003: 260). We are not living in 'the human age', for, as postcolonial and posthuman approaches may reflexively agree, there is no generalisable human characteristic, natural or otherwise – there never has been – to which we may point as the causal locus for our contemporary ills. There are, instead, very many peoples, cultures, ways of life, and modes of living, both within the densest urban agglomerations and within the thickest forests, exposed plains, or relatively unpopulated shores where care and reciprocity foster forms of flourishing that are not destructive to human and ecological futures. They may be in the minority, and they may be tenuous and difficult to isolate in the overwhelming clamour of the present, but their presence, tenacity, and resurgence (see e.g. Coulthard, 2014; Simpson, 2011), give the lie to totalising generalities that lay blame for the Anthropocene at the feet of a bipedal, upright, rogue primate.

The elision of the conditions of our global planetary predicament, caused as it is by Wynter's 'Man' as a generalised idea of human, is itself part of the problem. The Anthropocene, cast as a human problem with technical solutions or apocalyptic reverberations, is the work of 'Man' over-representing itself. Alternatives to the finality of the ostensible Anthropocene have been operative for tens of thousands of years, and they continue to be the case in multiple instances around the world.

Of course, many past and present human societies either destroyed themselves, or are in the process of destroying themselves. But what destroyed them were complex, specific sets of social, political, economic, and geographical circumstances that are not generalisable about humanity as such. Critical responsibility must be to these specificities, and not to false abstractions or universals that enable new means to assert unfoundable and unjust legitimisations. There are, in other words, many imaginaries of the human that are not Anthropocenic, nor Enlightenment humanist, nor capitalist, nor industrialist, etc. Wynter, again, notes:

> there was always a something else besides the dominant cultural logic going on, and that something *else* constituted another – but also *transgressive* – ground of understanding. So that the plot is not simply a sociodemographic location but the site both of a form of life and of possible critical intervention.
> (2000: 164, emphases in original; see also McKittrick, 2006: 123)

Our politics and ethics must be ones that address the multiplicities of these and many, as yet, unthought, unimaginable, interventions and forms of life, as much as they – or even more so – must also now meet a generalising planetary condition of one overbearing form of human expression (for a similar claim, see also Chakrabarty, 2011).

Priorities: Or, what is meant by politics?

At the conjuncture of examining genealogies of the human multiple, the postcolonial and the posthuman may agree and overlap. Decolonising our thinking about the human means, necessarily, engaging with others' precepts of the human, which either are or also share much with many posthumanist approaches. The implications for their agreement have, however, given rise to divisions and differences across the two diverse discourses. And, it is here in accounting for the seemingly incommensurable implications, that the attempt to characterise a shared conversation, or shared ambition, becomes much, much more complicated. A simple consideration illustrates a central tension.

I have invoked, thus far, Bruno Latour's now famous critical quip, emboldened in his title *We Have Never Been Modern* (1993) to suggest that both the immanent conditions of the postcolonial and the posthuman subtend the emergence of an Enlightenment story of the human that now holds normative epistemic and political sway. Counter-intuitively, Latour argues, we have never been modern because we have always been constituted by the very material complexities and inter-weavings that modernity repeatedly denies: nature is not separate from culture; modern humans are not removed from nature. I have extended the analogy to the recognition, made by others, that, as the postmodern precedes the modern (e.g. Lyotard, 1984), and as the posthuman precedes the human (e.g. Latour, 1993; Serres, 2007), so the postcolonial precedes the colonial (e.g. Pratt, 2008). It is also in this sense that, I argue, the postcolonial shares many conceptual affinities with decolonial refusal. The postcolonial is but one of the critical means to enable thinking about what is necessarily actively excluded, or ignored, by colonialism.

Via the reflexive work of critique, we may recoup and rethink these immanent exclusions in considering future means of flourishing together. Just as we may rhetorically argue that we have never been modern, so we might say that we have always been postcolonial because the conditions of postcoloniality precede the exercises of colonial reduction.

Of course, it does not follow from this that we have never been colonial! Nor that the many ravages of modernity are figments of political ideology. Such claims are precisely the apolitical and revisionist attitudes critique seeks to expose and transform. We must continue to mobilise accepted terms of political critique to challenge the many past and present injustices of colonialism: racism, genocide, inequality, exploitation, over-accumulation, cultural and ecological destruction, etc. The trouble lies in how we continue to do this and at the same time extend the boundaries, location, and resonances of human agency.

The problem is captured by the postcolonial historian Dipesh Chakrabarty's identification of the tension for political modernity between the universal and the particular. On the one hand, Chakrabarty argues, we need an analytical heritage that helps us to universalise and produce critical readings of social injustice. On the other hand, we also need to attend to how these universal claims depend on excluding local and specific ways of understanding and living differently (2000: 254–255). The result, for him, is an irresolvable tension, but a tension in which there is always an opening towards the critical spatialisation of reflexive critique.

The legitimacy of this critical space is framed, however, within a recognition of epistemic limits as defined by a critical, Enlightenment humanism. As intimated in the introduction, decentring a Eurocentric epistemological model of human exceptionalism raises possibilities for framing alterity, and the potential for politics and ethics, otherwise. Decentring epistemic limits as the ground of critique raises serious questions for how we account for the production of meaning, what counts as semiosis, who or what participates in the work of signification, and what the boundaries are around how we know sociality and political participation. One of the most important questions is how we privilege the grounds of political legitimacy, a problem Chakrabarty addresses with his perpetual tension between the needs of the universal and the needs of the particular.

What happens, however, if we shift how the universal and how the particular are meaningful categories: universal or particular from which specific viewpoint? For whom? Cui bono (who benefits)?

If we prioritise a posthuman redescription of the human as an ecology of diverse relational processes that enrol the technical, the geologic, the bacterial, and the chemical, do we then risk losing sight of the important structural injustices that have set in place the historical legacies of untold suffering in our cities, on our shorelines, and across our open seas? Some say we do (see e.g. Braun, 2015; Gilroy, 2015) and that we must be careful not to lose sight of the intellectual and political priorities at the heart of critique in the humanities and the social sciences. On the other hand, if we prioritise the critical concerns of humanity through known and responsive methodological strategies that target politics of representation, identity, economic distribution, production, and consumption, are we not also invoking a false generalisation and an imposed conception of the human that attempts to speak for others, many of whom, ironically, operate with quite different understandings of what constitutes the human? Do we not need, instead, to problematise the Enlightenment production of political narratives as much as their co-emergent narratives of humanism? Some argue this is a harder task (see e.g. Blaser, 2013a; de la Cadena, 2010; Descola, 2013a, 2013b; Glissant, 1997; Hunt, 2014; Kohn, 2013; Massumi, 2014; Povinelli, 2002, 2011; Sundberg, 2014; Todd, 2016; Watts, 2013), but one that is necessary if we want to avoid recolonising the horizons of critical possibility with 'the imperial concept of the Human' inherited from the universality of a European worldview (Mignolo, 2011: 242).

In other words, where the discourses of the postcolonial and the posthuman often do not see eye to eye is, in each instance, around what counts as political, and where political responsibility first lies. For postcolonial theory, political responsibility has typically been imagined and described, as indicated at the beginning, as people focused, because it plays itself through an epistemologically grounded notion of critique that emerges from an attention to human, cognitive fallibility, and the epistemic indeterminacies of representation. This has gone on to play itself out in important studies of identity, race, nationalism, cultural hybridity, and diaspora. And we do very much need to recognise, critique, and seek to overcome the many continuities of colonial epistemic and physical violence that continue to affect human life on the planet.

But, at the same time, we also need to recognise – and this is Latour's point, as much as it is those who seek to decolonise him (see Sundberg, 2014; Todd, 2016) – that what is often taken for granted as political, as just response, or as meriting moral legitimacy is itself a product of the same modern exclusions it seeks to condemn and overcome. As suggested above, what is important to recognise about posthumanism is that it is also seeking to rethink and redefine the terms of the political on grounds other than assumed epistemic models of human exceptionalism, as much as it is also about the descriptive ecologies of material imbrication in terms of which the human founds itself. Whereas human agency is wrapped up within the agencies of many other vitalities, so what counts as political legitimacy must also change. It is not simply the case that posthuman critical thought is 'a-' or 'post-political', but that the terms of how we define, perform, and enact the responsibility of being-with others (i.e. being political) also need to change. As Braidotti writes, posthuman critical thought recasts political agency 'in the direction of transversal alliances and relational ontology' (2016: 24).

The now well-known development of 'cosmopolitics' (see e.g. Stengers, 2010) goes some way to expanding this landscape of intra-relational legitimacy. Postcolonial critique has often either been slow to take up these more radical forms of material and agential re-orientation proposed by some forms of posthuman analysis, or it has, in a more reactionary vein, actively dismissed posthumanist critique as anti-humanist, overly theoretical, navel gazing that ignores the very real political primacies of human suffering that need to be privileged by critically responsible and socially engaged praxis (see e.g. Gilroy, 2015).

There is a deeper reason, however, for worrying about human political priorities and the turn to posthuman descriptions as problematically anti-humanist. The reason lies in epistemology: how we define the limits and purview of what counts as human cognition, and therefore the limits of human knowledge. In the next section, I turn to the link between epistemology and the inherited conceptual categories that define the recognition of the political in the exercise of anti-colonial practices.

Politics of difference: Eurocentrism is epistemological, not geographical

I argue that we must open the causal genealogy for divergent articulations of politics further than simply a debate about how anthropocentric priorities shape the conception of the political. At the crux of the worry around primacy lies an important epistemological dispute. The dispute circulates around three interrelated philosophical problems: i) what is knowable and sayable by a human subject?; ii) how do limits to knowability shape what counts as critique?; and, iii) which conceptual genealogies are necessary to argue for characterisations of human subjectivity?

A key component of the debate around political priorities emerges from the epistemological commitment to what is knowable and sayable. If we can only make epistemic claims about, and from, human experience (individually and collectively manifest), and human experience itself is defined as conscious self-awareness about how the structures of thought and language work, then both epistemic claims and the domains over which those claims have purchase (ethics, politics, aesthetics, etc.) are necessarily limited to how we define human exceptionalism. Politics, ethics, and aesthetics, let alone more general relations like society, language, symbolism, rationality, etc., would not be attributable to those creatures not deemed linguistic or self-reflexive. It was in precisely these terms that Kant, who argued for a 'Copernican Turn' to the critical capacities of conscious self-awareness as the purview of human reason, also argued that non-Northern Europeans were neither rational, nor, in the case of Africans or North Americans, fully human, and so not capable of thinking. Because they were not able to exercise themselves in rational self-reflection, i.e. critique (see, Kant, 1960: 97–116; Dabashi, 2015: xix–xx, 258–260; Jackson, 2016), Kant's 'Copernican Turn' enshrined human exceptionalism and human reason within a racist account of the heights of phenomenal reflexion. Further, the limits of thought and the sayable were defined by a critical self-understanding that bracketed the natural world, about which it was possible to have knowledge only through the phenomenal structures of thought. Kant, and the post-Kantian tradition, thus protects, as a feature of critique and therefore as a function of epistemological necessity, a fundamental distinction between nature and culture. Definitions of the human, and, of course, the reach of political, ethical, and aesthetic agency, are limited to what can be reasonably said of the knowable: self-aware subjects separated in their critical awareness (phenomena/culture) from the world in and of itself (*noumena*/nature).

One can see from this brief sketch of the Kantian epistemological (and thus political) limits of the sayable, and the notion of the human that emerges from it, an account that has been fundamental to critical thinking in the social sciences and humanities since the late eighteenth century, and that continues to be the touchstone for how critical reason is understood and defined today. To argue against it, one can either challenge the fact that the rational human is limited to a northern

European character, all the while holding on to Kant's deeply persuasive (but in my view ultimately incorrect) distinction between the phenomenal structures of thought and the world itself, or one could challenge this distinction itself, and problematise the nature-culture distinction upon which it depends. The former path expands the epistemic landscape beyond the gendered, racist characterisation invoked by Kant. The latter requires a more fundamental reappraisal of human experience, ontology, and the nature of epistemology. Under the influence of post-structuralism, historical materialism, hermeneutic analysis, psychoanalysis, and everyday political struggles with the rubrics of nationalism and independence, postcolonial theory has largely attempted the former task. Posthumanism, on the other hand, has attempted the latter, no less influenced by post-structuralist thinking. It has done so in ways that have sought a more radical challenge to the modernist critical phenomenalism that has been so influential from Kant onwards, but almost exclusively without insights drawn from non-European forms of thought. The consequence of the distinction is a division around the terms of epistemology, the reasons for human cognitive delimitation, and what is enabled politically by such epistemological commitments.

In order to unpack the terms of this epistemological argument for the political, we need to contextualise (all too) briefly the broad conceptual landscape of postcolonial theory and critique.

The discursive apparatus of postcolonial critique emerged from roughly two interrelated human efforts: historically and geographically diverse anti-colonial struggles for liberation from occupation and violent oppression; and the work of representing culture and difference through diverse texts (oral, written, visual, and architectural), many of which, of course, pre-dated occupation, but that were operationalised in resistance to colonialism as means of social expression, description, explanation, and, more problematically perhaps, cohesion. At the time of these anti-colonial struggles, global narratives of political and social organisation were increasingly imagined within the modern furnace of the nation-state, together with their commensurate Eurocentric (meta-) narratives of individualism, freedom, citizenship, the public and private spheres, property, and nationalism. Anti-colonial struggles were inspired by these modern social principles and, importantly, the contradiction of European colonial presence invoking the arts of statecraft in order to legitimise colonisation and protect colonial authority, and yet withholding the same rights and responsibilities from their colonial subjects. As Chakrabarty writes, the result of the contradiction is the inevitable failure of modern colonialism.

> The colonial state, we must remember, was not just the agency that brought the modular forms of the modern state to the colonies; it was also an agency that was destined never to fulfil the normalising mission of the modern state because the premise of its power was a rule of colonial difference, namely, the preservation of the alienness of the ruling group.
>
> (2000: 10)

At the same time, interpretation and criticism of the social practices and texts used to mobilise anti-colonial struggles revealed differences in how languages, social hierarchies, family structures, economies, and cultures articulated political legitimacy. Difference emerged as a key principle in articulating cultural rights, and contrasted those universalisms that instead came to reflect imposed colonial orders. The impossibility of grounding unitary narratives in certain epistemological terms was reflected in the resistance of texts to totalising or fixed interpretations, and further reflected the underlying commitment to the subjective, epistemic limits of the sayable. Texts, languages, systems of meaning, and people all resisted simple or definite characterisation and gave rise to the theoretical approaches that explored the fundamental and incessant contingencies of language, texts, and meaning.

The experience of these limits to meaning was also mirrored in the situated experiences of diverse subjects articulating their consciousness of their differences. Experience, like texts, revealed its own contingencies, its own limits, and its own histories. Phenomenological characterisations of human cognition, the positional limits to rational enquiry, and the historical contexts for speaking, all shaped an approach to knowledge production that favoured reflexively negotiating textually bound and historically situated epistemic representations of situated sociocultural experience. The postcolonial theoretical approaches that came to prominence in the late nineteenth and mid to late twentieth centuries reflected this phenomenological, textual, and historicist tradition of reflexive knowledge production by socially contingent self-conscious agents inaugurated from Kant. Political principles of individual freedom, citizenship differentially responsible to public and private spheres, and social identity reflecting core tenets of productive self-expression consequently became a central part of the motivating animus legitimising anti-colonial struggle. European philosophical inheritance, accordingly, was partly answerable to the principles that authorised freedom struggles, and a key point of reference. Characterisations of self-conscious limits to human cognition stretched back at least to Kant, phenomenological accounts of history and subject production to legacies derived from Hegel, Husserl, and Merleau-Ponty, and the material production of class and economic inequality under colonial capitalism to Marx. Legacies of political articulation, then, navigated both the inherited frameworks of subjective self-awareness from phenomenology and reflexive Kantian critique, and the playing out of rational self-awareness as an epistemic problem within the political narratives of the nation-state, rights, citizenship, and common cause, all of which rallied as the basis for humanism. What came to count, then, for human political agency was framed within a Eurocentric epistemological tradition that inscribed the human as a particular kind of limited, yet rational, self-conscious, language bound thing.

Phenomenological and textual limits to critical self-awareness also reflected a further important conceptual inheritance. Since the Enlightenment move to secularise the validity of knowledge production, humanism has been concerned with two broad formulations. The first formulation is to aspire to find some feature of human life that sits apart from nature or the supernatural. The second formulation, in light of the first, is to show how human uniqueness may be found wherever there are people, that is, that specific uniqueness which is a general characteristic of all human agents (see Bilgrami, 2004: x). The modern European philosophical tradition since Kant has tended to locate these poles of human uniqueness from nature in cultural capacities for representational self-reflection, self-consciousness, and the exercise of reason. Reason, however, as Kant famously argued, could legislate only over itself. Human cognition and self-consciousness was separated from the natural world, and while we could know things about the world, we could experience only the phenomenal representation of this world in our particular human cognitions. We were limited then to our cultural productions, some of which were more valuable than others (i.e. European ones). Some peoples' rational words represented the world, but they were not the world or the material realm itself. Further, only (some) humans used words, signs, and language. Culture was the domain, navigable and knowable by reason but fundamentally separate from non-human nature, in which we traverse phenomenal awareness. Distinctions and capacities for rational self-reflection and distance from nature then became arbiters for human description and commonality, and, of course, one of the key means by which to exclude others from legitimate participation.

Advocating for participation on the grounds of rational thought, within a context of opposition to imperial violence and colonial oppression, often meant that anti-colonial struggles needed to inherit the terms of debate from the epistemological contexts within which they fought. This was because the tools for modern independence and freedom struggles were the terms of freedom (liberté), rationality (rationalité), and equality (égalité) hypocritically withheld from colonial subjects by colonising structures. And, further, because brotherhood (fraternité) was operationalised through inherited imaginaries of the nation-state, the conceptual touchstones for imagining and mobilising political agency were also often inherited from Europe. Independence and autonomy meant not being dependent on the authority of others, and it meant being free to take responsibility, through self-aware reason, for oneself and one's wider community. It meant, in the political words of Kant, being 'Enlightened' (Kant, 1996: 58–64). Maturity, as Kant expressed in his famous essay in response to the question, 'what is enlightenment?' is acting within each moment of the present to throw off self-imposed forms of dependence (scared texts, monarchic rule, tradition, etc.) in the exercise of self-reliant reason. In nineteenth- and twentieth-century struggles, independence and freedom took the form of individually inflected liberalism, or it took the form of collective class struggle. What is important for our context here is the fact that what counted as the political or being politically aware in each instance, individual or collective, was inherited within a metaphysical and epistemological framework mobilised under the aegis of a European model of human cognitive agency.

And, yet at the same time, accounts of the epistemic subject sought to be distinctive. They drew on their local histories and contexts to imagine communities of distinct association, but hybridised these non-Western imaginaries through the political frameworks inherited as modular forms from Europe. The postcolonial political imaginary, as Chatterjee writes, is often caught, as such, in an asymmetrical modern tension between 'the Western claiming to be the universal and the national aspiring to be different (1997: 285).' How the aspiration to be different translates the terms of the epistemological and the political away from the colonisation of the imagination inherited in the modular forms of Europe (Chatterjee,

1999: 5) is a matter of negotiating between what he terms the 'spiritual' (Ibid.: 10) grounds of manifest difference (e.g. oral traditions, literature, the arts, vernaculars, stories, sacred practices, family rites, food practices, everyday forms of uncolonised life, etc.), and the inherited categories through which the exercises of national aspiration build new modernities. It is perhaps in unpicking what is meant by this rather vague and unsatisfactory reference to the 'spiritual' that we might be able to decolonise the term, and so extend the critical insight that the everyday materialities of lived practice deemed 'different' and so the source of the 'spiritual', is actually not 'spiritual' at all, but the sites of different semiotic productions. It is their incommensurability within an Enlightenment framework of critique, and so their unhelpful label 'spiritual', that renders them potential. They, perhaps, may be better understood as 'cosmopolitical'.

Provincialising the human

I have made the detour through a postcolonial critique of the nation-state to highlight analogous concerns over inherited, modular forms of imagining the human. Indeed, the detour is less analogical than it is intimately connected. For my argument is that the importance of defining political critique in postcolonial terms has both drawn from, and reinforced, certain modular assumptions about human epistemic agency. What counts as the political has also shaped what counts, and what is recognised as human by postcolonial critique. Just as postcolonial critique needs to challenge, as Chatterjee does, the failure of political theory to account for non-inherited modular forms of modern political association, and so derive new ones based on non-European modes of spiritual and material association, our theoretical language also needs to challenge the modular forms of humanist epistemic agency also inherited as the basis for political legitimacy.

If we examine the terms by which non-European 'Others' describe their embodiment, agency, and relation to human and non-human forms, we can see very different modalities – modularities – at play than those inherited from Europe. These modalities challenge even the basic assumptions at work in Eurocentric humanisms, assumptions like: nature-culture distinctions; reason and language as a basis for human exceptionalism; living and non-living differences; mind and world distinctions; image and matter; dream and reality. Examples are, of course, as numerous as the 'infinite incommensurabilities through which we struggle – perennially, precariously, but unavoidably – to "world the earth" in order to live within our different senses of ontic belonging' (Chakrabarty, 2000: 254). I indicate a few here from a range of sources, including indigenous scholars and scholarships, anthropology, religious studies, and politics. They include, for instance, indigenous accounts of the ontological inseparability of thought and matter (e.g. Atleo, 2004, 2011; Glowczewski, 2015, 2016; Graham, 2008; Kohn, 2013; Magowan, 2001a, 2001b; Muecke, 2004); accounts of personhood, law, and non-linguistic communication extending across, and disrupting, traditional Eurocentric confines of the living and non-living, animal as well as mineral (e.g. Borrows, 2010; Hallowell, 1975; Povinelli, 1995; Bird Rose, 1996); inversions of the one nature/many cultures divisions assumed by a European tradition, in favour of 'multinaturalist' perspectives (e.g. Descola, 2013a, 2013b; Vivieros de Castro, 2014, 2015); commensurate attempts to account for pluriversal ontologies and the cosmopolitics of multiple incommensurable worlds (e.g. Blaser, 2013a; de la Cadena, 2010, 2015); or, the alignment of purposive human ideals with an ontological commitment to the unity of cosmos and sacred practice, spirit and matter (e.g. Garfield, 2015; Shari'ati, 1979; Kopenawa and Albert, 2013; Singh, 2015).

Appealing with these, here simply indicative, references – and there are many, many more – invokes the importance of, in Chakrabarty's germinal words, 'provincializing Europe' (2000). Chakrabarty, to remind ourselves, details in his book of the same name a tensile critique of the inherited historicist time of political modernity. He emphasises the importance of drawing on both the indispensability and inadequacy of European thought 'to think through the experiences of political modernity in non-Western nations' (2000:17). Provincialising Europe, a conceptual project about history and historiography, aims to read excluded and subaltern histories alongside European to challenge the historicist production of Europe and its thought as the central template of political modernity. The book poses the self-consciously impossible 'question of how we create conjoined and disjunctive genealogies . . . of political modernity as we contemplate the necessarily fragmentary histories of human belonging that never constitute a one or a whole' (Ibid.: 255). In other words, provincialising as a method reads the legacies of Europe against the grain, but leaves unresolved the tension of a necessary universalist history inherited from the West (what he calls 'History 1') set against 'History 2', those infinite ontic multiplicities that 'in practice always modify and interrupt the totalising thrusts of History 1' (p. 254). Accounting for the diverse experiences of political modernity outside – and, today, within – the geographies of the West requires necessarily reading subaltern, local perspectives alongside analytic, universalising stories. But, for Chakrabarty, the political tension of the local and the universal is not one that it is possible to overcome. Provincialising simply writes 'into the history of modernity the ambivalences, contradictions, the use of force, and the tragedies and ironies that attend it' (p. 43); provincialising is, then, a 'politics of despair' (p. 45) which recognises *both* gratitude for the critiques of injustice made possible by a universalising political modernity, *and* the inescapable predicament that such critiques also 'assimilate to the projects of the modern state [i.e. politics] all other possibilities of human solidarity'(Ibid., emphasis added). Under this rubric, does 'human solidarity' presuppose an Amer-European nature-culture distinction? I think it does. In what sense, then, is this radically decolonising?

Chakrabarty's aporetic argument circulates around the political and epistemic tension between co-emergent modern drives, on the one hand, for the analytic, and, on the other for the interpretive. The analytic, figured for him through the work of Marx, articulates a universal and necessary history whose abstract intent is to ground solidarity in socially just practices of rational order and rule: citizenship, the state, civil society, the public sphere, etc., whether these be in the liberal capitalist form or in their more critical political economic forms. The interpretive, figured through the work of Heidegger and, of course, the non-European narratives of plural history and alterity, situates local particular expressions of being human that contest and disrupt universalising, analytic tropes. The latter, disrupting as it does the former, 'makes it impossible to sum up a present through any totalising principle' (Ibid.: 251). Yet, the ideals of the totalising principle are also necessary to energise the promise of a universally just humanity. Both notions pivot, however, around the definition of political responsibility for an idea of the human inherited as the agent to which politics and ethics, under the veil of Europe, are cast. The first, the analytical, is a characterisation of the universal responsibility of generalised humanity to itself. The second is about the interpretive limits of and for the human; the human is delimited only in its interpretive capacities and to its local historical contexts. What this necessary duality between analysis and interpretation sets up, for Chakrabarty, is, as I have noted already, an inevitable politics of despair and impossibility, one that must negotiate its tension honestly, but not with any conceptual hubris that the tension can be overcome, hence the argument being aporetic. But, for Chakrabarty, the tension between the analytic and the interpretive is framed within a vision of the human as epistemically torn between abstract, non-natural, rational (perhaps even idealising) capacities to universalise, and particular situations to which it is limited in its contexts, histories, languages, etc. The tension is the aporia: therein emerges a human responsibility to universalise, but also to take into account the critical responsibility of its epistemic, interpretive limits. These limits are conceived as human because of their epistemic nature, because of the historical subject needing to be aware of its critical, impossible tensions.

Must solidarity, however, be a necessarily epistemological question? Do we need to know the other, or, more importantly, do we need to know the impossibility of our knowledge of the other – in order to have political or ethical solidarity with its flourishing? Could we not found critique in different ways, as embodied, for instance, in shared affects? We may wonder whether the posthuman query of epistemological centrality, read alongside postcolonial and decolonising accounts of ontological worlding in an effort to provincialise normative, Eurocentric accounts of the human, might also work to challenge the logic of insurmountable tension that Chakrabarty diagnoses for our political present. In other words, what a decolonising posthuman perspective might be able to do is question the production of the human so that solidarity is not framed as emergent from the recognition that we are epistemically, and so impossibly, torn between the universal and the local. Rather, we could foster sensibilities of solidarity that emerge from the multiplicities that make up the relational possibilities of even asking the question of the human and thinking about its being-with others. These could include sensibilities of affect, sensibility, and emotion. We might understand these solidarities or shared embodiments as 'boundary practices'[1] that affect different living and non-living processes: attraction, repulsion, hunger, desire, pleasure, fear, care, reciprocity, sharing, greed, touch, temporality, etc. Boundary practices are those that are recognised across different interactive yet incommensurable worldings and that, though different, maintain enough identity that they can be understood between the different viewpoints. An example might be something like human parents foraging and sharing their food with their helpless young, and avian parents foraging and sharing their food with their helpless young. Important with the example of boundary practices is the adaptive recognition of sharing as a means to build solidarity and reciprocity between avian and human worlds. The question of whether avian life experiences sharing in a way similar to how humans experience sharing is moot, and beside the point of building a necessary solidarity between ecologically connected life-forms. Both bird and human may depend, for instance, on the health of the sea and its fish for their food. Solidarity between the two depends, in this instance, not on shared senses of knowing, but on the health of the sea as a site for shared food as embodied in the recognition of the other's sharing and flourishing. Reading the human's and the bird's flourishing through the assemblage of the sea or the forest connects them in ways that arguments about the difference between the two's epistemic capacities never could connect them, and that, not only being unanswerable (we will never know what it is like to be a bird or a bat), are irrelevant to possible solidarities around mutual flourishing. Yanomami shaman, Davi Kopenawa articulates such a relationship of cross-species kinship solidarity.

> Animals are just like humans. We eat our fill when our gardens are full of bananas and rasa si palm fruit and they eat their fill when the fruit of the forest trees is abundant. This is their food as it is ours, for the animals we hunt are the ghosts of our ancestors transformed into game in the beginning of time. Another group of the first human beings were hurled into the underworld, but these remained in the forest in which we too were created. We refer to them as game, but in fact we are all human. It is so.
>
> (Kopenawa and Albert, 2013: 149)

What a decolonising posthuman and postcolonial challenge to the boundaries of human responsibility thus provides, that a historically focused provincialising rhetoric about political modernity cannot (despite its undoubted importance for theorising political modernity) is the capacity to pluralise, and radically provincialise beyond Eurocentric productions of human critique, notions of intra-relational material solidarity to which we are already bound and enabled – the forest, the sea, other creatures. Such relations even intimate towards 'a modern form of the sacred' (Glissant, 1997: 16; see also, Alexander, 2005; Puig de la Bellacasa, 2015). Chakrabarty's concern in Provincialising Europe is not with this more radical posthuman attention, because he is interested to critique modern human political historicisms.

We must decolonise our politics further than the political imaginaries of modernity. This includes rearticulating the notion of the human through both more-than-human approaches, but also approaches that are incommensurable with either the modern human or the modern more-than-human. Numerous indigenous perspectives have always already been doing this, and continue to do so today. If history is also more than human history, if our history is also a function of the geo-historical and cosmo-ecological forces that shape the possibility of human and wider ecological imaginaries, geographical as well as political, cultural as well as economic (e.g. Clark, 2011; Dixon, 2009; Dixon, et. al., 2012; Povinelli, 2016; Yusoff, 2013, 2014), then human history is part of an imminent, enfolded cosmology whose intimate boundaries are felt as sensibilities across human and more-than-human worlds, and across incommensurate, sometimes inarticulable, worlds. Could we not use embodied, non-epistemic intimacies to forge solidarities of relation that do not begin with a model for the limits of self-understanding, within which the problem of the human is considered unique? Can we not see in others' relational capacities possibilities that emerge from their everyday interactions as much as they do for us? Could we not build on these so-called alternative sensibilities of embodiment for enabling less destructive forms of political and ethical solidarity? Doing so means beginning with different descriptions and assumptions of the human than those we inherit from the modernist syntax of traditional critical lexicons.

Indeed, as Kopenawa intimates above, alter-human solidarities and accountabilities have been recognised and described for millennia by subaltern and indigenous worldviews. Three short examples, amongst innumerable others, suggest themselves from three disparate geographies: isaak, the Nuu-chah-nulth word-as-practice meaning 'respect' from the North-West Coast of North America (Atleo, 2004: 15); dreams, which connect the material continuities of Amazonian human and non-human socialities (Kohn, 2013: 155; Descola, 1996: 103–119); and the cosmo-ontological association of 'people-as-places' in Yolngu (Northern Australia) philosophies, wherein kinetic resonance in song and landscape combine, in the colour, shape, and sound of place, as the co-substantive essence of ancestor and human (Magowan, 2001a: 23; see also Glowczewski, 2015, 2016).

The Nuu-chah-nulth scholar E. Richard Atleo writes that Isaak is a conceptual practice 'not of human origin' (2004: 16) that necessitates, in its conditions of practice, a recognition that all of creation is one. Because the recognition 'is extended to all life forms', isaak creates, in practiced awareness, 'a climate or environment for the practitioner where communication with other life forms is possible' (Ibid.). Communication across life forms is possible because the conditions for mutuality do not come from a human centred epistemology, with mind existing in humans, on the one hand, and, on the other, with matter and others either dumb or unmindful, or with the terms of communication understood solely as linguistic cognition. Rather, mutuality derives from a common conscious participation amidst ecologies, of which humans are a part. This is nothing new nor is it surprising. Humans communicate with non-humans all the time. Far beyond asking our dog to come, or recognising her contentment and sense of safety as she sleeps, we learn to respond to what birds and plants tell us of their health; we change soil compositions in response to signs worms leave or do not leave; we select and propagate plant species based on size and colour of fruit or flower; plants diversify and extend their genes through the interactive relationships afforded in our interactions with them – in other words, we are their agents, rather than them simply ours (Pollan, 2002).

This collapse of polarities and distinctions is reflected in a second example: dreams connecting the material socialities of humans and non-humans amongst the Quichua speaking Runa in Ecuador's Upper Amazon. For these forest dwellers, parallel human and non-human experiential worlds communicate through dreams (Kohn, 2013: 155), with semantic commensurability arising from a shared culture or '"I perspective" that all selves, human and non-human alike, inhabit' (Ibid.: 156). It is argued that the Western metaphysic that reads nature as one and culture as many is inverted in this Amazonian worldview (Vivieros de Castro, 1998, 2015). Instead, culture is one, of which many natures – people, jaguars, peccaries, monkeys, plants, etc. – participate differentially: many natures, one culture. Moreover, the claim is that this inversion from the West's enlightenment constructs of one nature/many cultures, wild/domesticated, human/non-human is not unique to the Amazon.

> It seems clear that in many regions of our planet, contrasting perceptions of beings and places, depending on their greater or lesser proximity to the world of humans, coincides hardly at all with the body of meanings and values that, in the West, have become attached to the two poles represented by 'the wild' and the 'domesticated'.
>
> (Descola, 2013a: 48)

The argument is, then, that 'understanding the relations that human beings establish between one another and with non-humans cannot be based upon a cosmology and an ontology that are as closely bound as ours [i.e. the West's] are to one particular context' (Ibid.: xviii). The particular context I am emphasising here is the Enlightenment epistemic situation of centring a particular idea of human cognition as the locus of human understanding and the means of self-reflection. Such a situated knowledge, one built on a cosmology in which the Copernican Turn spirals into an individualised human reason cognitively separated from mute matter and dumb nature, a human ontology that bears distinct traces of its Judeo-Christian genealogy, is very different from the third example, that is drawn from Yolngu cosmology, which 'places an implacable insistence that [Yolngu personal] being is distributed through the world' (Verran, 2015: 499). For the Yolngu, ancestral immanence and non-human alterity are constituted through the kinetic agencies that connect landscapes, seascapes, and songs. Songs and dancing are part of the resonances by which people enact the ecologies, the kinship genealogies, that make them possible: past, present, and future (Magowan, 2001b). Kinetic ritual is part of the 'constellation of intersubjective relations, not only between people but also with all the elements of the natural, social and technical environment' (Glowczewski, 2011: 11).

The point of the above examples is not to invoke multiple instances in an effort to prove, should enough examples be marshalled, a stronger argument for indigenous cosmologies than Western, Enlightenment ones. The point is twofold. First, it is to suggest that that the idea of 'human' cognitive reflexivity, often mobilised as epistemic critique, emerges from a particular, historical cosmology that emerges from European modernity. There are, also, multiple imaginaries of the human, many radically different from assumed epistemic forms of embodiment which entail their own forms of reflexivity, self-awareness, and self-questioning. One must not assume that in speaking of critique, often in very laudable ways against social, economic and environmental injustices, that one is also speaking with others' conceptions of reflexivity, self-awareness, or self-questioning.

Often we are not. We are often precisely 'perpetuating the exploitation of [other] peoples' (Todd, 2016: 16) and the non-human others we attempt to include in descriptive pluralisms.

Institutionalised Western posthumanisms are, of course, as guilty of this perpetuation as are traditional humanist meta-narratives. Colonial Eurocentrisms begin, as Mignolo writes (2011: 19) with epistemologies that deny the material relationships of place and ecology. He enshrines this critique in the phrase, 'I am where I think' (Ibid.: 117). Mignolo challenges the 'cogito ergo sum' for neglecting the ecological conditions that frame the possibility of cognitive awareness and eventual self-knowledge. In wondering whether one is deceived, even by an evil demon, the world and what makes such thinking materially possible are exempted, much to the detriment of the story we tell about thinking itself. Why, however, limit thought and writing about thought as self-awareness to a model that emerges from a particular moment in the history of European colonialism? Why limit it to the narratives about human and more-than-human interactions derived from this limited historiography? As the literary scholar Hamid Dabashi writes,

> Instead of the habitual mise-en-scène within which we [non-European thinkers] talk to them [European thinkers] as they talk to themselves, we need to change the whole architechtonics of this interlocution altogether, and address the only interlocutor that has been left to all of us: a fractured and self-destructing world. The European philosophers can only overcome what they consider their 'crisis of the subject' by avoiding the Kantian cul-de-sac that defines the knowing subject as the European knowing subject and designates us – the rest of the world – as their knowable realm. We are no longer (if we ever were) knowable to that European knowing subject. Because we no longer exist as they had fathomed in their process of self-centering subjection, so have they ceased to exist as our or any other kind of knowing subject. They don't and cannot know anymore.
>
> (2015: 22–23)

Debashi's is, thus, a refusal to translate non-European thinking into the terms of the European philosopher in order to make it commensurable with a self-referential epistemological 'architectonic' that conditions what is politically and ethically sayable.

It is also the case, second, that posthuman discourses derived within the institutional apparatuses and protocols of the intellectual traditions animating most, if not all, modern universities actively ignore numerous traditions of subaltern and indigenous critical scholarship (de la Cadena, 2015; Hunt, 2014; Muecke, 2004; Sundberg, 2014; Todd, 2016). For instance, indigenous critical philosophies have for decades sought to decolonise epistemological and ethical narratives via perspectives commensurable with what is now termed 'cosmopolitics'. These approaches have neither mobilised themselves in normative Eurocentric terms, nor in authorial genealogies frequently invoked by the Amer-European academy. Or, more worryingly, if perhaps more accurately, they have simply been ignored because they have been produced by non-white critics in languages other than those six vernaculars at the modern/colonial core of the Amer-European academy: English, French, Spanish, German, Portuguese, and Italian (see Mignolo, 2011: 19). In particular, indigenous philosophy and the scholarships mobilised under the aegis of critical indigenous studies (see e.g. Alfred, 2005; Deloria, 1999; de la Cadena and Starn, 2007; Simpson and Smith, 2014; Tuhiwai Smith, 2012), in addition to the wider purviews of African philosophy and critique (see e.g. Eze, 1997; Mudimbe, 1988, 1994; Thiong'o, 1986), Caribbean philosophy and critique (see e.g. Césaire, 2000; Glissant, 1997; Henry, 2000; Nesbitt, 2013), and, Southern American philosophy and critique (see e.g. Dussel, 1995; Kusch, 2010; Mendieta, 2002; Mignolo, 2011; Rivera Cusicanqui, 2012) have long pre-dated and anticipated conceptual developments now recognised under the critical rubric of 'posthuman'.[2]

In the next section, I move beyond some of the generalities thus far indulged in identifying kinships and differences between postcolonial and posthuman positionalities to examine a more specific point of tension and mutual critique. I focus on the way more recent examinations of humanist critique, open to the reflexive limits imposed by coloniality, go some way to recognising wider ecological relationships that problematise the reflexive subject. These developments and tensions within self-consciously humanist and postcolonial modes of critique anticipate some of conceptual questions that today are framed under the rubrics of posthumanism, decolonisation, and related discourses. More significantly, they are attempts to grapple with the present's global inter-human and ecological imbrication.

Humanist intimations on the implicate other: Said and Wynter

Said: words as 'reality itself'

I begin where the great postcolonial scholar and humanist critic Edward Said left off. In a series of lectures at the end of his life, posthumously published as Humanism and Democratic Criticism (2004), Said sought to defend the principles of humanism from what he saw as the bicameral constituencies of 'anti-humanism' and 'anti-intellectualism' (2004: 10 and 18). On the one hand, he was concerned to avoid the 'crippling limitation in those varieties of deconstructive . . . readings that end (as they began) in undecideability and uncertainty' (Ibid.: 66), which produced French theoretical anti-humanist tendencies. On the other hand, he worried about the 'marginalisation of the humanities' (Ibid.: 125) in the reactionary environment of post 9/11 polarising world-views, which tore themselves between both an isolationist academy and a largely uncritical public sphere, and between the falsely bifurcated civilizational ideologies of East and West.

Both 'anti-humanism' and 'anti-intellectualism' were simplistic and false divisions Said spent much of his career problematising. But at the end of his career, in these lectures, there are also interesting and important tensions that point to him struggling with the implications of humanist critique as openness to others, human and perhaps non: 'humanism [is] an unsettling adventure in difference, in alternative traditions, in texts that need new deciphering within much wider contexts than hitherto given them' (Ibid.: 55). I argue here that while it is premature to read Said as invoking for a more-than-humanist ontology, he does, in certain passages, intimate towards a more troubled word/world distinction than might be assumed. Place, situation, sociogeny, commitment, the need to recognise multiple worlds, and pre-cognitive sensibility all feature as characteristics of critique in his lectures. Read in the context of 'the contemporary imperative of our global interhuman and environmental situation', Wynter's present with which we began, Said might concur that humanism's responsibility now is to thinking the multiple possibilities of the wider contexts that give rise to thought, and which are also threatened by particular forms of exclusivist thought today. In the following, I briefly outline Said's remarks about the political and ethical significance of humanism, and emphasise the constitutive tensions in Said's arguments regarding critique that point to overcoming text versus world, and ideal versus real distinctions. These tensions are important, for they will become the stepping stones to a conceptual landscape of contemporary theorising which seeks to address the political problematics of our global interhuman and environmental present, problematics which demand theoretical bridges between epistemological and ontological characterisations of alterity, bridges this book and its authors seek to provide.

The problem Said set for himself in his 2000 Columbia lectures was not 'the human' as such, but humanism. In particular, he was animated by the relationships of humanism, as an ethos of questioning and critique, to the importance of the humanities for self-criticism. By 'humanities' he meant chiefly literature, but also music, the visual arts, and, more broadly, the questions posed by aesthetics, sensibilities he spent much of his life studying. Said saw in the humanities both the uniqueness of a generalisable human – only humans, he thought, reflected on themselves through art – and the modes of the reflective capacities necessary for humans to think about themselves: language, representation, recognition, and resistance. Humanism thus became aligned, for him, with the peculiarly human capacities for critique itself. '[S]elf-understanding and self-realisation' (Ibid.: 26) were the fundamental human exercises in terms of which we enable 'always restlessly self-clarifying' means to democratic 'freedom, enlightenment, more agency, and certainly not their opposites' (Ibid.: 73).

Said located the generative source of critique (i.e. humanism) in language and textuality. 'Language is where we start from as humanists' (Ibid.: 28). As such, he continued both a tradition of textual and hermeneutic criticism, and reflected his own theoretical formation within a phenomenological commitment to human reason delimited by the textually framed conditions of human recognition and representation. 'Language,' he quotes the literary critic Richard Poirier, 'is the only way to get around the obstruction of language' (Poirier, as quoted in Ibid.: 28). In supplying thought with its basic material, signs, language and texts, constituted, for Said, the immanent, creative evidence of three interrelated human capacities: the experience of time and history in the evolution of representative form; the experience of knowledge making as a reflective act of manufacture; and, the experience of the radical incompleteness of representation. To situate these claims, and to argue for a human centred ethos of reflective self-awareness of the limits and, thus, requirements for restless understanding (awareness of finitude becomes an infinite task), he drew on Giambattista Vico's notion of the 'verum-factum equation'.

Humans are reflective, finite knowers. We know what we make, and to know how something is made, is 'to see it from the point of view of its human maker' (Ibid.: 11). Truth (verum) lies in its being understood as made. But verum, referring in its noun form to 'fact' or 'reality' contains, in its adverbial form, the sense of a non-comparable factum: 'true, but even so', 'true, but', and 'true, however'. With this conditionality come two forms of awareness. The first lies in the constructed-ness or made-ness of knowledge. The second lies in the historical, and thus temporal context of its production. Hence, if knowledge is constructed, and what we can know is temporally bound to the historical world, then it is also the case that knowledge is 'always something that is radically incomplete, insufficient, provisional, disputable, and arguable' (Ibid.: 12). The humanist's responsibility, then, for Vico as for Said, is to recognise the intimate, personal, or 'subjective' (Ibid.) element of knowledge that provides it both its ethical force and its limit. In the irretrievably poetic, or intimate, subjective constructedness of attention, lies the 'essence of humanism . . . to understand human history as a continuous process of self-understanding and self-realisation' (Ibid.: 26). The differential demands the world impresses in the form of unending disclosure and self-criticism orients attentive knowledge production to the play of difference in irretrievably human, that is, produced language. Humanist responsibility is to 'understand, re-interpret and grapple with the products of language in history, other languages and other histories. . . . It is a means of questioning, upsetting and re-formulating' (Ibid.: 28). The restless concern to the language of human difference as critique requires that, in being honest to itself, it resist 'cliché', 'unthinking' and 'idées recues' (received ideas) (Ibid., 43) in order that it take account of what 'it either repressed or deliberately ignored' (Ibid.: 46).

Given Said's interest in navigating between the Scylla and Charybdis of hermeticism and ideological simplification, and in meeting the political demands of his present, he goes on to explore how the humanists and the humanities are now exploring ever

> more attuned ways than any before it to the non-European, genderized, decolonised, and de-centred energies and currents of our time. . . . Critique is a continuous practice of questioning and accumulating knowledge, that is open . . . to the constituent historical realities of the post-Cold War world, its early colonial formation, and the frighteningly global reach of the last remaining superpower of today.
>
> (Ibid.: 47)

He continues,

> The role of the humanist [therefore] is not to consolidate and affirm one tradition over all others. It is rather to open them all, or as many as possible, to each other, to question each of them for what it has done with the others, to show how one may cultivate peaceful co-existence – humanism provides little short of stubborn, and secular, intellectual resistance.
>
> (Ibid.: 49)

Nationalism, geopolitical exceptionalism, religious fanaticism, interventionism, essentialism, slavery, Eurocentrism, and a narrowing parochialism of depoliticised academic focus, concern Said, amongst others. They are the demands of the present that humanists and the humanities must recognise and resist with relentless questioning.

A few characteristic things stand out within Said's description of humanism and its importance for constituting critique. The first is a focus on language, texts, and the problematics of representation. Thinking is manifest within and bound by language, and the focus for thought is what is revealed by the mediation of the word. Second, the critical focus remains resolutely trained on the human form and conventional human politics. Questions of human violence, nationalist social expression, geopolitical intervention, and legacies of human inequality animate Said's curiosity for justice. Politics and ethics resonate, therefore, through the register of the human as inherited from an admittedly queried enlightenment legacy. Although the environment and what might be termed 'natural effects' of colonial and imperial legacies merit very brief mention, such non-human concerns are constructed by Said as separate from the constitution of the human and self-critical responsibility. Indeed, the self that is operationalised in 'self-reflection', 'self-criticism', and 'self-consciousness' throughout the text is one imagined within a phenomenalist or phenomenological tradition bequeathed as modern by a genealogy of Kantian critique. What renders the human unique is the capacity for, and responsibility to, thought, but thought limited by material conditions for its possibility: language, history, and reflexivity. Reason is responsible to itself and its limits because of difference and the fallibility of knowledge, and critique reaches its highest capacities in an intensely reflexive practice of ceaseless interpretation, exploration, and humility before itself and its historically and linguistically bound limits. These limits are trained on the conditions of their own exclusion in the self-aware production of, and attention to, alterity.

Said's humanistic imaginary, though critical of both liberalism and its post-structuralist critics, is set firmly within an Enlightenment horizon that orchestrates its analysis around a conventional human locus. This is a uniquely symbolic human framed as separate in its cultural productions from a non-semiotic nature, and thus exceptional. It is also a human so constituted that it shares this commonality in the symbolic throughout its diverse geographies. There is a commonality to human self-consciousness and it lies in the aesthetic capacity for critical reflexion on its own textual productions (including, of course, literature, art, story, buildings, ornament, etc.). Aesthetic self-attunement denotes critical affirmation, but also limits self-understanding and political expression to the transcendental conditions of the self, that is, to the conditions for the possibility of its thought: the making of signs, their recognition, and re-interpretation. Said's humanism, though tempered by Vico's historicism, is not far removed from a secular transcendental epistemology that privileges the centrality of human concerns as social, political, or ethical. What counts as political, or aesthetic is always already framed by an assumption of human exceptionalism that legitimates itself through the exclusivities of language, texts, and reflexive self-awareness.

Yet, at the same time, humanism, for Said, is never dogmatic about its beginnings and sources: 'humanism is constitutively in need of revision, re-thinking, and re-vitalisation' (Ibid.: 32). Today, we recognise in Said's requirement that 'humanism . . . is being required to take account of what, in its High Protestant mode, it had either repressed or deliberately ignored' (Ibid.: 46), that the epistemic question of human alterity also necessitates the ontological question of implicate materiality, a cosmopolitics, that takes us beyond the classical lineages of European hegemony. Said quotes Fanon: 'the Graeco-Roman statue [of humanism] is crumbling in the colonies' (Ibid.: 47) and suggests, further, that we must 'unsettle' humanism in an 'adventure of difference, in alternative traditions, in texts that need new deciphering within much wider contexts than hitherto given them' (Ibid.: 55). There is a sense here that collective human history – an open question – is imperilled by too narrow a focus on the politics of identity and its worries about representational limits or constructions. Words, he argues, are part of the ontological real itself; they are not separate from what we term nature, but enrolled in the materiality of thought itself. Perhaps it is too much to read Said as a monist here (he is not), but nevertheless it is telling that he grapples with a word/world distinction that is far from Kantian in its noumenal and phenomenal separation: 'words', he writes, 'are not passive markers or signifiers standing in unassumingly for a higher reality; they are, instead, an integral formative part of the reality itself' (Ibid.: 59). Language, he likens, through Emerson, to 'fossil poetry', 'traces . . . of that aboriginal power by which *we invent ourselves as a unique form* of nature' (Poirier, as quoted in Said, Ibid., emphasis added). We 'invent ourselves as a unique form'. We are not unique; we tell ourselves we are, but we're not. And language itself belies this assumption.

Here Said is toying with quite a radical reading of the human relationship with language, the implications of which he, unfortunately, did not have the time to work out. 'No author', he continues, 'is completely sovereign or above the time, place, and circumstances of his or her life' (Ibid.: 62). We are situated, emergent from the autochthony of our speaking, but also shaping it with the words we invoke. Language and words are given an ontological and material power to shape the real, and our 'irreducibly personal act of commitment includes . . . opening oneself to the text' (Ibid., p. 66) and 'readers must also extend their readings out into the various worlds each one of us resides in' (Ibid.: 76). Texts, for Said, are worlds, and worlds texts. There is little distinction for the humanist critique, whose humanism emerges from her commitment to her sociogenic worlds. 'It is especially appropriate for the contemporary humanist to cultivate that sense of multiple worlds and complex interaction traditions, that inevitable combination . . . of belonging and detachment, reception and resistance' (Ibid.).

While it might be tempting to read Said as arguing, at the end of his life, for a political ontology of humanism, such a reading, as I have suggested, would be premature. Rather, he equivocates on any kind of monist claim by retreating to the more familiar critical ground of word and world distinctions. Humans are uniquely aware, he argues, and in the frailty of their awareness of their own finitude and inability to speak, in our silence before the world, we encounter our fundamental, and perhaps, tragic aporia. 'Humanism . . . is the means, perhaps the consciousness we have for providing that kind of finally anti-nomian or oppositional analysis between the space of words and their various origins and deployments in physical and social place' (Ibid.: 83).

Wynter: The sociogenic principle

We might be able to understand the tension in Said's identification between word and world incommensurability, and the word's textuality being a part of 'reality itself' through 'sociogeny', the quasi-ontological bridging concept first proposed by Fanon, and later take up by Said's contemporary, Sylvia Wynter. 'Sociogeny' was famously proposed by Fanon in his introduction to *Black Skin, White Masks*. It was posited as an additional mode of explanatory understanding, which, he wrote, mediated the failures of 'phylogeny' and 'ontogeny' (2008: xv). 'Sociogeny' was an attempt by Fanon to recognise the means of the social and the collective ideational to shape inter-active human agency. Language, for Fanon, and, in his case, a language attentive to changing the force and legacy of racisms, needed to be recognised as symbiotic with flesh (Gordon, 2015: 25). Transforming language also transforms the users of language, for speakers are situated and constituted in a performative act of speaking and invoking the materiality, a more just materiality, than that invoked through colonialism. For Fanon, it was the black person's responsibility to take it upon him or herself to transform the world through a different language, and so instantiate a new world of living differently from coloniality's necrocracy (Ibid.).

While largely underdeveloped as a conceptual grammar in Fanon, sociogeny becomes a 'governing principle' in Sylvia Wynter's work (see, 1984, 1995, 2001, 2003, 2006). Wynter argues that sociogeny is a 'master code of symbolic life and death . . . that is constitutive of the multiple and varying genres of the human in the terms of which we can alone experience ourselves as human' (as quoted in Scott, 2000: 183). However, unlike a human essence or a reductive natural ground that can be attributed to all humans, sociogeny is an

> explanatory concept at the level [of] nature-culture . . . whose processes of functioning, while inseparable from the physical (that is, neuro-biological) processes which implement them, would, at the same time, be non-reducible, as the indispensable condition of what it is like to be human, to these processes alone and, therefore, to the laws of nature by which those processes are governed'.
>
> (2001: 32)

In other words, the sociogenic principle is an explanatory concept that accounts for the specific interaction of human cultural productions with its various and different physical worlds. 'Nature-culture', as quoted in the above, is a collapsed distinction; words and matter co-implicate one-another. Sociogeny is a capacity within different humans to differently constitute their worlds in light of the specific biophysical and ideational productions that make differential communication possible. '[T]he individual must filter the external through the mediation of what he/she is socialized to experience with reference to his/her culture specific identity' (Ibid.: 49). The sociogenic is the boundary space within which the physical and the ideational interact to produce the differential social and cultural ontologies experienced by both humans and non-humans. Language here, for Wynter, acts not unlike Said's intimation towards aggregating the human as, in language, a unique form of nature.

Two things, I argue, are important to recognise about how sociogeny explains a unique form, or political ontology, of nature. These two facets are not part of Wynter's schema, but are implications of her schema for the present work and argument, and, as such, I elaborate on them here.

First, sociogeny is not limited to humans. Humans, of course, are not the only social creatures. While we are familiar with such a claim when it comes to whales, dolphins, other primates, birds, ants, or bees, I would argue, however, that sociality is also a feature of any interactive, adaptive, or enduring assemblage. Plants, fungi, bacteria, viruses and the like – even cybernetic and physical systems – also communicate within creative and adaptive self-organising arrangements. I draw here on an analogous articulation from A. N. Whitehead, who argued that all enduring things are 'societies'. Sociality should be read not as a system of complexity whose apogee is the human, nor are systems social to the extent that they mirror or participate in certain human forms. Sociality is a feature of endurance and semiosis, and emerges in the implicate creation of the material contexts in which we find ourselves.

> The stubborn reality of the absolute self-attainment of each individual is bound up with a relativity which it issues from and issues into. The analysis of the various strands of relativity is the analysis of the social structure of the Universe.
>
> (1933: 376, see also, Halewood, 2011: 103)

Second, signs, and hence language as a sophisticated use of signs, are both the cause and the product of being in relationships with others. The source of sign making does not come from a mind projecting out onto a mute sign-less world, but emerges in relationships of becoming and unfolding between any enduring, sometimes self-organising, always already material, and therefore, social systems. Human sociogeny is a peculiar, and peculiarly complex, form of sociogenic life, but sociogeny – the experiential mediation between what there is and the biological capacities of the organism in question – means that each life form is produced in its own experiential world. The source of the signs, though, comes from the sociality of the world itself, rather than a hierarchy of life with complexity at the top and simplicity at the bottom, rationality at the top and irrationality at the bottom. Mind does not emerge the more complex cognition becomes. Mind is itself a feature of all self-organising systems to the extent that they respond to relationships with others. It is the relational interaction with others, or semiosis, that produces the emergent property we term 'mind' or mindfulness (Kohn, 2013: 34; Wynter, 2001: 48). As Wheeler (2016: 2) writes, 'mind is a relation. More specifically, mind is a sign relation. Environment is nature and culture.'

Mind, therefore, is not peculiar to humans, nor, one might even argue, living non-humans. Humans have a peculiar and diverse relationship with what semiosis makes possible across the assemblages of embodied interaction that make up different human experiences, but mind is a diverse and emergent property of the relational world for any self-organising, embodied interaction that is able to distinguish, react, and endure through social relationships, or 'societies'. A plant protecting itself from poison or insect attack participates in a semiotic interaction that constitutes distinctions of self and other, just as more familiar states and experiences might also be understood within such distinctions. An ant massaging a caterpillar for sugary excretions or rearing aphids for milk, a flower co-evolving to be pollinated by a specific butterfly's proboscis or specific bird's tongue, a bacteria responding to an anti-biotic agent. These may be explained as functions of 'biosemiotic' relationships (e.g. Hoffmeyer, 1996, 2008; Wheeler, 2006, 2016), that is, as the symbiotic effects of signifying interrelationships between living systems, rather than the beneficial effect of simply random emergence (Barbieri, 2008). Evolutionary and cultural emergence lies within the systematic relationships of material coding and information. As Wynter explains,

> what causes . . . specific neural firings to be activated in a specific modality is not a property of the brain itself (of ontogeny). Instead, it is a property of the [semiotic] codes in whose . . . systems of meaning we institute ourselves as specific genres of being human. So, we note here that the mind is not the brain. Since the causal source of the nature of our response does not lie in the neurophysiological mechanisms of the brain, which implement that response. It lies instead in the master code of the sociogenic principle. Since it is its meaning systems that determine *how* the mechanisms of the brain will implement our experience of being human, in the terms of each culture's specific conception. This therefore means that our aversive responses to ourselves, our reality, are *socialized*, rather than *natural* responses.
>
> (as quoted in Scott, 2000: 189, emphases in original).

The power of the sociogenic principle lies in its capacity to begin to break down a false distinction between internal mind and external matter. I say 'begin', because Fanon did not intend sociogeny to be an ontological claim. His concern with sociogeny was in the extent to which words and human cognitive structures, specifically, racist cognitive structures, shaped experiences of the real, and because they were performative could be overcome with new stories. With Wynter, sociogeny becomes a more thoroughgoing principle that explains a socio-materialist collapse in the mind-matter distinction.

The embodied dynamic of sociogeny is in perpetual change and adaptation across numerous semiotic relationships that come from the worlds of sensibility and so constitute possibilities of thought as particular enfoldings of semiotic potential. These worlds will, of course, be different in and for different material arrangements. Some worlds will share commonalities and overlaps, some more so than others, as in the case of, say, people from within different geographies. Others will share little, if anything, of their worlds, as, for instance, a bacteria in the gut of a sea snail, or the nematodes in the compost pile at the end of the garden. The cosmopolitical relationships that emerge from these materialities are pluriversal rather than universal. Multiple worlds are constituted in the different semiotic relationships that produce them. Rather than needing to explain the commensurability of all forms to one universal explanation, the political problematic becomes, instead, first recognising the multiverse, and second building from these plural relationships without making worlds commensurable (de la Cadena, 2010: 361).

What Wynter calls for with the sociogenic principle, then, is a 're-enchanted humanism' (Scott, 2000: 195), or a 'counterhumanism' (Wynter and McKittrick, 2015: 11; see also McKittrick, 2015). Hers is a conception that ameliorates anti-humanist and posthumanist positions, for the simple reason that it does not take the characterisation of the human as it emerges from a Eurocentric narrative as the normative standard through which to measure or understand the human as such. Neither does she locate the human in a reductive explanation of some essential natural component or process. In one sense, yes, it is anti-humanist in so far as it critiques a colonial or Enlightenment vision of humanism, but it does not seek to jettison thinking from a characterisation of human experiential and normative uniqueness. Nor does she refuse the 'posthuman' turn to a so-called more-than-human ontology. Re-imagining human uniqueness in the context of, say, a Maori ontology would entail thinking the kinship relations between, for instance, a river and a person (see e.g. Thomas, 2015). What she does do is recognise that human thought emerges as a dynamic and malleable sociogenic genre within numerous and ever-changing semiotic relationships that have their own particular histories. Sociogeny enables us to recognise how the human co-evolves, materially, within its numerous narrations of itself. The human does not precede its narration. It emerges in the immanence of its numerous forms of ideational, and hence also material, invention and description. '[T]he human-as-a-*homo-narrans*-species cannot pre-exist their hitherto always genre-specific or culture-specific representations of origin any more than. . . *bees can pre-exist their beehives*' (Wynter and McKittrick, 2015: 72, emphases in original).

The role of critique becomes as much creative as it does restrictive. Critique is fundamentally committed to being creative in that, in the question being raised by a world's signifying interactions, it demands re-imagination and compositional production. The question emerges in the possibilities suggested by the signs themselves: things, and so thought, could be otherwise. This otherwise demands a creative response, of which the differential human capacities entail different types of creative possibility. The human capacity, therefore, does not reside in being a uniquely aporetic or limiting one. To do so, for Wynter, means succumbing to a false and imposed characterisation of Man. As she notes, we must 'reimagine the human in the terms of a new history whose narrative will enable us to co-identify ourselves each with the other, whatever our local ethnos/ethnoi' (as quoted in Scott, 2000: 198). We must take

> our present history, as narrated by historians, as empirical data for the study of a specific cultural encoding of a history whose narration has, together with other such disciplinary narrations, given rise to the existential reality of our present Western world system – that is, to the reality of a system enacted about the ethno-class conception of the human Man, which represents itself as if it were the human, and in which we all now live.
>
> (Ibid.)

Art, literature, and science, and, of course, the multiple genealogies within which these numerous forms of aesthesis and Scientia emerge become fundamental to Wynter's sociogenic characterisation of word and world interactions. They are important because they implement options for cosmopolitical invention that draw from numerous, always already ecological practices and embodiments of relational possibility with deep and varied histories.

> What is needed . . . is a ceremony beyond the secular (the secular being of course an alibi produced from within the monotheistic episteme of Man) that can connect the project of re-naturalisation with one of re-enchantment – finding a new *cosmogony* that can institute our new *sociogenic* descriptive statement, our revolutionary collective praxis of being human beyond Man.
>
> (Cornell and Seely, 2016: 157, emphases in original)

New cosmogonies, new pasts, are necessary for new and different futures. Pluralising narrations, as Said argues critique must, also pluralises materialities, as Wynter argues stories must also do because the source of stories are the material relationships, the signs, of the particular world itself.

Said goes part way to recognising how 'words . . . are integral formative part[s] of reality itself' (2004: 59). Wynter carries forward Said's pluralism by reading humanism through sociogeny to argue for a 'a new hybrid science of the Word/Nature' (Wynter and McKittrick, 2015: 73). Neither thinker banishes the human as such, but aims to re-enchant it through a call both for plural textual narratives, and the materialities or political ontologies invoked by those very stories. The human resides, then, less in the claim, or in the attempt to claim, universality of some common elements, but in the variable capacities in each ecological culture from which it emerges, always 'a new creation, still damp from the vast firmament' (Baldwin, 1963: 286). The emphasis lies neither in defining what is proper to the human, nor in asserting that there is no essence to the human, but in refusing the work that answering the question, 'what is proper to humans?' is put. Humanism becomes, instead, concerned with the proliferation of multiplicities, and an interminable challenge to the attempt to put the human into a separate order (Vivieros de Castro, 2014: 44–45).

What then of politics and ethics? They are still there, of course. Only their source or ground is not the human, nor the proximity to a standard of humanness. The next section moves on to explain briefly how, for a relational ontology attentive to multiplicity, the concerning ground is not relativised, but is the condition for emergence and interdependence as such. The ontological ground is interdependence.

Relations: Conditions of care as political ontology

What is crucial in the counter-intuitive, and seemingly paradoxical, recognition of human multiplicity as an 'irreducible' (Latour, 1988: 158) category is to recognise that the capacity to 'be-in-the-world' does not come from something special about the cognitive complexity that is peculiar to humans, nor perhaps partially peculiar to how the almost-human is defined (i.e. the Muslim, the Jew, the Black, the Indian, the cognitively impaired, the primate, etc.). 'Human' is neither contract nor ideal. Being-in-the-world comes from the particular relation its particular worlding constitutes. Nothing is 'poor in the world'. Non-humans are not 'poor in the world'; they are only 'poor-in-the-world' when compared to the standard of a specific definition of cognitive 'wealth' which privileges a model as ideal, whether it be a classical model, an Enlightenment model, a modern model, a masculinist model, a Christian model, an Islamic model, etc. Every social system is in and of its world on its own terms, in other words, within the relational rapports of its constitution. The terms of constitution come from the material cultures, the ecologies, within which the relational process (i.e. thing, object, collective, idea, etc.) emerges: relations are more fundamental than entities.

As the reference to Latour above makes clear, the claim to irreducibility is not new. Including Latour, many social scientific and humanities scholars have argued for relational ontologies in late twentieth-century Amer-European thought, including well known contemporary theorists like Donna Haraway, N. Katherine Hayles, Isabelle Stengers, and Michel Serres. Earlier theorists like Gilles Deleuze, Felix Guattari, and Alfred North Whitehead, too, advocated similar relational and processual ontologies. Such figures are well known and even, today, almost canonical.

Less recognised or admitted by these theorists and the majority of their interlocutors is the fact that relational ontologies have numerous other ancient and non-European lineages (see e.g. Hunt, 2014; Sundberg, 2014; Todd, 2016). These lineages are almost never invoked on a par with the philosophical pretence of Europe. The effects of this ignorance, deliberate or otherwise, are several. First, as has long been argued, modern Europe is constructed as the locus of serious, critical thought; others' thinking, if it is recognised, is 'othered' as primitive, spiritualist, anthropologically interesting, non-modern, etc. (see Dabashi, 2015). Second, others' thinking gains its validity or legitimacy to the extent that it mirrors or corresponds to Eurocentric ideas and concepts. Often, this is translation or comparison is not deliberate, but a function of recognising similarity and consonance. The effect, however, is to place the European concept as the standard against which validity is measured, rarely, if ever, the other way around. As a result, third, the conceptual armature of Europe becomes the standard for explanatory significance and epistemic value. With that armature, fourth, ethical and political structures are also normalised. Finally, because knowledge gets constructed, institutionalised, and legitimated through the lens of a specific suite of geographies, a very simple but profoundly destructive attrition of worlds ensues. Languages disappear, practices erode, ecologies go untended and dissipate, ideas and stories, and, therefore, ways and matters of life, are irretrievably forgotten. These effects have long been the subject of postcolonialism's efforts to provincialise and decentre European modernity, and so also part and parcel of its efforts to decolonise knowledge production.

What would it mean if we were to begin, for instance, not with European modernity in the form of Spinoza, Deleuze, or Latour, but in another geography, another time, place, and context? Let us consider one simple, if relatively well known, example.

Buddhist thinking, for instance, is profoundly immersed within relational ontologies. For Gautama Buddha, what might be called 'self' is an emergent function of the inter- and intra-relations between the five skandhas. Skandhas simply means, in Sanskrit, 'groupings', 'aggregates', or 'assemblages' (Gier, 2000: 47). Within Mahayana Buddhism, the five skandhas from which the non-being of self emerges are: material relations (matter), feeling relations (affects), perceptual relations (percepts); conceptual relations (concepts); and consciousness (self-reflective experiences) (Garfield, 2015: 12). Each of these are relational assemblages, with no definable existence or essence outside the multiple relations themselves; hence, the assertion that the aggregates or assemblages are without independent existence. Relations precede entities and sociality is entailed by the multiplicities that make up the aggregates. As is well known, entities like 'self', 'identity', 'ego', etc. are simply, for the Buddhist, illusory chimera.

What becomes vital in this recognition, however, is neither the epistemic nor ontological privilege of the explanation from multiple aggregates, but what we do within, and from, our relational immersions. If interdependence is the condition of emergence for anything whatsoever, then attending to that interdependence becomes the condition from which action (i.e. interaction) arises. When interdependence is foregrounded, a generic concern for maintaining and enduring this ontological significance becomes the focus for acting in the world. In other words, care is the 'generic' condition of every social process – human and non-human – and its attempt to maintain, continue, and repair its worlds (Tronto, 1993: 103). Care and reciprocity are the conditions of endurance within processes, because they come from the relations themselves. Care is not something added to processes to render them valuable. 'Care is concomitant to the . . . continuation of more-than-human entanglements' and is neither an imperative imposed by a particular capacity, nor a duty of some rather than other processes (Puig de la Bellacasa, 2017: 70). An ant tends to its home, just as a plant attends to its futurity when it moves to avoid harm, just as humans attend to their young, just as body colouration and background preferences mimic one another as in a feeding moth's wings; ensures mutual continuity. Each of these forms of doing maintenance or enduring depends on the complexity of an interwoven web of pre-existing relationships: interdependency. It is one of the ontological conditions of emergence of life, which includes the cosmic, as well as the cellular and the ideational. Ideas and matter are, after all, historically immanent within one another. Care is not a normative imposition by a uniquely human, or nearly human, capacity, but a condition of material entanglement shared by all things, of which humans are also a part.

> When we think about what we care for: one moment it seems it would be easy to remove our care; the moment after we begin to realise that our care does not belong to us, and that that/whom we care for, somehow owns, us, we belong to it through the care that has attached us.
>
> (Ibid.: 167)

For a Buddhist ontology, the conditions of impermanence, interdependence, and the absence of intrinsic realities (i.e. there is nothing 'behind' relations) entail the fundamental ethical ontology of sympathetic recognition of others' impermanence and interdependence. Recognising care as the condition of things as processes enduring means committing oneself to sustaining the web of those interrelations. Ethics, ontology, and epistemology collapse into a tightly bound and enfolded relational plane (Garfield, 2015: 14). Mind is not distinct from the plane about which it thinks. It is the plane that does the thinking and caring.

I have briefly traced an ecology of mindful and caring immanence through a reference to a very old Buddhist ethics and ontology. It is, perhaps, more familiar to many readers. Yet one could also trace a similar onto-history through concepts drawn from multiple, different, perspectives, many of whose ethico-ontological lineages long pre-date the Buddhist example outlined above, for example: Kusch's reflections from América[3] (2010), Vivieros de Castro from the Amazonas (2015); or Rivera Cusicanqui (1984, 2012) and de la Cadena's (2010, 2015) more focused analyses from the Quechua and Aymara of the Andes; Redekopp (2014) and Kohn's (2013) analyses in nearby Aruna contexts, or Kopenawa and Albert's narrations from Yanomami worlds (2013); E. R. Atleo's work with the Nuu-chah-nulth (2004, 2011); Simpson's work from the Nishnaabeg (2011); Nadasdy (2003) with the Kluane, and, Cruickshank's nearby Tlingit and Athapaskan studies (2006); Descola with the Achuar (1996); Blaser with the Yshiro (2010); Stevenson (2014) and Cameron (2016) with the Inuit; Glissant (1997) and Frankétienne (2014) from Caribbean geo-poetics; Glowczewski with the Warlpiri (2015, 2016); Benterrak, Muecke, and Roe with the Yawur (2014) etc. These are, currently, amongst some of the more invoked treatments towards developing, what Western contemporary thinking might term, a 'relational ontology' of indigenous thought. Again, they are far from the only ones.

Indeed, the recent conceptual discourse, sometimes termed 'political ontology' (see e.g. Blaser, 2009a, 2009b, 2010, 2013a, 2013b; Escobar, 2008, 2016) attempts to characterise something of these relational and ontological sensibilities of interdependence for contemporary, academic, critical theory. Like the Buddhist collapse of ontology, epistemology, and ethics into a plane of thinking wherein these distinctions – derived as they are from a modern Western tradition – do not hold, political ontology also seeks to characterise planes of thought within which words and worlds are synonymous (Blaser, 2013b: 552). Just as the Buddhist seeks, performatively, to align action with the care of interdependence, so too do numerous – Descola (2013a) may argue 'all' – indigenous perspectives hold that words and stories (i.e. representations) 'partake in the performance of that which they narrate' (Blaser, 2013b: 552; see also Archibald, 2008). Political ontology, following Blaser's definition, refers both to the politics that shape particular worlds, and the material, and hence social, conflicts that ensue as different worlds interact, mingle, and sustain one another (2009a: 877). What is distinctive about an approach that seeks to articulate political immanence as sociogenesis is the recognition that political ontologies are neither epistemic nor aesthetic reflexions that precede the constitution of ordinary and abstract practices. Rather, as noted above, sociogenic forms (i.e. signs) are shaped through the practices, processes, and material interactions that are existence, that are word qua world, and vice versa. Ontologies perform themselves – they story themselves, they word themselves – as worlds (Ibid.; see also Jackson, 2014: 81).

Conclusion: Five questions

It is quite clear that in drawing this genealogy of word and world immanence through a select few, merely indicative, examples that traverse indigeneity, Buddhism, as well as European and non-European humanisms, that even within the often taken for granted humanist, representational frame that separates mind from matter (exemplified here with the example of Said) the questions both of pluralising knowledge, and consequently, of troubling so-called foundational distinctions between mind and matter, object and process, nature and culture, etc. are far from either given or settled. Today, what have become far more interesting than the simple claim to provincialise knowledge production (a claim that few critical social sciences or humanities discourses would rationally reject) are the implications of provincialising thought. Pluralising stories and narratives means, literally, pluralising the interdependent materialities that constitute the possibilities of speaking, and so also pluralising the possibilities words present for worlding anew. Critique 'is concerned with reality making, including its own participation in reality making' (Blaser, 2013b: 552). Critique is a compositional practice, not a reflection on, nor an operation within, ineluctable cognitive limits.

As I have tried to argue, and show, in this chapter, provincialising knowledge and thought is more disruptive than simply incorporating multiple perspectives, or decentring epistemic hegemonies. Provincialising also means re-conceptualising the landscape of the human, and thus, critique itself. For the human is always already the non-human or the more-than-human, as much as critique is always already the immanent planes that make up its possibility, the plural stories, human and non-human, that render careful interdependence the dynamic ground of continuity, endurance, and sociality. In other words, both the postcolonial appeal and the posthuman appeal entail one another.

As has become apparent, in doing so, their bridging as co-implication also raises, I think, a number of rather profound questions for contemporary thought. By way of conclusion, I would like to outline five such questions. These questions are posed so as to stage an agenda for thinking across the bridges, or rather, pre-existing ecologies that always already connect matters concerning postcoloniality and the posthuman.

Question one: What is a decolonised critique?

The overarching question that emerges for me from thinking the implications of the postcolonial and the posthuman is the status of critique under the sign of decoloniality. What is critique, if it is no longer a self-conscious description of the limits of a verifiable and universalisable human cognitive capacity? Can we speak of critique in the same ways as that story launched in the Enlightenment about daring to know one's own capacities through one's own limits? I rather think we cannot.

Querying the nature of critique as the reflexive base of critical thinking, however, is a thoroughly destabilising and deeply fraught endeavour. Destabilising it should not mean that anything goes; nor should it mean the abandonment of the necessary and liberatory potential of questioning. However, as we have seen, the figure of the human that operationalises critique for the contemporary academy, and critical thinking more generally, is a figure that emerges from a specific history whose narratives are themselves both onto-epistemically problematic (i.e. they depend on false distinctions like mind and matter, nature and culture, etc.), and which actively exclude histories and perspectives outside a European centre, including the many materials, worlds, and cosmopolitical entanglements that make up the divergent processes necessary for thought. If thought and care emerge from the world itself, and are not unique to humans, let alone non-humans, and critique is a creative compositional effort within these processes, what is the status of truth, legitimacy, and the like? Or, are these similarly universalising questions also doomed to fail? If we decolonise critique, do critique and the question similarly fragment within the pluriverses from which they emerge as irreducible multiplicities? Can we decolonise critique? Or, like ontology, aesthetics, etc., should we give up speaking of 'critique' as such? What, though, are the troubling implications of becoming 'uncritical'?

Question two: What is a sign?

Throughout the chapter, I have endeavoured to argue that in bridging the postcolonial and the posthuman, we inevitably also need to rethink the emergence of mind. Mind, as I have suggested by drawing on an increasingly mobilised, alternative, and radical body of work, is not peculiar to humans or complex cognitive apparatuses. Instead, it is an emergent property of semiotic processes. And semiotic processes are the relational dynamics of enduring material systems. 'Signs', as Kohn's epigraph which opens our chapter notes, 'don't come from the mind. Rather, it is the other way around. What we call mind, or self, is a product of semiosis' (2013: 34). The implications of this argument are similarly profound. If signs are a function of relational systems, and mind is a product of this semiosis, then it might be argued that mind is an embedded characteristic of all life, and, possibly, 'non-life'. Does mind, so constituted, subtend 'societies', in other words, all enduring relations? I rather think it does.

Again, the implications of this claim are similarly disruptive, if also, at the same time, not new. Such a claim about the dispersed relationships between what might be termed 'mind' and what might be termed 'signs' is certainly not new to non-Western world-views, including those often termed 'indigenous'. Recognising this within posthuman discourses requires extending the frames of reference for conceptual claims like 'mind', 'sign' etc., beyond the philosophical lexicons and genealogies derived from Eurocentric sources. On the other hand, the conceptual genealogies around semiosis for most orthodox postcolonial literatures are drawn from a tradition derived from European structuralism and post-structuralism. Traditional structuralist (de Saussure, Jakobson, Levi-Strauss, etc.) and most post-structuralist perspectives (Derrida, Foucault, Baudrillard, Lyotard, etc.) generate their arguments about language, signs, and the contingencies of meaning from particular assumptions about mind, specifically those derived from the post-Kantian tradition of critical phenomenalism and phenomenology. The sign's difference and historicity set the basis for a claim about knowledge production, the model of mind mobilised in the argument assumes a fundamental phenomenological distinction between noumena and phenomena, word and world. Indeed, it is that foundational difference that constitutes the possibility of thinking from difference. If, however, we were to be with a different semiological story, one that also assumed a different model of mind, wherein the world and word separation was neither taken for granted, nor characterised as unbridgeable, then both the thesis of mind and of signs, each uniquely located within a particular definition of humans would itself need to change. Of course, doing so brings with it a host of other philosophical concerns or problems. For instance, a simple problem might be, how would such an account preclude claims to magic? Casting spells assumes that words do a particular work in the world; nevertheless, words do do work in the world, often some of the most important work ('I love you'; 'I don't love you', etc . . .). They do have an effect on material systems like hormones, muscles, heart rates, etc. These systems are themselves ecologies of wider relationships that have profound collective implications. The point here is not to entertain the specifics of an alternative model of semiosis and mind. It is merely to recognise that thinking about the sources of epistemic legitimacy, and opening our analysis to both decolonised and more-than-human ecologies, requires *also* rethinking assumed orthodoxies of signs and minds. The emergent area of bio-semiotics holds much promise for new and previously under-appreciated ideas regarding the relationships of mind to more-than-human semiosis (e.g. Gregory Bateson, C. S. Peirce, A. N. Whitehead, G. Tarde) transforming many contemporary orthodoxies in critical social theory.

Question three: Why is the life/non-life distinction continually problematic?

If mind is a product of semiosis, and semioses are the actual material dynamics of enduring relations (what Whitehead calls 'societies'), then the problematics of living and non-living distinctions necessarily also devolve differently because mind and capacities with reactive self-organisation are described on an atavistic scale from more to less sophisticated. Humans are at the top of the scale of mind, cells and plants down at the bottom, or at least is the story of an Enlightenment modernity. In this story, the self-sustaining capacities of each, cells (e.g. bacteria) and massive cell clusters (e.g. humans) are designated as living to the extent that both participate in some form of reactive self-organisation. If, however, we designate mind and life differently, say, transversally or as a web of relations on a topological plane rather than in hierarchical scales, then this will necessarily have significant implications for ethical and political conceptions.

Transversal distinctions, non-hierarchical, or differently hierarchical framings, wherein humans are not at the apogee,[4] are not new. Indeed, differences between life and non-life are neither given nor natural nor self-evident, nor are characterisations of mindfulness. How we are given to understand the meaning of 'life', 'living', 'non-living', and 'mind' are themselves the product of historical contexts and situated understandings. Different people and philosophical perspectives consider these meanings differently. For example, for the Ojibwa, some stones are living (Halloway, 1975). But it is a mistake to assume that living in the Ojibwa context means the same as in a present modern scientific context. What we need to attend to, in such cases, is the work done with the claim that something is living or non-living, mindful or non-mindful. Claims about living and non-living legitimise difference and so action in the world. They also legitimise claims about sign-making and the recognition of signs in construing or bestowing value. If a rock is mute matter and devoid of life, then we may be legitimated in breaking it apart violently to extract some value from within. If, however, the rock is recognised as part of a wider sign system, and perhaps then, of a system of relations that constitutes mind or life in its dynamics of action, then action toward the rock and its relations may be legitimated quite differently. Thus, the ethical and political work in the world with that rock's interrelationship (or lake interrelationship, or river interrelationship, or mountain interrelationship, or moon interrelationship, or light interrelationship etc.) as cosmological, totemic, or ecological agency will entail different causes, responsibilities, and outcomes.

Consider care again within this context. Mindful and caring relationships of respect articulated by many traditions and practices mean attending to the relations of emergence given by the interactions of which the human is but a small part. Acting in a careful way towards these relations attends to the care and possibilities given by the relations themselves. Care here is not constituted as a recognised value that is imputed onto the relations that were previously valueless. Value emerges in the recognition of co-constitution of caring emergence. Thus, the life/non-life distinction is itself almost irrelevant in legitimating action or non-action. More important is not whether something is alive and therefore valuable, or not, and therefore valuable in a different way (e.g. as a resource, as a commodity, etc.) but that it is an occasion or event within a participatory whole.

Question four: Is matter good?

Does it follow from the recognition of careful interrelationships of interdependence that material relations, whether living or non-living, human or non-human, are good? Is the human capable only of goodness or moral action? Of course, it depends what you mean by 'matter' and by 'good'. Perhaps if we ask the question in a different way, it may not seem outlandish or irrational to think non-human ecological and social systems as good. The reason for this lies in the fact that when we ask whether an ecological system is good, we need to ask it not from the division between human and 'nature', culture and the 'not human', but from a context wherein we appreciate caring relations on their own terms. 'The meaningful division . . . is not between humans and 'nature' but between high levels of care and low levels of care' (Bird Rose, 2016: n.p.). We are all familiar with various levels of care shown between non-human animals and their young. Some cows are better than others at caring for their young. Indeed, all cows evidence particularities of character and care. Some you can trust; some you cannot. The same holds for other complex ecological systems (horses, goats, sheep, parrots, ants, and the like).[5]

The point in this somewhat weak thought experiment is not to discern intentionality and then follow its action and so adjudicate on its proximity to human intentionality and thus moral capacity. It is, rather, to look for the material, and thus semiotic, relationships of co-constitution as the conditions for the possibility of intention (see, e.g. Cruikshank, 2005), and to see in them relations of ontological reaction and endurance, and then to ascribe from them an understanding of nurturing endurance or violence. Violence is here defined as the deliberate refusal and destruction of co-constituting relations. A social or ecological system cannot be violent and endure because violence violates the relations of care that constitute it. One could be dominant and aggressive and endure, but not violent. A lion eating a gazelle is not violence. That's simply eating. A lion killing its young for no reason is violent. A human poisoning its seas with plastics and dioxins is violence. Each of the latter could be considered 'not good'.

Seeing in other traditions and ways of life the capacity to think the human otherwise, also requires reframing the sources of epistemic legitimacy. By seeing in others the means to understand material and ethical relationships differently, we may find, in their lifeways, the possibilities to speak to present productions of knowledge otherwise. This means embracing the disruptions the postcolonial and posthuman (broadly defined) can bring to creative thinking.

Question five: Can the university address these questions?

Yes and no. And, so in that tension of 'both/and' rather than 'either/or' there is much work to do. On a simple level, the university, as one of the key legitimating spaces of knowledge production, has always been asking these and related questions. Fundamental questions of life and non-life, mind and matter, good and violence, the sign, and critique have always been a part of the agora. But the agora, the public space of assembly and discussion, I would argue, is not unique to a European and classically derived geography. While much good has come from this tradition, much good has also been ignored or deemed tangential or, at worst, denied and deemed unworthy. If the university is to transform its horizons in the ways that it must to help address the many ongoing violences threatening much planetary endurance, it must decolonise its ways of doing things.

Initiatives to this end have been gaining pace for some time, and it is clear that the university, and related spaces of knowledge production, are facing enormous pressures and threats from revisionist, creationist, fundamentalist, enclosing, privatising, and other deeply problematic directions. But an effective response will not be to erect bulwarks and ramparts. It will be, instead, to transform what spaces of knowledge production are about, where they are located, and how they work. Seeing in knowledge production its fundamentally relational, and always other constituted, production will mean returning knowledge and questioning to the grove, the garden, the space of earthly flourishing, where new laboratories of thinking can create worlds of possibility from the signs generated by the multiplicities that make up thinking and doing, human and non-human, modern and non-modern, 'Western and non-Western'. Yes, we can address these questions, perhaps not in the effort to answer them, but to think with them in new and enduring ways. But to do so, we also need to think with and in new socio-spatial, and thus, newly materialising ecologies of thought.

Notes

1 I am adapting the notion of 'boundary practices' from Star and Griesemer's consideration of 'boundary objects', which they define as 'objects that are adaptable to different viewpoints, but robust enough to maintain identity across them' (1989, see also 2015: 171).

2 This is the simplest and most rudimentary list. It is not in any way meant to be representative or exhaustive. Here is not the place, nor would it be productive, to try to list authors or works outside the Amer-European academic machinery that circulate outside Europe or that actively 'provincialise Europe'. It is not the place, because space and critical integrity prohibits an adequate engagement with these vast bodies of thought. It is not productive because the pretense to representation and comprehension is doomed to fail. Instead, the few references mentioned here are constrained in three ways. First, they refer, as in the case of Alfred (2005), Cèsaire (2000), or Thiong'o (1986) to more polemical reflections on decolonising thinking via philosophical and poetic refusal. Second, they refer, as in the case of Eze (1997), Henry (2000), and Mendieta (2002), to surveys of, or edited volumes in, their respective fields. And third, they narrow to a point made about a conceptual affinity with what a dominant academic lens has learned, perhaps problematically, to call 'posthumanism', and the fact that 'others' have been saying similar things for years, sometimes thousands of years, but with terminologies not wrought from or responsive to a Eurocentric lens.

3 América with an accent is the term Kusch uses to signify 'a form of thinking that furnishes and connotes the authentic but suppressed experience of millions of people in their everyday lives . . . América keeps the reader with another optic, an alternative set of perceptions and understandings that the English reader must grope for, because América is not so readily within ones grasp' (Lugones and Price, 2010: lxix–xx).

4 'Apogee' is from the Greek apogaion, which is derived etymologically from the roots, 'apo-' meaning 'distance away from', and '+gaia' or gē meaning 'earth'.

5 I have some intimate knowledge of these living systems, having grown up and lived with them for long periods of my life, and for that reason I use them as examples here. Others talk similarly of fish, trees, other insects, reptiles, fungi, and bacterial cultures.

References

Alexander, M.J. 2005. *Pedagogies of Crossing: Meditations on Feminism, Sexual Politics, Memory, and the Sacred*. Durham and London: Duke University Press.

Alfred, T. 2005. *Wasáse: Indigenous. Pathways of Action and Freedom*. Toronto: University of Toronto Press.

Archibald, J-A. 2008. *Indigenous Storywork: Educating the Heart, Mind, Body, and Spirit*. Vancouver: University of British Columbia Press.

Atleo, E.R. (Umeek). 2004. *Tsawalk: A Nuu-chah-nulth Worldview*. Vancouver: University of British Columbia Press.
Atleo, E.R. (Umeek). 2011. *Principles of Tsawalk: An Indigenous Approach to Global Crisis*. Vancouver: University of British Columbia Press.
Baldwin, J. 1963. *Another Country*. New York and London: Penguin Classics.
Barbieri, M. 2008. 'Biosemiotics: A New Understanding of Life', *Naturwissenschaften*, vol. 95, no. 7, pp. 577–599.
Benterrak, K., Muecke, S. and Roe, P. 2014. *Reading the Country: Introduction to Nomadology*. Melbourne: Re Press.
Bilgrami, A. 2004. 'Foreword', In: E. Said, ed., *Humanism and Democratic Criticism*. New York: Columbia University Press, pp. i–xx.
Bird Rose, D. 1996. *Nourishing Terrains: Australian Aboriginal Views of Landscape and Wilderness*. Canberra: Australian Heritage Commission.
Bird Rose, D. 2016. 'Thinking Like a Mantis?' //deborahbirdrose.com/2016/02/15/thinking-like-a-mantis/ [Accessed 19.02.2016].
Blaser, M. 2009a. 'Political Ontology: Cultural Studies Without "Cultures?"' *Cultural Studies*, vol. 23, pp. 873–896.
Blaser, M. 2009b. 'The Threat of the Yrmo: The Political Ontology of a Sustainable Hunting Program', *American Anthropologist*, vol. 111, no. 1, pp. 10–20.
Blaser, M. 2010. *Storytelling Globalization From the Chaco and Beyond*. Durham and London: Duke University Press.
Blaser, M. 2013a. 'Notes Towards a Political Ontology of Environmental Conflicts', In: L. Green, ed., *Contested Ecologies: Dialogues in the South on Nature and Knowledge*. Cape Town: Human Sciences Research Council, pp. 13–27.
Blaser, M. 2013b. 'Ontological Conflicts and the Stories of People in Spite of Europe', *Current Anthropology*, vol. 54, no. 5, pp. 547–568.
Borrows, J. 2010. *Drawing Out Law: A Spirit's Guide*. Toronto: University of Toronto Press.
Braidotti, R. 2016. 'Posthuman Critical Theory', In: D. Banerji and M. R. Paranjape, eds., *Critical Posthumanism and Planetary Futures*. New Delhi: Springer India, pp. 13–32.
Braun, B. 2015. 'New Materialisms and Neoliberal Natures', *Antipode*, vol. 47, pp. 1–14.
Cameron, E. 2016. *Far Off Metal River: Inuit Lands, Settler Stories, and the Making of the Arctic*. Vancouver: University of British Columbia Press.
Césaire, A. 2000. *Discourse on Colonialism*. Trans. J. Pinkham. New York: Monthly Review Press.
Chakrabarty, D. 2000. *Provincialising Europe: Postcolonial Thought and Historical Difference*. Princeton: Princeton University Press.
Chakrabarty, D. 2011. 'Postcolonial Studies and the Challenge of Climate Change', *New Literary History*, vol. 43, no. 1, pp. 1–18.
Chatterjee, P. 1997. 'Our Modernity', In: *The Present History of West Bengal: Essays in Political Criticism*. New Delhi: Oxford University Press, pp. 197–223.
Chatterjee, P. 1999. *The Nation and Its Fragments*. New Delhi: Oxford University Press.
Clark, N. 2011. *Inhuman Nature: Sociable Life on a Dynamic Planet*. London and Thousand Oaks: Sage.
Cornell, D. and Seely, S.D. 2016. *The Spirit of Revolution: Beyond the Dead Ends of Man*. Cambridge: Polity Press.
Coulthard, G. 2014. *Red Skins, White Masks: Rejecting the Colonial Politics of Recognition*. Minneapolis: University of Minnesota Press.

Cruikshank, J. 2005. *Do Glaciers Listen? Local Knowledges, Colonial Encounters, and Social Imagination*. Vancouver: University of British Columbia Press.
Dabashi, H. 2015. *Can Non-Europeans Think?* London: Zed Books.
De la Cadena, M. 2010. 'Indigenous Cosmopolitics in the Andes: Conceptual Reflections Beyond "Politics" ', *Cultural Anthropology*, vol. 25, no. 2, pp. 334–370.
De la Cadena, M. 2015. *Earth Beings: Ecologies of Practice Across Andean Worlds*. Durham and London: Duke University Press.
De la Cadena, M. and Starn, O., eds. 2007. *Indigenous Experience Today*. Oxford and New York: Berg.
Deloria, V. 1999. *Spirit and Reason: The Vine Deloria Jr. Reader*. Golden, CO: Fulcrum.
Descola, P. 1996. *The Spears of Twilight*. Trans. J. Lloyd. New York: New Press.
Descola, P. 2013a. *Beyond Nature and Culture*. Trans. J. Lloyd. Chicago and London: University of Chicago Press.
Descola, P. 2013b. *The Ecology of Others*. Trans. G. Godbout and B.P. Luley. Chicago: Prickly Paradigm Press.
Dixon, D. 2009. 'Creating the Semi-Living: On Politics, Aesthetics and the More-Than-Human', *Transactions of the Institute of British Geographers*, vol. 34, no. 4, pp. 411–425.
Dixon, D., Hawkins, H. and Straughan, E. 2012. 'Of Human Birds and Living Rocks: Remaking Aesthetics for Post-Human Worlds', *Dialogues in Human Geography*, vol. 2, no. 3, pp. 249–270.
Dussel, E. 1995. *The Invention of the Americas: Eclipse of "The Other" and the Myth of Modernity*. New York: Continuum.
Escobar, A. 2008. *Territories of Difference: Place, Movements, Life, Redes*. Durham and London: Duke University Press.
Escobar, A. 2016. 'Thinking-Feeling With the Earth: Territorial Struggles and the Ontological Dimension of the Epistemologies of the South', *Revista de Antropología Iberoamericana*, vol. 11, no. 1, pp. 11–32.
Eze, E.C. 1997. *Postcolonial African Philosophy: A Critical Reader*. Oxford: Blackwell.
Fanon, F. 2008. *Black Skin, White Masks*. Trans. R. Philcox. New York: Grove Press.
Frankétienne. 2014. *Ready to Burst*. Trans. K.L. Glover. Brooklyn: Archipelago Books.
Garfield, J.L. 2015. *Engaging Buddhism: Why It Matters to Philosophy*. Oxford: Oxford University Press.
Gier, N.F. 2000. *Spiritual Titanism: Indian, Chinese and Western Perspectives*. Albany: State University of New York.
Gilroy, P. 2015. 'Offshore Humanism', The 2015 *Antipode* RGS-IBG Lecture. 2 September 2015. http://antipodefoundation.org/2015/12/10/paul-gilroy-offshore-humanism/ [Accessed 19.02.2016].
Glissant, É. 1997. *The Poetics of Relation*. Trans. B. Wing. Ann Arbor: University of Michigan Press.
Glowczewski, B. 2011. 'Between Spectacle and Politics: Indigenous Singularities', In: B. Glowczewski and R. Henry, eds., *The Challenge of Indigenous Peoples*. Oxford: Bardwell Press, pp. 1–23.
Glowczewski, B. 2015. *Totemic Becomings: Cosmopolitics of the Dreaming*. Helsinki and São Paolo: n-1 Publications.
Glowczewski, B. 2016. *Desert Dreamers*. Minneapolis: Univocal.
Gordon, L.R. 2015. *What Fanon Said: A Philosophical Introduction to His Life and Thought*. London: Hurst and Co.

Graham, M. 2008. 'Some Thoughts About the Philosophical Underpinnings of Aboriginal Worldviews', *Australian Humanities Review*, 45. www.australianhumanitiesreview.org/archive/Issue-November-2008/graham.html. [Accessed 27.07.2016].

Halewood, M. 2011. *A. N. Whitehead and Social Theory: Tracing a Culture of Thought*. New York and London: Anthem Press.

Hallowell, A.I. 1975 [1960]. 'Ojibwa Ontology, Behaviour, and World View', In: D. Tedlock and B. Tedlock, eds., *Teachings From the American Earth*. New York: Liveright, pp. 141–179.

Henry, P. 2000. *Caliban's Reason: Introducing Afro-Caribbean Philosophy*. Abingdon: Routledge.

Hoffmeyer, J. 1996. *Signs of Meaning in the Universe*. Bloomington: Indiana University Press.

Hoffmeyer, J. 2008. *Biosemiotics: An Examination into the Signs of Life and the Life of Signs*. Trans. J. Hoffmeyer and D. Favareau. Scranton: University of Scranton Press.

Hunt, S. 2014. 'Ontologies of Indigeneity: The Politics of Embodying a Concept', *Cultural Geographies*, vol. 21, no. 1, pp. 27–32.

Jackson, M. 2014. 'Composing Postcolonial Geographies: Postconstructivism, Ecology and Overcoming Ontologies of Critique', *Singapore Journal of Tropical Geography*, vol. 35, no. 1, pp. 72–87.

Jackson, M. 2016. 'Aesthetics, Politics, and Attunement: On Some Questions Brought by Alterity and Ontology', *GeoHumanities*, vol. 2, no. 1, pp. 8–23.

Kant, I. 1960. *Observations on the Feeling of the Beautiful and Sublime*. Trans. J. Goldthwait. Berkeley: University of California Press.

Kant, I. 1996. 'What Is Enlightenment?' In: J. Schmidt, ed., *What Is Enlightenment? Eighteenth Century Answers and Twentieth Century Questions*. Oakland: University of California Press, pp. 58–64.

King, T. 2003. *The Truth About Stories: A Native Narrative*. Toronto: Anansi Press.

Kohn, E. 2013. *How Forests Think: Towards and Anthropology Beyond the Human*. Oakland: University of California Press.

Kopenawa, D. and Albert, B. 2013. *The Falling Sky: Words of a Yanomami Shaman*. Trans. N. Elliott and A. Dundy. Cambridge, MA: Cambridge University Press.

Kusch, R. 2010. *Indigenous Popular Thinking in América*. Trans. M. Lugones and J. Price. Durham and London: Duke University Press.

Latour, B. 1988. *The Pasteurization of France*. Trans. A. Sheridan and J. Law. Cambridge, MA: Harvard University Press.

Latour, B. 1993. *We Have Never Been Modern*. Trans. C. Porter. Cambridge, MA: Harvard University Press.

Lyotard, J-F. 1984. *The Postmodern Condition: A Report on Knowledge*. Trans. G. Bennington and B. Massumi. Manchester: Manchester University Press.

Magowan, F. 2001a. 'Syncretism and Synchronicity: Remapping the Yolngu Feeling of Place', *The Australian Journal of Anthropology*, vol. 12, pp. 275–290.

Magowan, F. 2001b. *Telling Stories: Indigenous History in Australia and New Zealand*. Sydney: Allen and Unwin.

Massumi, B. 2014. *What Animals Teach Us About Politics*. Durham and London: Duke University Press.

McKittrick, K. 2006. *Demonic Grounds: Black Women and the Cartographies of Struggle*. Minneapolis: University of Minnesota Press.

McKittrick, K. 2015. 'Yours in the Intellectual Struggle: Sylvia Wynter and the Realisation of the Living', In: K. McKittrick, ed., *Sylvia Wynter: On Being Human as Praxis*. Durham and London: Duke University Press, pp. 1–9.
Mendieta, E., ed. 2002. *Latin American Philosophy: Currents, Issues, Debates*. Indianapolis: Indiana University Press.
Mignolo, W. 2011. *The Darker Side of Western Modernity: Global Futures, Decolonial Options*. Durham and London: Duke University Press.
Mudimbe, V.Y. 1988. *The Invention of Africa: Gnosis, Philosophy and the Order of Knowledge*. Indianapolis: Indiana University Press.
Mudimbe, V.Y. 1994. *The Idea of Africa: African Systems of Thought*. Indianapolis: Indiana University Press.
Muecke, S. 2004. *Ancient and Modern: Time, Culture, and Indigenous Philosophy*. Sydney: UNSW Press.
Nadasdy, P. 2003. *Hunters and Bureaucrats: Power, Knowledge, and Aboriginal-State Relations in the Southwest Yukon*. Vancouver: University of British Columbia Press.
Nesbitt, N. 2013. *Caribbean Critique: Antillean Critical Theory From Toussaint to Glissant*. Liverpool: Liverpool University Press.
NOAA. 2015. 'State of the Climate Report: November 2015'. www.ncdc.noaa.gov/sotc/ [Accessed 01.12.2016].
Pollan, M. 2002. *The Botany of Desire: A Plant's-Eye View of the World*. London: Bloomsbury.
Povinelli, E. 1995. 'Do Rocks Listen? The Cultural Politics of Apprehending Australian Aboriginal Law', *American Anthropologist*, vol. 97, no. 3, pp. 505–518.
Povinelli, E. 2002. *The Cunning of Recognition: Indigenous Alterities and the Making of Australian Multi-Culturalism*. Durham and London: Duke University Press.
Povinelli, E. 2011. *Economies of Abandonment: Social Belonging and Endurance in Late Liberalism*. Durham and London: Duke University Press.
Povinelli, E. 2016. *Geontologies: A Requiem to Late Liberalism*. Durham and London: Duke University Press.
Pratt, M.L. 2008. 'In the Neo-Colony: Destiny, Destination, and the Traffic of Meaning', In: M. Moraña, E. Dussel and C.A. Jáuregui, eds., *Coloniality at Large: Latin America and the Postcolonial Debate*. Durham and London: Duke University Press, pp. 459–475.
Puig de la Bellacasa, M. 2015. 'Ecological Thinking, Material Spirituality, and the Poetics of Infrastructure', In: G. C. Bowker, et al., eds., *Boundary Objects and Beyond: Working with Leigh Star*. Cambridge, MA, and London: The MIT Press, pp. 47–68.
Puig de la Bellacasa, M. 2017. *Matters of Care: Speculative Ethics in More Than Human Worlds*. Minnesota: University of Minnesota Press.
Redekopp, J. 2014. 'Thinking Across Worlds: Indigenous Thought, Relational Ontology, and the Politics of Nature: Or, If Only Nietzsche Could Meet A Yachaj', Unpublished PhD Dissertation, University of Western Ontario, London, ON. Canada.
Rivera Cusicanqui, S. 1984. *Oppressed But Not Defeated: Peasant Struggles Among the Aymara and Quechua in Bolivia, 1900–1980*. Ginebra: UNRISD.
Rivera Cusicanqui, S. 2012. 'Ch'ixinakax utxiwa: A Reflection on the Practices and Discourses of Decolonization', *South Atlantic Quarterly*, vol. 111, no. 1, pp. 95–109.
Said, E. 2004. *Humanism and Democratic Criticism*. New York: Columbia University Press.
Scott, D. 2000. 'The Re-Enchantment of Humanism: An Interview With Sylvia Wynter', *Small Axe*, vol. 8 (September), pp. 119–207.
Serres, M. 2007. *The Parasite*. Trans. L.S. Schehr. Minneapolis: University of Minnesota Press.
Shari'ati, A. 1979. *On the Sociology of Islam*. Trans. H. Algar. Berkeley: Mizan Press.

Simpson, A. and Smith, A., eds. 2014. *Theorising Native Studies*. Durham and London: Duke University Press.

Simpson, L. 2011. *Dancing on Our Turtle's Back: Stories of Nishnaabeg Re-Creation, Resurgence, and a New Emergence*. Winnipeg: ARP Books.

Singh, B. 2015. *Poverty and the Quest for Life*. Chicago: University of Chicago Press.

Star, S.L. and Griesemer, J.R. 1989. 'Institutional Ecology, "Translations" and Boundary Objects: Amateurs and Professionals in Berkeley's Museum of Vertebrate Zoology, 1907–1939', *Social Studies of Science*, vol. 19, no. 3, pp. 387–420.

Star, S.L. and Griesemer, J.R. 2015. 'Institutional Ecology, "Translations" and Boundary Objects: Amateurs and Professionals in Berkeley's Museum of Vertebrate Zoology, 1907–1939', In: G.C. Bowker et al., eds., *Boundary Objects and Beyond: Working With Leigh Star*. Cambridge, MA and London: MIT Press, pp. 171–200.

Stengers, I. 2010. *Cosmopolitics I*. Minneapolis: University of Minnesota Press.

Stevenson, L. 2014. *Life Beside Itself: Imagining Care in the Canadian Arctic*. Oakland: University of California Press.

Sundberg, J. 2014. 'Decolonizing Posthumanist Geographies', *Cultural Geographies*, vol. 21, no. 1, pp. 33–47.

Thomas, A.C. 2015. 'Indigenous More-Than-Humanisms: Relational Ethics With the Huruni River in Aotearoa, New Zealand', *Social and Cultural Geography*, vol. 16, no. 8, pp. 975–990.

Todd, Z. 2016. 'An Indigenous Feminist's Take on the Ontological Turn: "Ontology" Is Just Another Word for Colonialism', *Journal of Historical Sociology*, vol. 29, no. 1 (March), pp. 4–22.

Tronto, J.C. 1993. *Moral Boundaries: A Political Argument for an Ethic of Care*. New York: Routledge.

Tuhiwai-Smith, L. 2012. *Decolonizing Methodologies: Research and Indigenous Peoples*. 2nd ed. London: Zed Books.

UNHCR. 2016. 'Mediterranean Death Toll Soars, 2016 Is Deadliest Year Yet', 25 October 2016. www.unhcr.org/afr/news/latest/2016/10/580f3e684/mediterranean-death-toll-soars-2016-deadliest-year.html [Accessed 01.12.2016].

Verran, H. 2015. 'Afterword: On the Distributedness of Leigh', In: G.C. Bowker et al., eds., *Boundary Objects and Beyond: Working With Leigh Star*. Cambridge, MA and London: MIT Press, pp. 499–500.

Vivieros de Castro, E. 1998. 'Cosmological Deixis and Amazonian Perspectivism', *Journal of the Royal Anthropological Institute*, vol. 4, no. 3, pp. 469–488.

Vivieros de Castro, E. 2014. *Cannibal Metaphysics*. Ed. and trans. P. Skafish. Minneapolis: Univocal.

Vivieros de Castro, E. 2015. *The Relative Native: Essays on Indigenous Conceptual Worlds*. Chicago: HAU Books.

wa Thiong'o, N. 1986. *Decolonising the Mind: The Politics of Language in African Literature*. Portsmouth, NH: Heinemann.

Watts, V. 2013. 'Indigenous Place-Thought and Agency Amongst Humans and Non-Humans (First Woman and Sky Woman Go On a European Tour!)', *Decolonization: Indigeneity, Education and Society*, vol. 2, no. 1, pp. 20–34.

Wheeler, W. 2006. *The Whole Creature: Complexity, Biosemiotics, and the Evolution of Culture*. London: Lawrence and Wishart.

Wheeler, W. 2016. *Expecting the Earth: Life, Culture, Biosemiotics*. London: Lawrence and Wishart.

Whitehead, A.N. 1933. *Adventures of Ideas*. Cambridge: Cambridge University Press.

Wolfe, C. 2007. *What Is Posthumanism?* Minnesota: University of Minnesota Press.

Wynter, S. 1984. 'The Ceremony Must Be Found: After Humanism', *Boundary 2*, vol. 12, no. 3, pp. 19–70.

Wynter, S. 1995. '1492: A New World View', In: V.L. Hyatt and R. Nettleford, eds., *Race, Discourse, and the Origin of the Americas: A New World View*. Washington, DC: Smithsonian Institution Press, pp. 5–57.

Wynter, S. 2001. 'Towards the Sociogenic Principle: Fanon, Identity, the Puzzle of Conscious Experience, and What it is Like to be "Black"', In: M. Durán-Cogan and F. Gomez-Moriana, eds., *National Identities and Socio-Political Changes in Latin America*. New York and London: Routledge.

Wynter, S. 2003. 'Unsettling the Coloniality of Being/Power/Truth/Freedom: Towards the Human, After Man, Its Overrepresentation – An Argument', *CR: The New Centennial Review*, vol. 3, no. 3, pp. 257–337.

Wynter, S. 2006. 'On How We Mistook the Map for the Territory, and Re-Imprisoned Ourselves in Our Unbearable Wrongness of Being, of Désêtre: Black Studies Towards the Human Project', In: L.R. Gordon and J.A. Gordon, eds., *Not Only His Master's Tools: African-American Studies in Theory and Practice*. Boulder and London: Paradigm.

Wynter, S. and McKittrick, K. 2015. 'Unparalleled Catastrophe for Our Species? Or to Give Humanness a Different Future: Conversations', In: K. McKittrick, ed., *Sylvia Wynter: On Being Human as Praxis*. Durham and London: Duke University Press, pp. 9–89.

Wynter, Sylvia, and Scott, D. 2000. "The Re-Enchantment of Humanism: An Interview with Sylvia Wynter." *Small Axe* 8 (September 2000): 119–207.

Yusoff, K. 2013. 'Geological Life: Prehistory, Life, Futures in the Anthropocene', *Environment and Planning D: Society and Space*, vol. 31, no. 5, pp. 779–795.

Yusoff, K. 2014. 'Geologic Subjects: Nonhuman Origins, Geomorphic Aesthetics and the Art of Becoming Inhuman', *Cultural Geographies*, vol. 23, no. 3, pp. 383–407.

2 Anti-colonial ontologies

A dialogue

Angela Last

How do you engage with the question of ontology in postcolonial theory? In light of the proposal of the Anthropocene, which regards the human as a geophysical force, two contrasting arguments have recently been made with regard to postcoloniality and ontology. These critically examine the kind of relational ontologies expounded by many theorists of science and technology studies or social anthropology. On the one hand, it has been suggested that postcoloniality suffers from an uncritical uptake of the European division of nature and culture (Jackson, 2014). Although attention has been paid to the multiplicity of cultures, nature's multiplicity, or nature's involvement with culture, has largely been denied. As a consequence, postcolonial theory has run into a dead end, where, at a time of obvious nature-culture entanglements, it is in danger of becoming irrelevant (see Chakrabarty, 2009). As one of the pathways of redemption, the relational ontologies of authors such as Tim Ingold and Bruno Latour have been proposed as a strategy to give greater force to the anti-colonial project of postcolonial theory. The author Eduardo Viveiros de Castro has even argued for anthropology, with its new ontological preoccupation, to become 'the theory/practice of the permanent decolonization of thought' (2014: 17). A similar turn to rethinking of ontology has been put forward by authors such as Arturo Escobar (2008) and Walter Mignolo (2000) in their engagements with geopolitical hierarchies.

On the other hand, the theories of Bruno Latour, Tim Ingold, and others have been exposed as deeply colonial in their non-recognition or appropriation of indigenous ontologies (Ramos, 2012; Todd, 2014), as protecting Western science through unhelpful boundary-setting (Graeber, 2015), or as perpetuating an image of human homogeneity that takes the white, privileged subject as a starting point and model organism (Sundberg, 2014; Weheliye, 2014). With this, most new 'materialisms', from ANT to posthumanism, have come under fire for being unable to recognise human difference or the political problems this blind spot entails (Fuller, 2000: 11, 23; Harding, 2008: 122). Likewise, their authors have been taken to account for looking towards other white authors as discourse participants, while potential non-white or 'indigenous' participants remain empirical examples rather than voices in their own right (see Tuhiwai Smith, 2012; Ko, 2015). This has led to calls for an overall decolonisation of academic discourse, directed at both knowledge-makers and knowledge itself.

Both positions, in my opinion, have valid points. Postcolonial theory, as framed by the majority of the Euro-American academy, has a well-known and justified aversion to nature, materiality, and ontologies that fail to recognise hierarchy or difference. In a similar manner to Jackson, the Martinican poet and theorist Edouard Glissant accuses postcolonial theory of choosing 'reason' over messier connections and registers (see Diawara, 2015). In defence, many postcolonial authors draw attention to the negative experiences of being 'indigenised' by the coloniser in a fashion that placed them outside culture and, with this, outside of humanity. Nature was not only something to be fought against as a category, but also played a part in the performance of punishments such as lynchings (Finney, 2014). Maryse Condé notes how, in the 1930s and 1940s, Martinican négritude writer Suzanne Césaire rallied against the association of black Caribbeans with nature, including their own identification with it in literature and cultural production in general (1998a). Although the négritude writers did not condemn associations with nature outright – it seems that they mainly fought against exoticisation and envaironmental determinism – Condé finds that they often privileged 'ideologisation over description' (1998): nature was identified as culture. This sensitivity towards the utilisation of 'nature' against the colonised continues to echo through postcolonial discourse, where I have witnessed many instances of hostility towards any form of 'materialisation', particularly the utilisation of the term 'nonhuman' (Gilroy, 2015). Often, it seems accepted to deconstruct naturalisations of coloniality, but not necessarily to reimagine or reappropriate them. As Arun Saldanha notes, matter is often seen as human prison or violent distortion (2007: 206). This fear of matter is, of course, not exclusive to postcolonial theory.

The second position highlights how epistemological violence translates into ontological violence. Even in the recent past, literary scholars who wanted to study literature from former colonies were told to change to Anthropology (Price and Price, 2013). Today, white European/Euro-American writers still dominate academic discourse around the world. As Arturo Escobar writes on the subject of ontology: 'The fact that dominant modern ontologies are connected to these other ontologies asymmetrically through the very same projects means that the latter are almost inevitably refunctionalised at the service at the former' (2008: 15). Academics themselves have drawn attention to the ongoing reinforcement of colonial imaginaries and practices inside universities (Bhambra, 2015; Coleman, 2014; Escobar, 2008; Law and Lin, 2015; Shilliam, 2014), including the habit to relegate questions of colonialism and race are almost exclusively to modules on postcolonialism. This position also brings out a problem that has been noted by other authors such as Patricia Noxolo and Marika Preziuso (2013), namely, that the discourse not only around materiality, but also around postcoloniality, in the Euro-American academy, consists of a very narrow canon. This privileging of a few authors that serve to reference 'postcolonial theory', tends to exclude the many authors that already offer alternative or even explicitly decolonial approaches to both culture and nature, especially in the form of literature or poetry. Even when materialism or relationality are proposed as a deconstructive path, it is rarely authors from former colonies, who explicitly constructed their materialisms as decolonial tools, that are brought to the discourse.

One could say that, in light of the aforementioned problems of both postcolonial theory and new materialism, such anti-colonial materialist or relational proposals generally pass under the radar of the anticipated in both directions. Yet an attention to human equality and difference is not mutually exclusive with a focus on materialism and vice versa. In fact, many authors have advocated a focus on 'nature' or 'matter' to rethink issues such as race, class, sovereignty and even gender, or have even insisted that radical rethinking of ontology should be seen as a key feature of decolonial thought and practice (Blaser, 2009; Haraway, 1991: 173; Noxolo, 2009, 60; Verran, 2001: 33; Weheliye, 2014: 32). Likewise, Patricia Noxolo reminds us that 'postcolonial geography can also be understood as a struggle for materiality' (Noxolo, 2009: 60). Achille Mbembe (2015), for instance, drawing on Ngugi's call to 'see ourselves clearly', notes how 'Black thought' has a strong tradition of theorising relationality (he cites the example of Edouard Glissant) that should continue to be mobilised, especially as imperialist practices continue to threaten life for everyone on this planet:

> We are called upon to see ourselves clearly, not as an act of secession from the rest of the humanity, but in relation to ourselves and to other selves with whom we share the universe. And the term "other selves" is open ended enough to include, in this Age of the Anthropocene, all sorts of living species and objects, including the biosphere itself. (. . .)
>
> In other words, a new understanding of ontology, epistemology, ethics and politics has to be achieved. It can only be achieved by overcoming anthropocentrism and humanism, the split between nature and culture.

Given that a rich literature on materiality and decoloniality already exists in both in the postcolonial and STS discourse, how else could a potential dialogue on the deconstruction of the nature-culture boundary look like, other than a story of appropriation or correction?

In this chapter, I am proposing a dialogue between postcolonial and feminist STS based engagements with the nature-culture boundary. The latter has paid close attention to the decolonial dimension of their deconstructions of 'Man' (or 'Scientist' as in the case of Isabelle Stengers), reflected in a greater sensitivity to the relationship between human-nonhuman and inter-human violence. The work of Donna Haraway, Helen Verran, Isabelle Stengers, and Sandra Harding, for instance, grapples with questions of racism, sexism, colonialism and other 'rationalisations' of epistemological, ontological and physical violence. It is striking that feminist and postcolonial challenges here have a strong overlap of terminology and concepts, such as Edouard Glissant's and Helen Verran's 'imaginary', Isabelle Stengers' and Aimé Césaire's respective 'cosmopolitics', Donna Haraway's and Sylvia Wynter's contestations of 'Man', Daniel Maximin's and Donna Haraway's differing proposals of human-nonhuman hybridity, to name but a few examples.

What makes these overlaps interesting is their emergence from, as Isabelle Stengers puts it, opposite sides of the divide (2012). While both postcolonial critiques and feminist/postcolonial STS critiques are critical of Western science and the nature-culture imaginary, they perform different positions, not just because of their different disciplines – as scientists/philosophers of science and poets/literary scholars respectively – but frequently also because of their different subject positions. Feminist philosophy of science places great emphasis on the situatedness (often of the writer) on the colonising side of a divide (through whiteness, through science), and does not make assumptions that all humans are comfortably united on the 'culture' side of the nature-culture boundary. This situating is an acknowledgement of privilege and implication in ongoing violence, which has implications for the way white academics can participate in decolonisation and related forms of activism. As Stengers writes in her article on 'reclaiming animism': 'I must accept to not feel free to speak and speculate in a way that would situate others. Rather, I must acknowledge the fact that my own practice and tradition situate me on one side of the divide, the side that characterised 'others' as animists' (2012). Further, she notes how the various demands and separations placed upon herself and others by her tradition prompt her to remain on her side of the divide. It becomes obvious to her that 'there is work do be done on this side' (2012).

Postcolonial engagements with the nature-culture boundary, on the other hand, often speak from the 'other side of the divide', from the 'interruption of Western discourse', to paraphrase Homi Bhabha (1994: 199). Just how much this was an 'other side' was rendered to me rather strikingly when I received the 1979 German translation (Kindler Verlag) of Edouard Glissant's novel 'La Lézarde', originally published in 1958 (I was curious how Glissant was made to sound in my native language). I was shocked to read that the cover blurb declared the book 'an unusual example of modern negro literature' ('moderne Negerliteratur'). Further, while 'La Lézarde' is a highly political attempt at destabilising the holy grail of European subjectivity by challenging nature-culture boundaries, the blurb declared it a documentation of the 'transformation of negro awareness from the natural and instinctive, to the rational and conscious'. This racist misreading renders apparent how European knowledge criteria, imposed on the authors through colonialism, situated them both within and without. Any divergence from or experimentation with European norms was initially equated with racial and cultural inferiority. It was not registered that the constant racist boundary policing engendered a variety of contestations, including the complete rejection of science or theory as system (Césaire, 1978), the mockery of the European obsession with classification and measurement (Maximin, 1995), or a critique of how European norms and imaginaries have become accepted in Caribbean societies (Césaire, 2012; Condé, 1998b).

While authors such as Glissant may have lost the 'negro' disclaimer in the past decades and have become embedded in the Euro-American knowledge economy (Hallward, 2001), their theoretical contributions continue to remain less visible in disciplines such as geography, where 'new materialism', for instance, begins with Latour and Haraway, or with Deleuze and Guattari. This chapter, then, has three purposes: first of all, to contribute to the existing project of postcolonial attempts to challenge the hegemonic European construction of nature. Too often, during discussion in geography, have I encountered the statement: there is no postcolonial literature on the environment, and I strongly disagree that this is the case.

Second, to contribute to highlighting how a postcolonial sensitivity can shape, from the positions we write from, our capacity to stop participating in the 'the continuing subordination of other worlds' (Blaser, 2009: 18). Thirdly, to continue dialogue across the divide, while acknowledging that different kinds of work need to be done from different subject positions. Having entered the postcolonial discourse, as a white academic, through science and technology studies as well as science itself (physics), I found the diversity of anti-colonial materialist proposals thought-provoking, as well as alarming in their absence in what generally counts as materialist discourse (for instance, 'humanity' is often unproblematically presented opposite 'nonhumanity'). Specifically, I became interested in the conversations between generations of French Caribbean writers around materiality and Caribbean identity, which involves discussions of Caribbean nature-culture constructions. Authors include Suzanne and Aimé Césaire, Edouard Glissant, Maryse Condé and Daniel Maximin (I apologise in advance to readers from literature studies, as these authors and their concepts will be familiar). For me, these authors draw out very sharply the far-reaching repercussions of colonialism that, in their destructiveness, move across nature and culture.

In the following pages, I would like to construct one possible dialogue across the divide, around challenges to universality, one of the key issues that runs across ontology and epistemology. I have organised these dialogues around the concepts of 'cosmopolitics', which, at its core, addresses the problem of universality as exclusion, and 'imaginaries', which addresses the cultural and material situatedness of rationality. These dialogues have helped me to deepen my own understanding of the divide and ways of speaking across the divide. I use these examples as a reminder that discussing 'postcolonial ontologies' necessarily involves an attention to one's own situatedness within the tension between existing and proposed ontologies.

Cosmopolitics

The universal as exclusion has been attacked by both postcolonial and STS scholars. Particularly in the STS discourse, a great degree of hostility has been directed at the 'cosmos' as a universalising imaginary: 'The problem we face now is that it's precisely this 'one cosmos', what I call mononaturalism, that has disappeared. It's impossible for us now to inherit the beautiful idea of cosmopolitanism since what we lack is just what our prestigious ancestors possessed: a cosmos' (Latour, 2004b: 453). At the same time, the cosmos has become re-introduced as a provocation by authors such as Isabelle Stengers, who does not associate it with Latour's 'mononaturalism', but with an impossibility to delineate 'what is human':

> The prefix 'cosmo-' indicates the impossibility of appropriating or representing 'what is human in man' and should not be confused with what we call the universal. The universal is a question within the tradition that has invented it as a requirement and also as a way of disqualifying those who do not refer to it. The cosmos has nothing to do with this universal or with the universe as an object of science. But neither should the 'cosmo' of cosmopolitical be

> confused with a speculative definition of the cosmos, capable of establishing a 'cosmopolitics'. The prefix makes present, helps resonate, the unknown affecting our questions that our political tradition is at significant risk of disqualifying.
>
> (Stengers, 2011: 355)

While Stengers' 'cosmopolitics' 'expresses the insistence of the 'cosmos' within politics' (2011: 415), in the sense that it allows traffic between 'politics' and that which is excluded from it, it remains suspicious of a material or ideological basis for universalism. What is or unites humanity? We don't, and may never, know, and thus maybe shouldn't look for it. Instead, we should focus on dismantling existing hierarchies. What is matter? We don't know either, but we shouldn't exclude it and how 'our' actions and decisions are entangled with it (2010: 40).

The theme of the cosmopolitical has also been played a part in anti-colonial writing. During the decolonisation struggle in the mid-twentieth century, many anti-colonial writers sought to contest the prevailing, exclusionary kind of universality, while it remained unquestioned by the coloniser. In fact, as Shepard (2006) argues, decolonisation became a strategy for the coloniser to reaffirm their European values against those of the newly independent countries. The aim of these anti-colonial writers was not only to question European universalism(s), but to question the aspiration towards them. Partly, this was achieved by drawing on materialist theories. These included insights from both historical materialism and from the natural sciences. Matter was regarded as a battleground for various reasons. The key ideological tension of the time was between matter as an alternative universality and matter as a difference engine. While imperialists and fascists came to regard material difference as a negative form of biological determination, decolonisation activists embraced the creative potentiality of both universality and difference to rewrite hierarchies.

A key theme of this period was the cosmos, which gave rise to a great diversity of 'cosmic materialisms' that explored how to think with this vast expanse of space, time and matter. For many writers, especially of the interwar period, 'the body becomes a way of accessing the universe, a medium that unites world and consciousness' (Weil, 1956: 592). Thinking cosmically meant resituating the human in terms of significance, as well as in terms of common origins, vulnerability and potential. The 'cosmological universalism' (Wilder, 2015: 236) that suggested itself, however, was not simply an attempt to level human cultural difference in the face of a universal nature. Instead, it acknowledged the role of the material or natural in the construction of difference. Importantly, such anti-colonial materialisms highlighted how 'natural difference' is in constant operation to increase the coloniser's power, but can also be put to constructive use to counter such destructive naturalisations. The Martinican writers Aimé and Suzanne Césaire, for instance, employed natural imagery to protest and inverse colonial stereotypes. The closing sentences of Suzanne Césaire's 1942 essay 'Poetic Destitution' exemplify this strategy: 'To hell with hibiscus, frangipane, bougainvillea. Martinican literature will be cannibal, or it will not be' (2012: 27).

As co-founders of the négritude movement, their aim was to try to 'unthink' colonial logics about nature and culture, and create an alternative that could be described as a counter-universal emergent from a unification of the excluded. Their deliberate deployment of monstrous and tropical imagery thus went beyond mere inversion to liberate the colonised from their relegation to exploitable nature. It experimented with the liberating potential of a self-selected alignment with 'other natures' ('nonhumans', including those humans excluded from 'humanity'), partly through a cosmic deep-time perspective where boundaries between humans, flora, fauna and geology remain blurred, where nature is not (just) external, but internal. 'In us [resides] the mankind of the ages. In us, all men. In us, the animal, the vegetable, the mineral. Man is not merely man. He is universe' (Césaire, 1978: 162).

This search for new alliances, again, worked in two directions: first, it addressed the colonised's alienation from nature, during colonisation and post-colonisation (Dash, 1985: 15). The Césaires and others recognised that 'nature' has to be reclaimed and its boundary shifted, redistributed or dissolved. Second, it tapped into the widespread European colonial anxiety of becoming 'a different kind of organisation of matter and thus a different kind of being' in non-native climates (Allewaert, 2013: 4). Alliances across other species, other matter, were threatening, precisely because they disturbed nature-culture boundaries. In her writing, Suzanne Césaire mocks the boundary that the white Europeans have placed between themselves and the rest of humanity, and the fear it creates about boundary crossing, for example, through interracial procreation (Césaire, 2012: 43). Since this fear not only affected the coloniser, the Césaires proposed that conscious and new alliances across the nature-culture boundary could be empowering for those who were being dehumanised. In this respect, the cosmopolitics of the Césaires are perhaps closer to those of Donna Haraway's cyborg imaginary. As beings that transgress human-animal-machine boundaries, cyborgs do not seek to be part of the cosmos of the currently agreed 'universal': 'cyborgs are not reverent; they do not remember the cosmos. They are wary of holism, but needy for connection' (2004: 9–10).

Connection is a key theme. Reading the Césaires, Stengers/Haraway together, one notices a theme of solidarity building. Négritude became such a strong provocation because it attempted to 'speak the universal from the colonial periphery' (Garraway, 2010: 84). This alternative universality was as much about undermining hegemonic notions of racial hierarchies as it was about building new solidarities across the racialised or otherwise negatively materialised. This was partly to be achieved through surrealist poetics and their exploration of the unconscious. By contrast, Isabelle Stengers also advocates audacity of imagination, but by way of laziness as an energiser: 'Today, to fight for the right to laziness, means to regain the strength not to make common cause with the enemy, not to give in to the resentment that is instilled into us, not to accept the slave songs we are required to sing' (2013). With a strong anti-capitalist, rather than explicit anti-colonial slant, she calls for disenfranchised publics to self-organise against current notions of 'progress' and 'growth' (2009). Donna Haraway, on the other hand, emphasises the intersection of struggles and imaginaries. She takes care to point out that cyborg imagery has been used across genres, fields and struggles to contest the equally fictional 'whole' of nature or of Man. Drawing on Audre Lorde's 'Sister Outsider', Haraway points to a diversity of already existing cyborgs, such as 'women of colour' who constitute a cyborg identity due to their 'fusions of outsider identities (women/of colour), and because of their economic role as preferred workers in a supposedly white phallocentric technological universe (2004: 32). Cyborgs are intersectional beings that not only have a potential for multiple oppression but multiple solidarities.

For both directions, the cosmos serves as a provocation: all authors expose that most imaginaries of cosmopolitanism and universality fail to perform already at a superficial level. With their counter-imagery, they show how the universal, so far, has been about control of privilege and not about a desire for inclusion. In this context, the cosmic or cosmopolitical can be seen as an attempt at building solidarity without dictating a particular alliance/universality. Rather, it is intended as a reversal of the violence of hierarchical materialisation that also moves beyond the human. Both directions also call attention to the fact that building new solidarities does not just involve thinking new relations, but also thinking about the terms on which these relations are built. Both sets of authors call not so much for a 'multinaturalism', but for an enquiry into the conditions of such a multinaturalism – as with 'multiculturalism', the insistence on 'multinaturalism' can be used to avoid dealing with certain questions. These include hierarchy, material dynamics, desire for universality and the situatedness of authors. They remind us that it is not just a matter of diversifying, but of ensuring that existing hierarchies and privileges are not replicated. Here, authors such as Aimé and Suzanne Césaire remind us that, while common or co-constituted materiality remains a strong imaginary, the embracing of this kind of imaginary is less problematic for those who already operate from within the space of universality (including the space of appropriated land). The step or gesture to a 'parliament of things' (Latour, 2004a) or a different utopian/political material multiplicity is not that far. From outside this space, the alliance with materiality is taken as given and frequently remains overlooked in its radical conceptualisation. While materiality itself may be colour blind, humans are not (Ferreira da Silva, 2007: 9).

Imaginaries

The attempts of the négritude movement to build solidarities among the excluded were at once welcomed and rejected. The rejection occurred for two reasons: because of the desire for assimilation into universality; and, because of the impression of essentialism (the association of blackness with common characteristics or aims). Despite this critique, the idea of a provocative cosmic materiality has retained its appeal. As Edouard Glissant, a known critic of négritude phrases it: '[t]he being of the world cannot be divided from the being of the universe and whatever imagines itself suspended in this whole. This suspension is not primarily spatial' (Glissant, 1997: 187). In his own work, Glissant also experimented with the decolonisation of the landscape, albeit with different redescriptions. This also included his own version of relationality, the 'Poetics of Relation' (2010 [1990]).

Like Aimé Césaire, Glissant insists on poetics as a means of building new imaginaries, because of a disillusionment that things can be changed through politics (or politics alone). Poetics, he argued, may be, on the one hand, totally ineffective against oppression, but that also makes them powerful – they are outside of the system and within reach of those outside it (2009: 12–13, 19–20).

At a conference on Edouard Glissant's work, Maryse Condé referred back to Suzanne Césaire's angry attack on Martinicans writers' colonial use of 'nature'. While pointing out that nature kept being subversively used in the Caribbean, including by the Césaires themselves (1998), she particularly credits Glissant with reintroducing 'nature' as an acceptable topic for Caribbean literature and as a vital part of the decolonial project. In his critique of colonial binaries, Glissant attacked not only cultural divisions, but the nature/culture division as well (1997: 14). In his 1958 novel, 'La Lézarde', entities and phenomena normally relegated to 'nature' permeate nearly every sentence, every human action that is described. Even the courtroom is not free from the heat of the sun or from issues surrounding nonhumans, such as the theft of an animal (1979: 195–196). Within this constant permeation of the 'cultural' with the 'natural', the meaning of these 'natural' events and entities remain deeply unstable. Things are shaped by context, by each other – the sea, for instance, can be a source of balance or threat. With this oscillation between nature and culture, Glissant illustrates how the two categories are constantly (re)shaped by one another – no clear cut can be made.

This indeterminacy extends to humans and human decisions. In 'La Lézarde', this is illustrated through the murder of a mulatto landowner, who intends to privatise the land around the river. To the reader, it is never clear who killed him. Thaël, the 'country boy' who gets co-opted by urban radicals to deflect attention from the group? His radical 'friends' who insist on co-responsibility? The river that pulls the man underwater and tears him away during the fight? The landowner himself, though his strange actions or emerging conscience? The values or environment he was shaped by? Glissant takes care to place and leave the reader in this space of indeterminacy. Various proposals are made throughout the book, but no position is taken other than that of a determined exposure of relationality and a commitment to the 'inexhaustible diversity in reality' (Dash, 1989: 609). Even the narrator seems apologetic and intervenes in less than helpful ways in the search for answers. Glissant's translator Michael Dash describes the author's aim as a dismantling of the individual and the 'heroics of self-formulation' (Dash, 1989: 609), through situating the mind in the body, and the body in relation to the world. With this, Glissant also warns against the exclusion of the 'natural agency' from both identity formation and land formation. As Condé puts it, Glissant insisted that 'self-knowledge is impossible without appropriation of the geographical universe that it is embedded in' (Condé, 1998a). Or, in the words of Michael Dash, Glissant focused on how the land was open to being shaped in people's imagination, but also on the land's 'shaping power on those who inhabit it' (Dash, 1985: 7).

It seems as if, with every publication, Glissant has sought to further destabilise our perspective of the world, later adding the relations between the local and the global in his conceptualisations of the 'tout-monde' (literally 'all-world'; often translated as 'whole world', but maybe 'world entirety', 'world-whole' or 'world-totality' would be more appropriate) and the 'chaos-monde' ('chaos-world'). This tout-monde advances a positive re-reading of totality: 'what is totality, once again, and through return, if not the relation of each matter to all others? (2009: 16?). Glissant's tout-monde attempts to replace 'totalitarian modes of thought' with a focus on openness and process (Burns, 2012: 123). Within this totality, relation gives rise to difference as newness, rather than exclusion (Burns, 2012: 82; Dash, 1995: 22). This later work has often been criticised as 'apolitical', for its extreme destabilisation (leading to what?) and its reliance on chaos theory's dictum that small events in remote places can have big consequences somewhere else (Hallward, 2001: 67–68). Its particular negotiation of the micro and the macro can be read both as an affirmation of the non-futility of small actions, but also as an excuse to 'not think big'. Glissant himself states that his 'Relation' 'does not have morals' (2009: 73). Instead, what his Relations prompt is to consider how, that which we ignore or don't know about the world in our moral and political knowledge and decision-making, continues to affect us, and others, beyond our present (2009: 74).

At the same time, Glissant insists that Relation is not passive participation (Glissant, 1997: 14), not the equivalent of 'valium', as Chris Bongie puts it (2008), to endure globalisation. It is about developing a sensitivity to relations in a way that renders one unable to participate in non-relational imaginaries (Glissant in Bongie, 2008: 336–337). Celia Britton, writing about Glissant's Relation, sides with the inherent political position that is required of adopting an extreme relational vision: because it is such a different conception of the world, 'Relation is an anti-imperialist project' (1999: 12). It is difficult to imagine or put into practice by the imperialist 'West', because the West is 'anxiously determined to cling to their conception of a world in which they dominate and control rather than 'a world in which one is, quite simply, one agrees to be, with and among others' (Glissant, 2010: 128). The question 'where are we located?' thus results in a productive confusion, since anyone or anything is located across the material and non-material, across geographical boundaries. Here, Glissant's work highlights how everyone has to navigate the colonial/decolonial boundary at all times, whether consciously or unconsciously, because any object contains the whole world in the form of relations.

The idea of relation as an anti-colonial strategy has also been discussed by the science studies scholar Helen Verran, who, like Glissant, uses the term 'imaginary' to highlight the consequences of the Western exclusion of images, and relations, especially the distributed or indeterminate nature of 'the mind'. For Verran, 'the imaginary does not involve the mind, and certainly is not located there' (2001: 37). Situating it close to cultural theorist Michael Carter's definition of the imaginary as 'not a 'thing' of the mind, but an over-arching relation' (Carter, cited in Verran, 2001: 37), she emphasises that it is a product of 'doing' – action being a nexus of many non-rational influences that move across space, time and 'matter'. Thinking with an 'imaginary' 'eschews the picture of a given foundation and its symbolising, instead of having worlds, with their objects and subjects' (2001: 37). For Glissant, an imaginary performs a similar role: it destabilises boundaries between thought and world:

> Thinking thought usually amounts to withdrawing into a dimensionless place in which the idea of thought alone persists. But thought in reality spaces itself out into the world. It informs the imaginary of peoples, their varied poetics, which it then transforms, meaning, in them its risk becomes realized.
>
> (1997: 1)

To recognise Western knowledge-making practices as culturally situated and not the pinnacle of development, because they are apparently freed from 'imaginaries', in turn, challenges the notion of the kind of 'development' that continues to be imposed on 'developing' countries, not only in terms of their economies, but also their knowledge-making practices (arguably, also part of their economies). Part of Verran's strategy is to highlight how all cultures have built such imaginaries, and that the way forward is to 'learn to celebrate rather than deny the ontic/epistemic imaginaries involved in modern knowledge traditions' (Verran, 1998: 249).

The focus on mind/thought and world in both Glissant and Verran brings out the importance of thinking about links between representation and nonrepresentation, especially in terms of privilege:

> Transgression of the boundaries of the given categories of world, knower, and knowledge is perhaps the most serious category mistake in foundationism. It is sometimes claimed that 'primitives' transgress the boundary between knowledge and the world in literalising – constructing fetishes by having symbols and objects as the same thing. This 'mistake' subverts all claims that 'fetishising primitives' might make for the certainty of their explanations of the world.
>
> (Verran, 2001: 34)

Here, a similar situation occurs as with alliances with the 'natural'. Who can cross what boundaries and remain heard or registered? Who can use this as a means? Glissant seems to remain more optimistic than Verran in his insistence on accessibility. In his Philosophy of Relations, he highlights how relations exist not only between material entities, but between 'ideas, identities, intuitions' (Glissant, 2009: 72): the common world runs across the material and the virtual. Although privilege makes both domains more accessible, everyone technically has access to both. One means for the underprivileged is to resort to 'opacity' – to refuse total transparency. Erica Violet Lee (2015), on her blog 'Moontime Warrior', gives one such example of anti-imperialist opacity:

> Here's the thing: even if Indigenous people were to spend our entire lives trying to explain our philosophies to settlers (. . .), they still might not get it. Not only was the classroom content I described above inaccurate, but silly, as well, demonstrating a basic lack of understanding regarding the ways our communities operate. As Thurman Lee Hester Jr. and Dennis McPherson write in "The Euro-American Philosophical Tradition and its Ability to Examine Indigenous Philosophy", ". . . to operate within the paradigm of

> Euro-American philosophy would mean that you are necessarily cut off from any real understanding of Indigenous thought."
>
> This is the worst accusation of all for a white male philosopher: the suggestion that he cannot know something, and further, to point out that others have knowledge he might never be able to have. Not as a result of biology (!!!), but as the result of a redefinition of knowledge, giving weight to embodied experience; a recognition of Indigenous epistemic privilege.

While opacity is an anti-imperialist practice, for Glissant, opacity does not prevent, but rather constitutes, a condition for relationality – relationality does not necessitate or imply transparency (Mbom, 1999). Opacity, in a sense, makes relations possible, or, as Marilyn Strathern would put it, relations are created through separation (2004: 111). Opacity is that which cannot be communicated, that which remains unintelligible (especially to the West that is in need for transparency). Glissant and Verran also insist on the link between language and opacity (Glissant, 2009: 70; Verran, 2001: 234). For them, language (and place) shapes a worldview, causes things to be untranslatable, nonrepresentable to a person from another 'world' (even the concepts of representation and relation are culturally specific). For Verran, difference even 'exists outside worlds', because when one world is translated into another, it needs to be reshaped to that world to be intelligible (2001: 151) – this is how universality naturally reproduces.

In a similar manner to Lee, Verran also refers to opacity as part of anti-colonial resistance, giving the example of teachers who refuse to entirely subject to a European curriculum and teaching style:

> Mr Ojeniyi's lesson might be alternatively explained as a heroic resistance. It could be said that Mr Ojeniyi had engaged understandings of a logic of Yoruba numbering that he shares with the children, but which is entirely opaque to the òyinbó (white) lecturer. A story could be told of Mr Ojeniyi and the children joining forces to offer a heroic resistance to the European colonising inherent in the accredited Nigerian mathematics syllabus.
>
> (Verran, 2001: 14)

Both Verran and Lee continue to point to the power dynamics surrounding opacity. Teachers who want to teach a non-colonial curriculum are taking risks in terms of their job security, while the adoption of the colonial curriculum might result in strained relationships with pupils. Lee points to the coloniser's desire for authority, no matter whether something is understood or not, and the tendency to sideline indigenous colleagues (2015; see also Ko, 2015). With this, they suggest that decisions on the nature of reality are intimately tied to negotiating privilege. The building of a 'worldmentality'[1] (mondialité) (Diawara, 2015) as a counter-imaginary to universality thus needs to take into account that, in a sense, everything (including one's own position; across nature/culture) is other. That the other cannot be fully known must not be used as an excuse to sideline communication or ideas, or to cling to the performance of cultural stereotypes, but should translate into struggle for an abandonment of hierarchies. Such a movement against hierarchy would involve an openness to being changed by the other (Glissant, 2009: 66), rather than a continuation of the drive to universalise in ways that create further inequalities.

Openings

In her book 'Towards a Global Theory of Race', Denise Ferreira da Silva diagnoses that 'the proliferating (racial, ethical, postcolonial) 'subaltern ontologies and epistemologies' are an indicator that the subject-object division is still very much alive and something to be contested' (2007: xix). To this, I would like to add the call to arms 'don't conclude – populate!' of Maria Puig de la Bellacasa (2014). In her book she calls for an expansion of the repertory of visions, which, in the context of this chapter, could translate as a need to create a growing body of examples as a resource against the (self)destructive system in which we currently seem forced to participate. It is useful to think of theory as a tool that can aid in unsettling of the status quo. As Anne-Marie Mol recently put it: 'My strategy has always been to strengthen the margins, not criticise the mainstream' (2015). With this chapter, I have tried to add a further example that also highlights the importance of promoting dialogue across different kinds of projects and resources. To use Donna Haraway's phrase, the question may be not just a matter of 'exploding binaries' by ourselves (2004: 116), but to combine such explosions in the most effective way. The need to talk to one another (even if in angry voices) is especially important for academics who, as Richa Nagar points out, tend to 'become locked into pure theoretical positions and loyalties', which, in turn, affects their capacity for solidarity (2014: 2). I feel that dialogues and debates, such as the one around postcoloniality and ontology (are we colonising or decolonising, or rather, who is?), can help all participants, but especially people on my (the colonising) side of the divide, to 'change with'.

Such dialogues can also help to remind us, regardless of position, that we must not lose sight of what we are exploding binaries for, both in theory and practice. Where (and how) do we want to be? This question also carries resonances of a related debate, that around knowledge making practices. At present, calls for a de-universalisation of European knowledge and knowledge production appear to grow in strength. In universities across Europe and its former colonies (that includes those in North America), calls for the diversification of the curriculum have started to become integrated into strategic institutional aims. One could be cynical about this development – the problematic relation between mondialisation and modialité (globalisation and 'worldmentality') – since such efforts are often paired with calls for internationalisation, often meaning the attraction of lucrative overseas students and funds (a similar argument can be made for publishing). At the same time, attempts to re-examine the dominance of European (and often male dominated) knowledge have belatedly led to some deeper revisioning efforts. These have raised questions about the way knowledge-making should proceed, and especially around questions of inequality. Here, the attention to the lack of boundaries between the material world and the world of ideas, ontology and epistemology, can be a useful tool for keeping track of uncomfortable relations.

Further, writers such as Vandana Shiva or Richa Nagar remind us to always think or go beyond mere theorisations. While the two authors differ substantially in their theoretical set-up and rhetorical style, they insist, through their writing as well as actions, on both theoretical and physical means of decolonisation. Although Shiva's work seems to have gone out of fashion – she is frequently dismissed as essentialist, reactionary and un-feminist for speaking of protecting 'Mother Nature' (for an overview of the debate see Agarwal, 1992) – her analyses of the relationships between natural and societal degradation remain disconcertingly pertinent (1989). In her texts and talks she constantly moves between material and representational violence and offers means for different population groups to address this violence (2013). Nagar's work offers such means, too, and especially problematises the frequent distinction between academic 'abstract thinking' and activist 'concrete doing' (2014: 2). Despite their differences, is worth bearing in mind how both authors not only embody the intersection of academia and activism, but also the intersection between postcolonial theory and STS sensitivities in the way they attend to both theory and empirics. With this, they also show the difficulty and risk of embodying such a position. Rejection – whether from journals, discourses, hiring committees, activists or whoever defends 'pure' positions – is one such risk, but, as Nagar emphasises, 'muddying theories and genres' has never been more necessary in order to build relations across discourses that share the same goal (2014: 2).

Shared goals can also function as something to build towards when solidarities across the colonial divide remain problematic for obvious reasons. The most difficult issue here, perhaps, is the often forgotten materiality of land in postcolonial discourse, which is intimately tied to issues of knowledge, reason and subjecthood, whether directly or by proxy. In his later texts, Glissant accused postcolonial discourse not only of siding with reason, but also of participating in the 'destruction of lives and the environment' much the same ways as former and ongoing colonisers (Diawara, 2015). Tuck and Yang stress how the use of 'decolonisiation' as a metaphor in recent discourse (mostly by 'settlers'), empties it of any possibilities for claims to land and other reparations (2012: 7). Such examples of land grabs and physical violence render apparent how ontologies – and ontological hierarchies – have very real consequences. At present, given the ongoing imperialist violence, it seems ironic to talk about postcolonial ontologies. The desire for postcolonial ontologies, however, might be one step towards recognising the profundity of this enterprise and, for the colonising side of the divide, to recognise other ontologies – not as quick fixes for our own problems, but as a call to support decolonisation across the mental and the material world.

Acknowledgements

I would like to thank Uli Beisel, Anja Kanngieser, Kathryn Yusoff, Heather McLean, Tahani Nadim, Gurminder Bhambra, and Mark Jackson for reading suggestions and discussions. I would also like to thank Maria Puig de la Bellacasa for clarification of her book conclusion.

Note

1 I like Manthia Diawara's (2015) translation of mondialité 'worldmentality', because of its blurring of the mental and material. Another possibility would have 'worldliness' (Diawara uses that, too) or 'worlding', both of which come with too much Arendtian and Heideggerian baggage attached.

References

Agarwal, B. 1992. 'The Gender and Environment Debate: Lessons From India', *Feminist Studies*, vol. 18, no. 1, pp. 119–158.

Allewaert, M. 2013. *Ariel's Ecology: Plantations, Personhood, and Colonialism in the American Tropics*. Minneapolis: University of Minnesota Press.

Bhabha, H. 1994. *The Location of Culture*. London: Routledge.

Bhambra, G. 2015. *Connected Sociologies*. London: Bloomsbury.

Blaser, M. 2009. 'The Threat of the Yrmo: The Political Ontology of a Sustainable Hunting Program', *American Anthropologist*, vol. 111, no. 1, pp. 10–20.

Bongie, C. 2008. *Friends and Enemies: The Scribal Politics of Post/Colonial Literature*. Liverpool: Liverpool University Press.

Britton, C. 1999. *Edouard Glissant and Postcolonial Theory: Strategies of Language and Resistance*. Charlottesville and London: University Press of Virginia.

Burns, L. 2012. *Contemporary Caribbean Writing and Deleuze: Literature Between Postcolonialism and Post-Continental Philosophy*. New York: Continuum.

Césaire, A. 1978. 'Poésie et Connaissance', In: A. Césaire and R. Ménil, eds., *Tropiques 1941–1945, Collection Complète*. Paris: Jean-Michel Place.

Césaire, S.R. 2012. *The Great Camouflage: Writings of Dissent (1941–1945)*. Ed. D. Maximin. Middletown, CT: Wesleyan University Press.

Chakrabarty, D. 2009. 'The Climate of History: Four Theses', *Critical Inquiry*, vol. 35, no. 2, pp. 197–222.

Coleman, N.A.T. 2014. 'Philosophy Is Dead White – and Dead Wrong', *Times Higher Education*, 20 March 2014. www.timeshighereducation.com/comment/opinion/philosophy-is-deadwhite-and-dead-wrong/2012122.article

Condé, M. 1998a. 'Edouard Glissant, ou les Antilles repossédées', Presentation at the colloquium 'Edouard Glissant: de la pensée archipélique au Tout-Monde, New York.

Condé, M. 1998b. 'Suzanne Césaire and the Construct of a Caribbean Identity', In: A.S. Newson and L. Strong-Leek, eds., *Winds of Change: The Transforming Voices of Caribbean Women Writers and Scholars*. New York: Peter Lang, pp. 61–66.

Da Silva, D.F. 2007. *Towards a Global Idea of Race*. Minneapolis: University of Minnesota Press.

Dash, J.M. 1985. 'Introduction', In: E. Glissant, ed., *The Ripening*. London: Heinemann.

Dash, J.M. 1989. 'Writing the Body: Edouard Glissant's Poetics of Re-Membering', *World Literature Today*, vol. 63, no. 4, pp. 609–612.

Diawara, M. 2015. 'Edouard Glissant's Worldmentality: An Introduction to One World in Relation', *South as a State of Mind Journal #6 (documenta 14 #1)*. Kassel: documenta und Museum Fridericianum. Online Version. www.documenta14.de/en/south/34_douard_glissant_s_worldmentality_an_introduction_to_one_world_in_relation

Escobar, A. 2008. *Territories of Difference: Place, Movement, Life, Redes*. Durham: Duke University Press.

Finney, C. 2014. *Black Faces, White Spaces*. Chapel Hill: The University of North Carolina Press.

Fuller, S. 2000. *Thomas Kuhn: A Political History of Our Time*. Chicago: Chicago University Press.

Garraway, D.L. 2010. '"What Is Mine": Césairean Negritude Between the Particular and the Universal', *Research in African Literatures*, vol. 41, no. 1, pp. 71–86.

Gilroy, P. 2015. 'Offshore Humanism', The 2015 *Antipode* Lecture at the RGS-IBG Annual Conference, 2 September 2015. http://antipodefoundation.org/2015/12/10/paul-gilroy-offshore-humanism/

Glissant, É. 1979. *Sturzflut*. Munich: Kindler-Verlag.

Glissant, É. 1985. *The Ripening*. London: Heinemann.

Glissant, É. 1997. *The Poetics of Relation*. Trans. B. Wing. Ann Arbor: University of Michigan Press.

Glissant, É. 2009. *Philosophie de la relation: poésie en entendue*. Paris: Gallimard.

Glissant, É. 2010. *Poetics of Relation*. Ann Arbor: University of Michigan Press.

Graeber, D. 2015. 'Radical Alterity Is Just Another Way of Saying "Reality": A Reply to Eduardo Viveiros de Castro', *HAU Journal of Ethnographic Theory*, vol. 5, no. 2. www.haujournal.org/index.php/hau/article/view/hau5.2.003/1978

Hallward, P. 2001. *Absolutely Postcolonial: Writing Between the Singular and the Specific*. Manchester: Manchester University Press.

Haraway, D. 1991. *Simians, Cyborgs, and Women: The Reinvention of Nature*. New York: Routledge.

Haraway, D. 2004. *The Haraway Reader*. New York: Routledge.

Harding, S. 2008. *Sciences from Below: Feminisms, Postcolonialities, and Modernities*. Durham and London: Duke University Press.

Jackson, M.S. 2014. 'Composing Postcolonial Geographies: Postconstructivism, Ecology and Overcoming Ontologies of Critique', *Singapore Journal of Tropical Geography*, vol. 35, no. 1, pp. 72–87.

Ko, A. 2015. 'Aph's Guide to the Revolutionary, Decolonizing Praxis of "Moving Over"', *Aphro-Ism Weblog*, 5 October 2015. http://aphro-ism.com/2015/10/05/aphs-guide-to-the-revolutionary-decolonizing-praxis-of-moving-over/

Latour, B. 2004a. 'Whose Cosmos, Which Cosmopolitics?' *Common Knowledge*, vol. 10, no. 3, pp. 450–462.

Latour, B. 2004b. *Politics of Nature: How to Bring the Sciences into Democracy*. Cambridge, MA: Harvard University Press.

Law, J. and Lin, W.Y. 2015. 'Provincialising STS: Postcoloniality, Symmetry and Method', Presented at the Bernal Prize plenary at the Society for Social Studies of Science Annual Meeting, Denver. www.heterogeneities.net/publications/LawLin2015ProvincialisingSTS.pdf

Lee, E.V. 2015. 'Indigenizing the Academy Without Indigenous People: Who Can Teach Our Stories? Moontime Warrior Weblog', 9 November 2015. http://moontimewarrior.com/2015/11/09/who-can-teach-indigenous-philosophy/

Maximin, D. 1995. *L'île et une nuit*. Paris: Editions Seuil.

Mbembe, A. 2015. 'Decolonizing Knowledge and the Question of the Archive', Presentation at the University of Witwatersrand, Johannesburg, 22 April 2015. http://wiser.wits.ac.za/content/achille-mbembe-decolonizing-knowledge-and-question-archive-12054

Mbom, C. 1999. 'Edouard Glissant, De l'opacité à la relation', In: J. Chevrier, ed., *Poétiques d'Edouard Glissant: actes du colloque international "Poétiques d'Edouard Glissant":* Paris-Sorbonne, 11–13 mars 1998.

Mignolo, W. 2000. *Local Histories/Global Designs: Coloniality, Subaltern Knowledges, and Border Thinking*. Princeton: Princeton University Press.

Mol, A. 2015. 'Presentation at Nature in Modern Society, Now and in the Future – A Philosophers "Debate" '. http://themasites.pbl.nl/natureoutlook/2016/news-2/nature-in-modern-society-now-and-in-the-future-a-philosophers-debate

Nagar, R. 2014. *Muddying the Waters: Co-Authoring Feminisms Across Scholarship & Activism*. Urbana, Chicago and Springfield: University of Illinois Press.

Noxolo, P. 2009. ' "My Paper, My Paper": Reflections on the Embodied Production of Postcolonial Geographical Responsibility in Academic Writing', *Geoforum*, vol. 40, pp. 55–65.

Noxolo, P. and Preziuso, M. 2013. 'Approaching a "Fictionable" World Through the Novels of Maryse Condé and Wilson Harris', *Annals of the Association of American Geographers*, vol. 103, no. 1, pp. 163–179.

Price, R. and Price, S. 2013. 'Suriname Literary Geography: The Changing Same', In: M. C. Fumagalli, P. Hulme, O. Robinson and L. Wylie, eds., *Surveying the American Tropics: A Literary Geography from New York to Rio*. Liverpool: Liverpool University Press, pp. 285–312.

Puig de la Bellacasa, M. 2014. *Science et épistémologies féministes: Les savoirs situés de Sandra Harding et Donna Haraway*. Paris: Harmattan.

Ramos, A.R. 2012. 'The Politics of Perspectivism', *Annual Review of Anthropology*, vol. 41, pp. 481–494.

Saldanha, A. 2007. *Psychedelic White: Goa Trance and the Viscosity of Race*. Minneapolis: University of Minnesota Press.

Shepard, T. 2006. *The Invention of Decolonisation: The Algerian War and the Making of France*. Ithaca, NY and London: Cornell University Press.

Shilliam, R. 2014. ' "Open the Gates Mek We Repatriate": Caribbean Slavery, Constructivism and Hermeneutic Tensions', *International Theory*, vol. 6, no. 2, pp. 349–372.

Shiva, V. 1989. *Staying Alive: Women, Ecology and Development*. London: Zed Books.

Stengers, I. 2009. *Au temps des catastrophes: Résister à la barbarie qui vient*. Paris: La Découverte.

Stengers, I. 2010. *Cosmopolitics I*. Minneapolis: University of Minnesota Press.

Stengers, I. 2011. *Cosmopolitics II*. Minneapolis: University of Minnesota Press.

Stengers, I. 2012. 'Reclaiming Animism', *E-Flux #36*. www.e-flux.com/journal/reclaiming-animism/

Stengers, I. 2013. 'The Right to Laziness, an Urgent Claim: De Buren Weblog', 28 March 2013. www.deburen.eu/en/blog/detail/the-right-to-laziness-an-urgent-claim

Strathern, M. 2004. *Partial Connections*. Walnut Creek, CA: Altamira Press and Rowman & Littlefield.

Sundberg, J. 2014. 'Decolonizing Posthumanist Geographies', *Cultural Geographies*, vol. 21, no. 1, pp. 33–47.

Todd, Z. 2014. 'An Indigenous Feminist's Take on the Ontological Turn: "Ontology" Is Just Another Word for Colonialism, Urbane Adventurer: Amiskwacî', 24 October 2014. https://zoeandthecity.wordpress.com/2014/10/24/an-indigenous-feminists-take-on-the-ontological-turn-ontology-is-just-another-word-for-colonialism/

Tuck, E. and Yang, K.W. 2012. 'Decolonization Is Not a Metaphor', *Decolonization: Indigeneity, Education & Society*, vol. 1, no. 1, pp. 1–40.

Tuhiwai Smith, L. 2012. *Decolonizing Methodologies: Research and Indigenous Peoples*. London: Zed Books.

Verran, H. 1998. 'Re-Imagining Land Ownership in Australia', *Postcolonial Studies*, vol. 1, no. 2, pp. 237–254.

Verran, H. 2001. *Science and an African Logic*. Chicago and London: University of Chicago Press.

Vivieros de Castro, E. 2014. *Cannibal Metaphyics*. Ed. and trans. by P. Skafish. Minneapolis: Univocal Publishing.

Weheliye, A.G. 2014. *Habeas Viscus: Racializing Assemblages, Biopolitics, and Theories of the Human*. Durham and London: Duke University Press.

Weil, S. 1956. *The Notebooks of Simone Weil: Volumes 1 & 2*. London: Routledge and Kegan Paul.

Wilder, G. 2015. *Freedom Time: Négritude, Decolonisation, and the Future of the World*. Durham and London: Duke University Press.

3 Chronic carriers

Creole pigs, postplantation politics, and disturbing agrarian ontologies in Haiti

Sophie Moore

Introduction

In 1983, 2000 pink-skinned, North American piglets arrived in Haiti, their passage paid by the United States Agency for International Development (USAID). These so-called sentinel pigs were castrated males, products of a Hampshire and Duroc cross, and had been raised in a sterile environment, thus qualifying as SPF – specific pathogen free (Allison, 1982: 6–7). When the sentinels tested negative for African Swine Fever (ASF) some months later, 452 of their kin were imported for breeding at a handful of swine multiplication centers (SMCs) around the country, part of a program called the Interim Swine Repopulation Project (ISRP). Why did Haiti, a predominantly rural society in which pigs had long been the backbone of the peasant economy, have need of a swine repopulation program? The answer, this chapter proposes, is entangled with transnational agrarian politics, and in particular, the life and death of a small black animal called the creole pig, or kochon kreyol.

The creole pig, *Sus scrofa domesticus*, was likely the offspring of Spanish boars introduced in the Columbian exchange and the landrace of wild pigs native to the Caribbean. By 1984, however, the creole pig was officially defunct in Haiti. In 1982, the Interamericano Instituto de Ciencias Agricola (IICA) had led the slaughter of 384,391 Haitian pigs, executing the eradication portion of an endeavor called the "Program to Eradicate African Swine Fever and to Develop Pig Raising" (PEPPADEP, following from the French acronym).[1] While the project remains a frequently cited example of the tragedies wrought by the advent of overzealous neoliberal development in 1980s Haiti (see, e.g. Trouillot, 1990; Diederich, 1985; Taluy, 2015)), the political futures set in motion by PEPPADEP are not so easily reduced to either an endorsement or a critique of neoliberal globalization. With grassroots, state, and civil society actors alike each claiming a measure of success in the set of events that I here gather under the loose title of the Creole Pig Affair (CPA), PEPPADEP's afterlives extend well beyond the project scope. The CPA was and is, I suggest, entangled with the broader political, ecological, and economic processes that continue to define Haiti's political horizons, and in particular with the country's so-called, unfinished democratic transition between 1986 and 1990.

The CPA ultimately makes clear the stakes of postcolonial nation building, throwing into relief how transnational scientific and political practices converge to enact make or unmake particular human and nonhuman bodies. If we are, as David Scott advises, to learn to ask new questions of the postcolonial (2004: 7), we must take seriously the emergence of decolonial politics in unexpected places. To do so means to attend not only to the epistemology of coloniality – how the colonial project has shaped ways of knowing, but also to its ontology – the ways of being that coloniality makes possible or forecloses. Of critical concern is how to disarticulate such questions from the immanent ontological violence of the modern West, a ceaseless reproduction of the colonial present through economic, scientific, and political practices. Haiti's agrarian worlds are one site of perpetual disturbance to this order – what Anna Tsing would call "a hair in the flour" (2005: 206). This chapter focuses in particular on Afro-descendant smallholders' claims to political sovereignty, and examines how state, international, and civil society actors enact political and economic transformation through agrarian relations. The stakes are mortal, as they have always been for the seekers of Black freedom in the Americas.

While Haiti's Revolution remains a touchstone for scholarship on anti-colonial and anti-racist struggle in the Americas, the continuing decolonial project of the descendants of those enslaved persons who won their freedom and their land has been less thoroughly examined. I extend anthropologist and medical doctor Paul Farmer's examination of the body – its existence or nonexistence – as the indispensable terrain for the making of political claims (Farmer, 2006) in arguing that political bodies are not only human, but also collective projects made across more-than-human relations. Hence, neither the familiar humanist nor postcolonial frameworks (nor their postmodern critiques) are sufficient for understanding the emergent political horizons of the postplantation Americas. Caribbean scholars have made this provocation to postcolonial studies since the inception of the field, noting the particular virulence of antiblack violence foundational to the Americas (Fanon, 2008; Wynter, 2003; Glissant, 1981). The aim of this chapter, then, is to extend a uniquely Caribbean vein of postcolonial critique to consider the more than human relations at work in postplantation politics. A question worth asking at the present conjuncture, I propose, is how we might materially and theoretically claim ground for political ontologies otherwise, that is such that we begin to cultivate Black political vitality rather than to hasten its extinction.

In this chapter, I pursue three goals:

1 To situate the CPA as one key conjuncture of political, economic, and agrarian transformation in Haiti's democratic transition.
2 To untangle how the CPA's neoliberal and neocolonial logics shaped the material futures of Haiti's smallholders by enacting specific sets of scientific, political, and economic practices.
3 To examine how state, civil society, and intergovernmental (IGO) actors strategically enacted a zone of indistinction around the creole pig.

The Creole Pig Affair: 1978–1988

Although PEPPADEP was launched in 1982, the CPA actually begins in 1978, when an outbreak of African Swine Fever (ASF) was reported in the Western region of the Dominican Republic, near the Haitian border. Following the directive of the Dominican Republic's government, Haitian army personnel and Department of Agriculture (DDA) agents were sent to slaughter all the pigs in the eradication area, a 15-km swathe on each side of the border. Records report 20,671 pigs killed between July and September 1978 (Toussaint, 1989: 5). While the Ministry of Agriculture and Natural Resources (MARNDR) claims to have kept records of proprietors, the army did not, and compensation for those proprietors who could be identified did not begin until almost four years after they had lost their pigs (Alexander, 1992: 254). The 1978 endeavor was apparently equally unsuccessful in containing ASF. Later that year, 30,000 pigs were reported to have died in the fertile Artibonite Valley; serum tests revealed ASF as the likely cause of death. By 1979, large numbers of pig deaths were reported around the country, perhaps as many as 300,000, and panic was beginning to set in as peasants found no market for their meat, their livestock holdings severely depleted, and rising prices. Pigs were perishing so quickly that they could not be buried, their bodies stacked on irrigation dikes. Between December 1978 and December 1979, Haiti's pig population was cut at least in half. Out of 1368 serum samples taken from dead pigs in the Artibonite, however, only 7% tested positive for ASF in 1978 (Alexander, 1992: 253; Toussaint, 1989: 6).

The coast of Florida is only about 600 miles from Haiti's shores. After an urgent 1980 meeting about the ASF outbreak, an animal health specialist working on the project, remarkably, "obtained a declaration from the Acting U.S. Secretary of State with support from both outgoing and incoming administrations, that the presence of ASF in Haiti constituted an emergency for the United States," thus making U.S. funds available (Alexander, 1992: 253). Haiti's neighbors were closely monitoring the ongoing outbreak, and the IICA convened another "urgent" meeting in 1981 to establish a strategy for controlling ASF. After 10 months of negotiations, the IICA established the PEPPADEP protocol, which would have two phases, eradication and repopulation, entailing four steps: a public education campaign; slaughter and compensation; raking, cleaning, and disinfection; and, sentinelization and repopulation (MARNDR, 1982a: 1, 4). President Jean-Claude Duvalier issued the eradication decree for PEPPADEP in 1982, and over 380,000 pigs were slaughtered that year (Toussaint, 1989: 8)

Although repopulation constituted the essential second phase of PEPPADEP, the IICA's financing did not extend beyond eradication, which was completed in 1984. The second phase was to be financed by the International Development Bank (IDB) and the Canadian government (MARNDR, 1982b: 1). When eradication was complete, however, the IDB was not ready to release the funds, and so in 1983 USAID agreed to fund the "Interim Swine Repopulation Project" (ISRP), which would ostensibly hasten the distribution of pigs to the peasants who had suffered the most from the eradication campaign. Ultimately, the ISRP ran concurrently with PEPPADEP's repopulation campaign, terminating in 1987 and 1989, respectively. While the ISRP defined its beneficiaries as the peasants most affected by eradication, PEPPADEP did away with even this nominal attention to the poorest farmers, imposing prohibitive conditions for the procurement of "improved" piglets. As anthropologist Michel-Rolph Trouillot notes, pig raising rapidly became a bourgeois affair in Haiti (1990: 216), and the swine population would never again reach pre-CPA levels. The lack of rural infrastructure or agricultural extension services ensured that few of the "white" pigs would make it deep into the provinces, and rumors began to swirl of their poor adaptation to the tropical climate. Project documents dispute this claim, but belatedly, asserting in 1987 that the first generation of pigs from the swine multiplication centers (SMCs) moved readily into homeostasis with local conditions (USAID, 1987: 5). Peasants' groups, many of which were at this time operating clandestinely from the cover of the Church, soon joined the critiques of the project, pointing out that the program conditions for receiving a new, improved piglet required the construction of a piggery significantly more luxurious than peasants' own lodgings, not to mention the purchase of expensive imported bagged feed. Peasants' groups, most prominently the Mouvman Peyizan Papay (MPP), a large and powerful group on Haiti's Central Plateau that had long opposed the dictatorship, launched their own repopulation campaign, using consanguine pigs from Jamaica in an effort to re-produce the kochon kreyol.

Project documents term both the ISRP and the IICA repopulation efforts a success; MPP likewise claims their repopulation campaign as a victory. PEP-PADEP was not officially terminated until 1989, as Jean-Bertrand Aristide, who in 1990 would become Haiti's first democratically elected president, was rising to power. By that time, the creole pig had regained a measure of both its symbolic and material power, coming to serve as a symbol of the surge in grassroots civil society that helped to topple the dictatorship. Nonetheless, peasants' futures were significantly less secure in 1988 than they had been in 1978. It is important to recognize the extent to which U.S. involvement sought to create political back doors by which to exert control over Haiti's shifting political terrain. The CPA is one crucial example of how supposedly apolitical scientific discourses in fact generated the material conditions of possibility for a limited democratic transition in which the United States would maintain significant influence. The CPA likewise demonstrates how civil society actors sought to mobilize the radical excess expelled from, and dangerous to, the neoliberal hemispheric order. The significance of the CPA lies less in the intent of the actors involved, however, than in the expected and unexpected outcomes of their friction. The long afterlives of the CPA make clear that the constitutive relations – political, economic, and environmental – of a neoliberal order were actively debated and materially negotiated not only by state actors, but also by the supposed beneficiaries of liberalizing markets.

Neoliberal pigs

The market

When president-for-life Jean-Claude "Baby Doc" Duvalier came to power in 1973, he inherited a deeply polarized political landscape. The state's function was largely extractive and disciplinary, with a tiny political class sustained by the extraction of agrarian labor and resources from an immiserated and terrorized peasant majority. Whereas his father, François "Papa Doc" Duvalier, had pushed forward an authoritarian nationalist project that consolidated and centralized state power in Port-au-Prince, his son's tastes ran more to extravagant cosmopolitan luxuries. Jean-Claude's economic policy highlighted light industry and tourism, seeking to lure business to Haiti with a plentiful supply of cheap labor recently migrated to urban centers from the provinces. This labor force was made up largely of peasants forced from their land by increasing rural precarity throughout the late 1970s and early 1980s. Six years after Jean-Claude came to power, the number of U.S. firms in Haiti had nearly doubled; cost of living and food prices had increased dramatically, the food index doubling between 1972 and 1973 as it rode an "inflationary wave." The profits of Jean-Claude's "economic revolution" tended to exit the country at a rapid rate (Trouillot, 1990: 201, 241, 216; 1994: 130). Baby Doc's economic development strategy ultimately only heightened an impending political crisis, as Trouillot points out: "The light industry strategy was destined to misfire primarily because it ignored the impasse posed by the relations of production in the agrarian world. The crisis inherited by François Duvalier and exacerbated by his regime and that of his son had its roots in the contradictions of a peasant country ruled by an unholy alliance of merchants and political profiteers, an alliance cemented by the state" (1990: 209–210). By 1986, when Baby Doc and Duvalierism were dechouke (uprooted), the gulf between Haiti's rich and poor was wider than ever before.

Accordingly, as Trouillot notes, after the CPA, pig raising "becomes a bourgeois venture, and the price of all meat had increased greatly" (1990: 216). Haiti's deeply entrenched political and economic inequality likewise facilitated the movement of entrepreneurs and business people more directly into the political arena. It is important to understand the CPA as a crucial turning point in the final years of the Duvalier dictatorship, at which volatile political power was directly transferred by economic means to a set of national and international elites with little concern for or connection to the agrarian sector, at the time the majority of the country's population. While the entities in charge of PEPPADEP were ostensibly apolitical IGOs, the forced liberalization of the market nonetheless helped to create the conditions for a democratic transition that would protect American interests in Haiti. Development was one means by which neoliberal globalization inserted international interests into Haiti's "unholy alliance" between the state, the private sector, and the political class. Already heavily indebted to international lenders, Haiti's external debt grew in proportion with the steady flow of cash separated from the national coffers during the dictatorship.

Even the project documents admit that peasants' situation was worse after PEPPADEP than before. The central logic of PEPPADEP was one of "creative destruction" (Harvey, 2007: 3), that is, concretely, the transformation of subsistence-oriented household pig raising into an export-focused pork industry. It would be a grave mistake to ignore either the violent nature of this transformation or the ontological work that it performs. PEPPADEP heralded the kinds of transformations that agrarian systems across Latin America would undergo as market liberalization converged with structural adjustment policy in the 1980s, although the particular futures of Afro-descendant smallholders are perhaps nowhere more starkly at stake than in Haiti. The CPA makes the logic of substitution that drove the project abundantly clear, articulating the conditions for belonging in a neoliberal agrarian world in policy, and enacting them in practice.

Neoliberal structural adjustment quite literally transformed the conditions of possibility for agrarian life in rural Haiti. The science behind PEPPADEP reproduced its neoliberal logic both materially and epistemologically. The three phases of the CPA – contamination, eradication, and repopulation – each evidence the emergence of a scientific logic that aligns tightly with the reproduction of globalized neoliberal agrarian worlds. In this sense, the project was indeed a resounding success. Solipsistic ally, USAID's, 1987 ISRP report commends PEPPADEP for project management and administration, while nevertheless asserting that "The original project goal to raise peasant farmers' standard of living solely on the basis of providing them with increased numbers of pigs, in the absence of appropriate supporting infrastructure (e.g. marketing, credit, easy access to feed, vitamins, medication, etc.), is not realistic" (USAID, 1987: 4). It is indeed unrealistic, to say the least, to expect a short-term project to remake agrarian livelihoods solely on the basis of the eradication of one species and the multiplication of another. The material evidence of this gap was abundant: feed shortages, unstable markets and high prices, vacant breeding centers, and of course an overwhelmingly unsatisfied demand for pigs coupled with a decrease in pork consumption. Yet, the project equivocates. Abundant U.S. agricultural surpluses could fill at least one gap in Haiti's deficient agrarian economy:

> Wheat shorts and rice bran will never solve the feed problem and therefore, it is suggested that the GOH [Government of Haiti] and USAID discuss the possibility of importing feed sorghum for swine and poultry. Importing soy meal might be another solution.
>
> (17)

Soy meal, however, proved to be an equally unsatisfactory answer, at least for the peasants who could not afford to purchase any bagged feed at all. PEPPADEP demonstrates that the metric by which the project measured success depended on a logic of substitution that proved deadly to its target beneficiaries.

Deadly substitution

On 11 January 1979, MARNDR declared Haiti entirely infected with ASF (Toussaint, 1989: 6). MARNDR considered four possible responses: define the contamination zone and kill infected pigs within it; establish a cordon to control movement of people and pigs, and enforce quarantine; let the disease run its course; or slaughter, disinfection, and repopulation. MARNDR based their assessment on 1368 blood serum samples, 7% of which tested positive for ASF (Alexander, 1992: 253). The virus had likely entered the Dominican Republic when pigs near the capital were fed food scraps from a flight originating in Spain, and soon crossed the porous border with Haiti. How did state and international actors' efforts come to entail the eradication of Haiti's entire swine population, and what motivated this decision? An animal health specialist who worked on the project references outbreaks in Spain, Portugal, Malta, Sardinia, and Brazil since 1978, but highlights the urgency of the outbreak in Hispaniola: "The outbreaks in Brazil were of much less concern to us as meat trade with South American countries was limited because of foot-and-mouth disease. The Spanish-speaking Dominican Republic also had little contact but when ASF spread to Haiti, the alarm bells started to ring because of the island's proximity and the interchange of informal traders" (Alexander, 1992: 252). Why did a global ASF outbreak set off "alarm bells" at this particular juncture? At the time of the outbreak, Haiti's pork exports were insignificant, and production was concentrated in urban and peri-urban industrial piggeries. The rapid adoption of eradication as an appropriate response stemmed from a scientific logic that could not help but reproduce ontological violence.

The commentator above highlights not the actual threat of ASF, but the threat of future losses that it represents. The ungovernability of Haiti's agrarian life, its unlikely endurance in remote, rugged geographies farmed by fugitive slaves and forgotten peasants, has disturbed America's capitalist ontologies since the plantation. Slaves' and forced laborers' refusal to work under the plantation system, their preservation of African foodways and cultures, their revolt against occupation and exploitation all demonstrate Afro-descendant smallholders' ontological commitment to a world in which Black freedom is not only possible, but in the making (Trouillot, 1990; Fischer, 2004; Carney and Rosomoff, 2011; Hallward, 2007). The scientific orthodoxies of rural development do the work of dematerializing these decolonial commitments in agrarian ecologies across the global south. Decolonization is thus caught in the balance of a material struggle for the ontological territory of Black freedom. The CPA demonstrates how the terrain for making political claims, whether hegemonic or radical, emerges from agrarian relations. Reflecting the settler colonial logic which so quickly exterminated Hispaniola's native population, PEPPADEP's eradication-repopulation cycle replaces one swine world with another, the tidiness of this substitutive logic accounting little for agrarian lives in its violent ontological transformation.

There is material evidence for this deadly logic. The failure of the state to fulfill its obligation to compensate peasants for their slaughtered pigs was both a cause and an effect of its hollowing out under Baby Doc's rule. One of the conditions for the launch of PEPPADEP in 1982 was that MARNDR would begin paying compensation for the pigs slaughtered in 1978 (MARNDR claimed to have kept records of proprietors, while the army made no such claim) (Alexander, 1992: 254). Drawing from international coffers to remedy state violence, the project thus set in place an economic model of development that would come to dominate the conditional aid provided by the IGOs and NGOs that would flood the country after the end of the dictatorship. The prioritization of U.S. interests over local demands, even to the extent of clear internal contradictions, characterized the project at the biopolitical as well as the economic level. As he reflects on the early stages of PEPPADEP, a project veterinarian is perplexed: "with the history of ASF in the Dominican Republic and the border slaughter that occurred," he "cannot explain why inoculations against hog cholera preceded steps to contain [the] massive outbreak in the Artibonite" in 1978, wherein 30,000 swine perished (Alexander, 1992: 253).

It would seem that the curtailing of ASF was secondary to the more diffuse preparation of the terrain for the emergence of a "modern" system of pig raising. The project documents, improbably, cast peasants as docile subjects patiently awaiting this gift of development. In a 1982 report, for example, MARNDR recounts how proprietors soon realized the futility of their local efforts to repopulate the creole pig from the few creole pigs that had been successfully hidden during eradication, and "simply awaited the return of better times and a means of making pig raising profitable again" (MARNDR, 1982a: 6). While the project accounts for the possibility of fugitive creole pigs, it assumes that the docility of the peasantry will provide an appropriate check on the covert multiplication of potential ASF vectors. Although a grassroots effort did eventually return a version of the creole pig to Haiti's peasants, this effort was superseded by the state and IGO-led repopulation efforts, which were closely aligned with the production of the kind of agrarian worlds best suited to the absorption of Northern surplus and labor demands. This aim is explicit in the project documents, which state that in service of their own interests, other Western countries must assist Haiti in eradicating the menace of ASF (MARNDR, 1982a: 7).

The end of PEPPADEP's eradication phase coincided with a funding gap in the IICA's plan, which USAID soon stepped in to fill, though not soon enough to prevent a definitive break between industrial and small farm pig raising. Some of the delay was perhaps attributable to the project's lofty goal of effecting a wholesale transformation in Haiti's pork production industry. The pervasive contradictions of PEPPADEP indicate the circularity of the project's substitutive logic, in which the project reproduced or even created the very same conditions it was intended to remedy, not least through the management of a delicate equilibrium between vulnerability to disease and economic growth. A project veterinarian reflects that Haiti offers a "big opportunity" that other countries do not have: to "uproot" an array of porcine maladies like cholera, pseudorabies, and atrophic rhinitis. He emphasizes that in the first three years of the program, when peasants do not yet have access to any pigs, it is essential to "confront the reality" of a swine world whose entanglement with disease poses a perpetual threat (Allison, 1982: 40–41). Improbably, project documents propose a "general cleaning" of all areas where pigs were raised, reinforcing the project's dictum that a "pure" pig population is possible – but not, it seems, for peasants, who were given only castrated males, under suspicion of harboring fugitive creole pigs (1982: 16, 41). Eventually, "less desirable" piglets would be sold (not given) to peasants, but not until several generations had been distributed from the SMCs to larger producers and appropriately democratically minded civil society groups (Allison, 1982: 41).

This new national swine herd, however, was no less vulnerable to disease than the now-defunct one. Perhaps it was more vulnerable, for the creole pig had demonstrated resistance to ASF. Confusingly, the ISRP both touts its production of a "pure" ASF-free swine herd, and acknowledges the existence of a fugitive herd, host to a perpetual contamination threat. A veterinarian writes that the

> ISRP must acknowledge that swine being raised outside the realm of the project are in close association with Project swine and are of equal importance to the individual peasant farmer regardless of their origin. Consequently, a

> national swine herd now exists which is ever expanding in size and very susceptible to disease.
>
> (17)

Now that Haiti's pig production was sufficiently modern, a new set of diseases threatened the herd. The veterinarian advises that

> the Haitians can triumph and be proud of having one of the least disease ridden swine herds in the world. However, the real goal now is to maintain this health status knowing full well that the herd is fully susceptible to [the] common infectious swine diseases [. . .] which plague swine herds in more industrial countries.
>
> (USAID, 1987: 90)

He notes the spread of infectious diseases of unknown etiology throughout PEPPADEP's herd, suggesting a criticism of the IICA's assumption of tidy substitution between one pig and another (USAID, 1987: 95).

The ISRP validates its own urgent intervention and the necessity of rapidly getting pigs to the poorest farmers, claiming that "a philosophy of noninvolvement could result in an accelerated decline of the fragile environment in which the [creole pig's] culture is preserved" (USAID, 1987: 74). Indeed, this is exactly what happened. When peasants were unable to "bank" on creole pig futures, they turned to tree cutting for charcoal production and small commerce to satisfy a demand for cash, which in turn produced lasting economic and environmental effects. For the state and the private sector, however, the project's success rapidly exceeded expectations. An ISRP expert writes,

> The performance of the imported improved breeding stock has been good across all production units [. . .], which provides a good base for a sound national swine program. [. . .] The project has assisted, in varying degrees, in the establishment of 440 SMCs, thus exceeding the 55–70 planned target by an overwhelming margin.
>
> (USAID, 1987: 5)

Repopulation thus led to the proliferation of both pigs and sites of pig production, the sudden increase of which put mounting pressure both on the market and fragile rural infrastructure. USAID's commitments to "making as small an investment as possible, [and focusing on] infrastructure and institution-building" (1987: 13) seemed to work in opposition to each other, as the private sector eagerly expanded the repopulation efforts beyond the agrarian economy's capacity to absorb the transformation. The neat equivalence of the eradication-compensation-distribution cycle was soon steered significantly off course by the rapid liberalization of the market, saturated with an excess of new and improved pigs.

Too many pigs

It was apparent that the repopulation effort had been all too successful. Although the project logic depended on the supposition of a self-regulating market, surplus pigs – a fleshly excess – began to affect the market as well as animal health. The private sector had been importing its own pigs – non-SPF, healthy pigs from the U.S. (USAID, 1987: 32). The USAID, 1987 ISRP report predicted that "equity issues" would arise because of the proliferation of entities engaging in repopulation efforts (77), and indeed the repopulation phase is a testament to the capacity of free market relations to increase economic polarization and reinforce existing structures of inequity. The equivalence between pig, money, and meat established during the eradication phase was quickly disrupted. A project veterinarian relates his observations of the PEPPADEP killing sites:

> the routine was for the owner to present the pig and have both its ears tagged. The pig was killed and bled, and one ear was removed. The owner took the dead pig, cleaned it, returned to the paymaster with the other tagged ear, which was matched, and then received compensation, all within an hour. In addition to the payment, the owner also received the carcass. In a short time, killing sites became lively trading posts for fresh pork.
>
> (Alexander, 1992: 255)

This accounting makes clear that bodies – flesh, human and animal – are mortally at stake in the CPA. While the project's stated objective was to create a productive pork industry, carnal matters posed a continual problem.

In the most concrete possible sense, the project produced a weighty material excess, for which there existed no structural outlets. The conditions for managing this excess were envisioned as emerging organically from the project, sometime in an undefined future. The liberalization of the market that accompanied the CPA both augmented the production of fleshly excess and installed a solution for its management – a now-classic paradigm of neoliberal development (see Mitchell, 2002, e.g.). In the break between total eradication and robust pork futures there emerged an excess of pigs, their living and dead flesh constituting the ontological terrain for Haiti's neoliberal futures. Although no hard data are available on pork consumption at the time, observers attest to a reduced consumption of pork and a subsequent drop in the market during the initial outbreak. During this chaotic phase, market relations emerged that put not only the pigs' but also peasants' lives in the balance. The project tried to channel this excess into the market, allowing peasants six to nine months to dispose of their swine herds "naturally" on the local market. During this period, pig breeding was forbidden, and education ostensibly provided on the benefits of improved breeding stock (MARNDR, 1982a: 15). Nonetheless, when eradication began, the equivalency between pig and cash was established as separate from market relations. Again, demands excessive to the scope of the project resulted. Requests for compensation increased dramatically, and USDA personnel felt the agency was "being cheated." "There was no way of anticipating the large increase in the number of pigs," a project veterinarian remembers, and although "auditors kept pouring into PEPPADEP headquarters," few irregularities were found in this unexpected wave of demand (Alexander, 1992: 255).

The situation only worsened with Phase II of PEPPADEP and the mostly concurrent ISRP repopulation program. The SMCs – the secondary multiplication centers at which the first generation of crossed pigs were bred – were soon drawn fully into a rapidly liberalizing market, proliferating as the private sector got in on the game. The first, "better," project SMCs received their first crossed pigs between late 1984 and early 1985. Both SMCs and the new pigs proliferated over the course of the following year. While the distribution of the first generation of piglets generated an excess of privatized SMCs, the distribution of second-generation piglets further concentrated industrial pork production and flooded the market with surplus pork, while simultaneously creating barriers to peasants' successful market entry. The market was soon flooded not only with piglets peasants could not afford to buy, but also with bagged feed and the meat of the new, larger, pigs. Some SMCs were closed because of feed shortfalls, as the IICA had anticipated. (USAID, 1987: 77). Ironically, the project had to "decentralize" pigs to peasants to raise because of feed shortfalls, leaving many SMCs thus lying vacant. The project "was constrained in time and financing [. . .]. Therefore, all efforts were to focus on producing the maximum number of pigs" (USAID, 1987: 30). The project was working exactly as intended.

Slippery ontologies

Uncertain futures

In Agamben's (1998) " 'uncertain and nameless terrain' where life and death enter 'zones of indistinction,' " Elizabeth Povinelli locates "a space of potentiality where new forms of life can emerge" in the oscillation between "striving to persevere and any actual idea or action that emerges from this striving" (2011: 9).[2] That zone of indistinction, this chapter argues, is a site of more-than-human encounter, where political possibilities emerge from agrarian relations. The CPA comprised multiple efforts to "capacitate an alternative set of human and nonhuman worlds" (Povinelli, 2011: 7), which in turn created opening for an array of actors to materialize their political commitments within or against broader processes of agrarian change. The production and management of uncertainty and indistinction was central to the CPA, contributing to and reflecting the emergence of a particularly unstable configuration of the state, civil society, and the agrarian means of production at the end of the Duvalier dictatorship. Despite the abundant ironies of the "successful" termination of PEPPADEP and the ISRP in 1989, the relativist claim that the project was successful for donors and catastrophic for beneficiaries, accounts neither for the multiple and immanent failures within the project itself, nor its remarkable ability to disavow its own collateral damage. Empirically, such contradictions are borne out in the data gathered and deployed under the auspices of the CPA, as well as equivocations that throw into doubt the viability of the futures therein. The theoretical basis for the project is likewise entangled with contradictory political ontologies, to which the actors involved in the CPA make strategic claims. The section that follows examines what kinds of political futures emerge from such uncertain ontological terrain, focusing on how the project's outcomes trouble orthodoxies both of scientific knowledge and of political agency in processes of agrarian change.

The CPA represents the emergence of a set of contradictory political conditions that would allow success and failure, endurance and extinction, anticolonial revolution and the reproduction of Northern hegemony, to occupy the same ontological territory. Concretely, these contradictions would manifest in Haiti as a set of symptoms constitutive of a failing or weak state: massive economic equality and high indebtedness, political polarization, violence, and corruption, and dependence on foreign aid. Understanding the ontological grounds for the emergence of these conditions is central to understanding how Haiti's democratic transition has afforded widely divergent political horizons for the state and for its citizens. The CPA demonstrates the real material effects of ontological misapprehension in politics, whether by design or by omission, with the political and material vitality of agrarian ecosystems in the balance. We might call the CPA, then, a case study in the politics of slippery ontologies.

Swine worlds: Old pig/new pig

Like different salmon "done" in different sets of constitutive relations (Law and Lien, 2012: 3), the ontology of the creole pig is slippery, an unstable axis on which turned an array of political and material futures. In what follows, I focus on two key sets of relations in which smallholders' futures were held in the balance: that of data (what is the creole pig?), and of responsibility (who will account for the pigs' life and death, and how?). In a 1987 report, USAID congratulates itself for the success of the ISRP, and in particular the improvements wrought by the substitution of a "new" Haitian pig for the "old" one:

> It is worthy to note that the selective cross breeding which has occurred in the project has generated stock which will ultimately result in a new Haitian pig with characteristics of the three imported breeds. It is estimated that many pigs in succeeding generations will be partially or wholly black due to the Hampshire breed input.
>
> (USAID, 1987: 70)

I highlight the qualities that USAID attaches to the future Haitian pig: its creoleness (both in the sense of 'born in the New World,' and of mixed heritage), and its blackness. USAID is confident that the new pig will, sometime in the future, achieve an acceptable standard both of indigeneity and of pigmentation. That is to say, the ISRP elides ontological difference through the principle of (colonial) mimicry – the new pigs are the same but not quite, or, quite literally, the same but not white (Bhabha, 1984). It is remarkable not only how closely this formulation effects reflects a broader colonial logic of white supremacy and Northern hegemony, but also the extent to which its material and political effects play out in the CPA itself. Attending to empirical ontology offers a way in to understanding the shifting relations between the old and new creole pig. Law and Lien write:

> We have two sets of practices here, and those practices are doing different salmon because the relations that they enact are themselves different. [. . .]

> A salmon is not general but specific. It depends on how 'it' is being done in practice. [. . .] it follows that since those practices aren't the same, different and multiple salmon subsist in different and multiple worlds.
>
> (2012: 4)

To extend Law and Lien's argument, I propose that the sets of practices that made up the CPA enacted different pigs, divergent swine worlds that could not, and were never intended to, sustain equivalent forms of life.

These multiple pig ontologies were rooted in scientific practices of categorization and analysis that were themselves slippery. An animal called the 'creole pig' exists today across the Greater Antilles, marking the islands where Spanish colonists first landed. A constellation of consanguineous species maps the Spanish colonial encounter onto the Caribbean archipelago. Its taxonomic indistinction means that the creole pig is identified by its traits – its hardiness and indiscriminate appetite, its small size, lean flanks, long snout, and ability to fatten quickly when needed. For Haitians, the creole pig is likewise distinguished by its particular eco-social role, a vital "savings account" that also aerates garden soil with sharp hooves and rooting snout, and indiscriminately devours kitchen scraps and agricultural surplus. Although the PEPPADEP and ISRP documents assert repeatedly that the new pig is moving into homeostasis with local conditions during the project's repopulation phase (USAID, 1987: 5), data nonetheless indicate that the new pig's adaptation is, rather, to a modern pork industry in which peasants' agrarian worlds matter little. In other words, these pigs' ontologies are incommensurable. Yet the policy that shaped the project assumed that these pigs occupied the same ontological territory, with an obligatory temporal separation that established their linear progression to an ideally modern pig. For the IICA and USAID, one pig could not only be replaced by another (the logic of substitution), but could also transform into another, given the appropriate conditions. Taking the slipperiness of the pig's ontology seriously – and resisting the assumption that the pigs' difference is a matter of perspective – is crucial to the imagining of agrarian futures otherwise.

Empirically, neither the genetic composition nor the population and distribution of swine herds were accurately determined before, during, or after PEPPADEP. The project data itself demonstrate this uncertainty, suggesting the problematic endeavor of counting pigs when the criteria for belonging to the countable category remain in flux. We do not know how many creole pigs were in Haiti before the CPA began, nor do we know if any survived the eradication. We likewise do not know how the new pigs fared after the third generation, nor do we have general animal health statistics over time, although launching a national animal health plan was one of the project's goals (MARNDR, 1982a, 1982b; Alexander, 1992). Such lacunae played a crucial role in the CPA's production of destabilizing effects felt well beyond the project. The justification for the eradication program, for example, depended on a sampling protocol that qualified the threshold of tolerance for ASF in a rapidly globalizing pork industry. A project veterinarian writes,

> Before the slaughter program a serological survey was undertaken by the USDA with technicians from the University of Georgia. Sera from 1295

> swine of 83 different areas showed 77% of the areas had pigs positive for ASF. Overall prevalence [. . .] was 20%.
>
> (Alexander, 1992: 254)

Did the presence of ASF in Haiti necessarily warrant eradication? It was a technique that had been used in Malta, but was not undertaken in the Dominican Republic, for example, or in Brazil, not to mention in Spain, the presumed source of the Caribbean ASF outbreak. It matters how you measure, Marilyn Strathern writes (1999). In the CPA, the consequence of measuring pigs-as-vectors was that the creole pig was deemed eminently killable.

The CPA's sampling protocol thus depended on the assumption of a critical threshold for action, whether for eradication or for repopulation. Any peripheral effects – the collapse of rural pork markets, or the spread of other porcine diseases, for example – were collateral damage, conveniently confined to a geographically separate location. Attending to the creole pig's ontological slipperiness allows us to understand that the misapprehension that created such divergent outcomes, in such unequal distribution, was not epistemological – project operatives knew what the creole pig meant to peasants – but ontological, in that it generated contradictory visions of how pig futures would and should be enacted. A pig that had the qualities of the "original" creole pig – itself a New World hybrid – eventually emerged, whether by adaptation, selective breeding, forgetting, or a mixture of all three. The trove of data available on the creole pig and its replacement spans 1978–1989, and then falls into silence. Before concluding this chapter, I examine how responsibility for Haiti's agrarian futures emerged across the more than human relations constituting the creole pig.

Intra-action: Responsibility

The neoliberalization of Caribbean economies marks a new turn in the political horizons of international development. The entanglement of international actors like the IICA and USAID with processes of agrarian change, state-building, and democratization in Latin America and the Caribbean spreads political agency across more-than-human networks still entangled with colonial histories. Just as we can say that the slippery ontologies of the creole pig were multiplied and strategically deployed in the CPA, the kinds of ontological claims that could be made by grassroots, national, and international actors were likewise defined through a period of political transition in postcolonial hemispheric relations. Through the lens of political ontology, then, the entities engaged in the CPA were constituted in action and touch (Law and Lien, 2012; Barad, 2012). I turn to the politics of interaction across more than human relations in order to understand how we might make agrarian worlds matter differently, answering the critical demand from decolonial political ontology to take seriously the distribution of responsibility across more than human relations.[3] Karen Barad writes,

> Intra-action (in contrast to the usual 'interaction' which presumes the prior existence of independent entities or relata) represents a profound conceptual

> shift. It is through specific agential intra-actions that the boundaries and properties of the components of phenomena become determinate and that particular concepts (that is, material articulations of the world) become meaningful.
>
> (Barad, 2007: 139)

Barad articulates why taking political ontologies seriously matters, offering by extension a useful means of understanding Black political life in the Americas, a condition of justice-to-come that remains excessive to the always colonial nation-state. She writes,

> What is needed is a starting point for analysis that does not presume that the terms on either side of equivalent relations are given, but instead directly takes up the matter of the cuts that produce distinctions between 'humans' and 'nonhumans' [. . .] That is, what is needed is accountability for the cuts that are made and the constitutive entanglements that are effected. In particular, the 'posthumanist' point is [. . .] to understand the materializing effects of particular ways of drawing boundaries between 'humans' and 'nonhumans.'
>
> (2011: 123)

To extend Barad's point, it is thus not sufficient to critique the actors involved in the CPA for misunderstanding the political and material cultures surrounding the creole pig. It can indeed be taken as a given that IGOs such as the IICA and USAID sought to reproduce capitalist logics of Northern hegemony. What is important, however, is to understand how those logics are enacted and themselves enact particular ontological terrain – that is, very concretely, how they make and unmake worlds depending on how one measures "pig." The CPA shows us what globalized agrarian capitalism looks like when it is working exactly as it should. In the CPA, USAID cemented its material role in the neocolonial transformation of Haiti's agrarian worlds, while simultaneously disavowing its political role in Haiti's democratic transition. The CPA's liberation of pork futures from smallholders' agrarian worlds depended on the dislocation of democratization from the political vitality of the peasantry. This dislocation was effected by the manipulation of scientific data, of agrarian markets, and of the bodies of pigs and peasants themselves. How might we imagine such political futures otherwise? Barad insists,

> Responsibility entails providing opportunities for the organism to respond. [. . .] Understanding that the objects of investigation are effects rather than causes does not settle once and for all the matter of causation as one of acausality or of no causal relationship whatsoever in the space vacated by determinism. [. . .] The responsible practice of science is simultaneously a matter of good scientific practice (epistemologically sound science) and justice-to-come.
>
> (2011: 136–137)

While Barad is writing about responsible laboratory practices, the conditions she articulates for the emergence of justice-to-come resonate across the more than human relations that constitute agrarian worlds, too. Attending to political ontology thus allows for a more expansive understanding of the political terrain of late liberal postcoloniality, including not only evolving notions of the emergence of new political actors, but also of their constitutive role in the material futures of precarious life.

Chronic carriers: Black freedom and radical excess

In "Concerning Violence," Frantz Fanon reminds us that decolonization is always ontological, for it entails the replacement of one 'species' or kind of man with another. He compares the moral work of colonization to the eradication of yellow fever, the application of D.D.T. to proselytizing Christianity (1967). The taxonomic and biopolitical implications of Fanon's claim are significant, in particular with respect to Afro-descendant populations in the Caribbean. In Haiti, the price for Black life has been not only on the horrors of plantation slavery, but also a swift and brutal Native genocide. Neither settler nor native, Black smallholders work the land that their ancestors won in the break between an ongoing colonial project whose outcome is always extinction, and the imagining of Black freedom. That freedom had always been and continues to be in radical excess to the political ontology of the modern West. And yet, precarious agrarian worlds endure at the edge of the impossible.

Trouillot likewise frames the unthinkability of the Haitian Revolution as ontological,

> based not so much on empirical evidence as on an ontology, an implicit organization of the world and its inhabitants. [. . .] Although it left room for variations, none of these variations included the possibility of a revolutionary uprising in the slave plantation, let alone a successful one leading to the creation of an independent state. The Haitian Revolution thus entered history with the peculiar characteristic of being unthinkable even as it happened. [. . .] How does one write a history of the impossible?
>
> (1995/2012: 73)

The endurance of Haiti's peasantry is testament both to the remarkable persistence of decolonial agrarian worlds, and the ceaseless reproduction of the foundational violence of the Americas. As becomes clear through the CPA, Haiti's smallholders have continually sought to materialize an agrarian politics otherwise, one that accounts for the endurance of Black life within and against the postplantation politics that both need and neutralize it. To conclude, this chapter offers a brief reflection on the methodological implications of attending to the slippery ontologies of postcolonial agrarian worlds.

A PEPPADEP veterinarian notes, "At the end of 1979, the swine population was estimated at only 600,000. [ASF] had evidently spread all over the country; Haiti was considered enzootic for the disease with chronic carriers of the ASF virus" (Alexander, 1992: 253). He notes, in passing, that many of these carriers were asymptomatic; the threat of a transnational disease outbreak, presumably, trumps scientific interest in the pigs' resistance. These chronic carriers of a threat to American pork futures were thus marked for eradication. The question we should be asking is not about resilience – how have the creole pig, or a precarious peasantry, managed to survive in this world – but rather about refusal – what other worlds we are not seeing, not attending to at all, that might sustain life otherwise? PEPPADEP's claim to success encompasses the production of appropriately docile political citizens who await only the stabilizing effects of a modern, democratic state, always to come. The peasantry's resistance to PEPPADEP shows their unwilling pacification into Western political ontologies, never having been particularly concerned with the structures of the nation state, in any case. A posthuman approach to this emergent postcolonial North-South horizon allows us to see how political relations constituting the state, civil society, or their remainders, have emerged from the material work of neoliberal world-making.

The neoliberal logic driving the eradication-repopulation cycle thus sought to create appropriate political subjects in conjunction with opening markets. In so doing, the CPA performed ontological work, installing an agrarian world in which peasants could meet the conditions of market entry by becoming appropriate citizen-subjects. Pig distribution recipients were asked to "perform community projects" in exchange for receiving a pig: to build a latrine or a piggery, to do 25 hours of community labor, to read and write their names, or to give three females and one male from the first four farrowings. Twenty-five more hours of community service were required to receive a second pig (USAID, 1987: 80). Ultimately, these conditions proved too stringent for their widespread adoption by the peasantry, and new swine worlds emerged, attached to multiple iterations of the creole pig. It is of the utmost importance to take the ontological disturbances posed by the endurance of Black life in the postplantation seriously, not simply as resistance, but as the materialization of agrarian politics otherwise. As Sybille Fischer notes, we can now answer a new demand for "read[ing beliefs and attributed beliefs] against the grain, with an ear for improbable distortions, so that they might reveal the shadow of other futures, or projects not realized and ideas rarely remembered" (Fischer, 2004: 23).

It is against the grain and in the break that we can read the "shadows of other futures," the possibility of postplantation worlds otherwise (Moten, 2003; Fischer, 2004: 23; McKittrick, 2011, 2013). I draw on Caribbean scholars' long tradition of antiracist and anticolonial critiques of humanism to address the present conjuncture, at which decolonial horizons must contend with new political, ecological, and material demands. Finally, I echo Juanita Sundberg's insistence that decolonizing posthumanist geographies in fact "implies making political choices about the worlds we wish to enact, choices for some ways of living together over others" (2014: 42). The precarious lives of Haiti's peasants, whose ancestors' blood was sacrificed to capital and to the modern state, demand of us that we look beyond what David Scott calls the longing for anticolonial revolution (2004: 7). In seeking to disarticulate a collective decolonial project from the political ontology of the postplantation, let us pose an interdisciplinary challenge to the ruins of the postplantation – it could be otherwise (Whatmore, 2006; Woolgar and Lezaun, 2013).

Notes

1 Throughout, I refer to project documents originally published in French. All translations are my own.
2 Agamben writes, "The term *sacer* "indicates . . . a life that may be killed by anyone – an object of violence that exceeds the sphere both of law and of sacrifice. This double excess opens the zone of indistinction between and beyond the profane and the religious. [. . .] From this perspective, many of the apparent contradictions of the term "sacred" dissolve. Thus the Latins called pigs *pure* if they were held to be fit for sacrifice ten days after their birth. But Varro (*De re rustica*, 2.4.16) relates that in ancient times the pigs fit for sacrifice were called *sacres*. Far from contradicting the unsacrificeability of *homo sacer*, here the term gestures toward an originary zone of indistinction in which *sacer* simply meant a life that could be killed" (1998: 86).
3 Kim et al. (2012) remind us that despite the sub-field's critical commitment to examining the production of inequality under relations of power, nature, and capital, political ecologists have tended to reproduce the claim that there is only one way to be radical in geography. Along with Sundberg (2014), their call for a decolonial political ecology resonates with the emerging commitments of a posthuman political ontology.

References

Agamben, G. 1998. *Homo Sacer: Sovereign Power and Bare Life*. Trans. D. Heller-Roazen. Palo Alto: Stanford University Press. www.sup.org/books/title/?id=2003

Alexander, F. 1992. 'Experiences With African Swine Fever in Haiti', *Annals of the New York Academy of Sciences*, vol. 653, no. 1, pp. 251–256.

Allison, J. 1982. *Rapport préliminaire de la repopulation du cochon pour Haiti*. Port-au-Prince: MARNDR.

Barad, K. 2007. *Meeting the Universe Halfway: Quantum Physics and the Entanglement of Matter and Meaning*. Durham: Duke University Press.

Barad, K. 2011. 'Nature's Queer Performativity', *Qui Parle: Critical Humanities and Social Sciences*, vol. 19, no. 2, pp. 121–158.

Barad, K. 2012. 'On Touching—The Inhuman That Therefore I Am', *differences: a Journal of Feminist Cultural Studies*, vol. 23, no. 3, pp. 206–223.

Bhabha, H. 1984. 'Of Mimicry and Man: The Ambivalence of Colonial Discourse', *October*, vol. 28, no. 1, pp. 125–133.

Carney, J.A. and Rosomoff, R.N. 2011. *In the Shadow of Slavery: Africa's Botanical Legacy in the Atlantic World*. Berkeley: University of California Press.

Dejean, M. 1987. *Situation de l'elevage porcin: Strategie de repeuplement et gestion d'elevage*. Port-au-Prince: MARNDR.

Diederich, B. 1985. 'Swine Fever Ironies: The Slaughter of the Haitian Black Pig', *Caribbean Review*, vol. 14, no. 1, p. 16.

Fanon, F. 1967. *The Wretched of the Earth*. Trans. C. Farrington. New York: Penguin.

Fanon, F. 2008. *Black Skin, White Masks*. Trans. R. Philcox. New York: Grove Press.

Farmer, P. 2006. *AIDS and Accusation: Haiti and the Geography of Blame*. Berkeley: University of California Press.

Fischer, S. 2004. *Modernity Disavowed: Haiti and the Cultures of Slavery in the Age of Revolution*. Durham: Duke University Press.

Glissant, É. 1981. *Le discours antillais*. Paris: Seuil.

Hallward, P. 2007. *Damming the Flood: Haiti, Aristide, and the Politics of Containment*. London: Verso.

Harvey, D. 2007. *A Brief History of Neoliberalism*. Oxford: Oxford University Press.

Kim, S., Ojo, G.U., Zaidi, R.Z. and Bryant, R.L. 2012. 'Bringing the Other into Political Ecology: Reflecting on Preoccupations in a Research Field', *Singapore Journal of Tropical Geography*, vol. 33, no. 1, pp. 34–48.

Law, J. and Lien, M. 2012. 'Slippery: Field Notes on Empirical Ontology', *Social Studies of Science*, vol. 43, no. 3, pp. 363–378.

MARNDR. 1982a. 'Projet d'éradication de la peste porcine africaine et de developpement de l'elevage porcin en Haiti: Document de Base', MARNDR.

MARNDR. 1982b. 'Projet d'eradication de la peste porcine africaine et de developpement de l'elevage porcin en Haiti: Rapport No. 1', MARNDR.

McKittrick, K. 2011. 'On Plantations, Prisons, and a Black Sense of Place', *Social & Cultural Geography*, vol. 12, no. 8, pp. 947–963.

McKittrick, K. 2013. 'Plantation Futures', *Small Axe*, vol. 17, 3_42, pp. 1–15.

Mitchell, T. 2002. *Rule of Experts: Egypt, Techno-Politics, Modernity*. Berkeley: University of California Press.

Moten, F. 2003. *In the Break*. Minneapolis: University of Minnesota Press.

Povinelli, E.A. 2011. *Economies of Abandonment: Social Belonging and Endurance in Late Liberalism*. Durham: Duke University Press.

Rolland, B. 1987. *Présentation d'une enquete sur l'élevage porcin Haitien issu du repeuplement*. Institut d'Elevage et de Médécine Vetérinaire des Pays Tropicaux. Maisons-Alfort: CIRAD-IEMVT

Scott, D. 2004. *Conscripts of Modernity: The Tragedy of Colonial Enlightenment*. Durham: Duke University Press.

Strathern, M. 1999. *Property, Substance, and Effect: Anthropological Essays on Persons and Things*. London: Athlone Press.

Sundberg, J. 2014. 'Decolonizing Posthumanist Geographies', *Cultural Geographies*, vol. 21, no. 1, pp. 33–47.

Taluy, S. 2015. 'USA's Crusade on the Haitian Black Pig Population & Its Toll on Haitian Peasantry and Agriculture', *Caribbean Quilt*, vol. 3, no. 1, p. 51.

Toussaint, J.J. 1989. 'Eradication de la Peste Porcine Africaine et Développement de l'Elevage Porcin en Haiti', *Bulletin Agricole*, vol. 54, pp. 5–9.

Trouillot, M-R. 1990. *Haiti, State Against Nation: The Origins and Legacy of Duvalierism*. New York: Monthly Review Press.

Trouillot, M-R. 1994. 'Culture, Color, and Politics in Haiti', *Race*, pp. 146–174.

Tsing, A.L. 2005. *Friction: An Ethnography of Global Connection*. Princeton: Princeton University Press.

USAID. 1987. *USAID/Haiti Evaluation Summary Part I*. United States Agency for International Development.

Whatmore, S. 2006. 'Materialist Returns: Practising Cultural Geography in and for a More-Than-Human World', *Cultural Geographies*, vol. 13, no. 4, pp. 600–609.

Woolgar, S. and Lezaun, J. 2013. 'The Wrong Bin Bag: A Turn to Ontology in Science and Technology Studies?' *Social Studies of Science*, vol. 43, no. 3, pp. 321–340.

Wynter, S. 2003. 'Unsettling the Coloniality of Being/Power/Truth/Freedom: Towards the Human, After Man, Its Overrepresentation – An Argument', *CR: The New Centennial Review*, vol. 3, no. 3, pp. 257–337.

4 Terra plena

Revisiting contemporary agrarian struggles in Central America through a "full earth" perspective

Naomi Millner

Introduction

To say the earth we share with approximately seven billion other human beings, who speak 6,500 different languages, as well as 8.7 million species of animal and plant life, is populated by many 'worlds' is an understatement. The earth teems with microbial communities in perpetual interaction; with the remains of once-living things in de- and re-composition; with disintegrating rock and crystalline minerals; with slow-growing root-systems and, between them, burrowing macro-organisms. We would not have food without all of these interrelating elemental processes. Nor would we have food without the millennia of experimental know-how for adapting soils, seeds, and water-systems that underpin modern agriculture.

It is easy to reduce planetary diversities when conceiving of social practices like food production. This is partly necessary: epistemic reduction enables coping with complexity and allows for the creation of knowledge about specific components within these processes, such as the soil. Yet there remain serious contradictions in the ways that complexity is managed to create authoritative forms of environmental knowledge that inform environmental governance. For example, despite the thousands of years of soil management, observation, and classification in places intimately related to the major centres of plant domestication, local soil knowledge has barely been reflected in soil science research (Barrera-Bassols and Zinck, 2003). This is important partly because such knowledge is elsewhere being enclosed for private profit (Hayden, 2003), but also because everyday conservation practices reflect modalities of 'worlding' nonhuman environments that may offer principles for more sustainable forms of inhabitation (Rist and Dahdouh-Guebas, 2006).

Indeed, the very notion of 'worlding,' increasingly applied within anthropological accounts of indigenous knowledge, reaches to grasp the singular ways that the animacy of nonhuman (or 'more-than-human') ecologies are culturally articulated. According to Western convention, modern science helps divest 'myth' of false claims by establishing principles of experimental methodology. We learn, thus, to appreciate that there are many cultures interpreting one given and complex nature. However, while empirical testing unsettles many unhelpful 'truths' about the past and the nature of the world, recent scholarship troubles the notion that only one notion of nature exists. Long before pre-ecological thinker Alexander von Humboldt devised his innovative graphic maps to depict the 'kosmological' inter-connections of the biophysical world in the early nineteenth century, the earth was invested with agency, spirit and connectivity through diverse mythological and cosmological narratives (de la Cadena, 2015; Gemein, 2016). Meanwhile, based in depth explorations of particular cosmologies, many critical anthropologists increasingly use the term 'multinaturalism' to emphasise that there are plural coherent ways of knowing and relating to the holisms of nonhuman environments. Eduardo Viveiros de Castro's (2004) work on Amerindian ontologies laid the foundation for this work, showing how Amazonian cosmologies presume nondifferentiation between the internal forms of humans and nonhuman beings; both are understood to have been 'human' originally. Meanwhile, in other ontologies, species ecologies are organised per their correlation with planetary bodies and ancestral narratives that invest nonhumans with dreams, memories, and other forms of agency (Descola, 2005; Kohn, 2013). This research has initiated a reconsideration of political concepts, for example, cosmopolitanism, which traditionally acknowledges cultural difference, but does not acknowledge the richness of plural 'natures.'

There is a risk of wrongly applying multinaturalist theory as reinscribing indigenous cultures as more authentic because they are 'closer to nature.' The central point, however, is to understand how the historical enclosure of land and livelihoods depends on a powerful knowledge politics that centres on 'ontological' forms of violence.[1] Euro-american articulations of 'nature' (all that is, without human intervention) are premised in a fundamental separation from 'culture' (all that is produced by humans, through the application of reason), and the belief that all beings are radically separated by their internal lives (Descola, 2005). Of course, this binary is not straightforward, as, since the Romantic period, nature has been used in Western thought both to support the capitalist theory of value, and to undermine it (Morton, 2007: 19). However, the violence of an Euro-American worldview is performed through the denial of the specificity of this cosmological narrative, which relegates all other forms of knowing to myth. Such contradictions are often compounded in articulations of indigeneity. For example, in environmental protests defending the forests of Clayoquot Sound in British Columbia, Canada, white Western activists attempted to speak 'in the name' of the forests and First Nation forest inhabitants in order to protect them against industrial interests. In doing so, they situated both in the (silent) primitive past (Braun, 2002).

Such knowledge politics can be associated with the concept of terra nullius: the designation of land managed through subsistence technologies or for common use as 'empty' or 'wild,' as a precursor to enclosure. In cultural terms, it is possible to say that inhabited land is empty only if you can portray existing practices as 'natural' (and thus in need of civilisation) or 'cultural' (and largely influenced by mythic, false belief. Through this logic contemporary 'land-grabbing' scenarios, as well as the 'new' enclosures of environmental knowledge and genetic material are shown to recapitulate a pattern laid down by the enclosure of commons in sixteenth and seventeenth century England (Makki, 2014). The designation of land as empty and uncultivated forms the basis for claiming it as private property, or selling it, and thus of extracting profit to be accumulated elsewhere (Harvey, 2006).

Mario Blaser (2014) devises the term 'political ontology' to capture two different ways that the terminology of ontology has been used to articulate the violence of terra nullius. While critical anthropologists trace multiple ontologies – a multiplicity of worlds animated in different ways – geographers increasingly use the language of relational ontologies to emphasise that whatever we know as 'reality' is an always emergent assemblage of humans and nonhumans that cannot be separated into autonomous parts.[2] This difference is important: in the first, emphasis falls on ontologies as diverse, coherent entities; whereas, the second points to the inevitably partial, fluid, and hybrid aspects of knowing. Where both forms of scholarship agree is in recognising the violence of 'imposed universals' wherein the extension of highly specific notions of ethics or freedom (including liberalism and human rights) as regimes of truth are made to count for all (Tsing, 2005). Multinaturalist ethics demand attention to moments of 'translation' or 'transliteration' wherein one claim or concept is moved to another site, taking with it a host of other linked meanings and associations. This is underlined by a number of recent anthropological studies that bring together these two senses of ontology by following the piecemeal way that heterogeneous elements of species life and human practice, each characterised by its own sign-based (semiotic) communication systems, become connected into 'assemblages' (Tsing, 2015; Li, 2014; de la Cadena, 2015). An assemblage has its own coherence, rules, and margins, but it exists only by connecting with other coherent systems by means of gatekeepers who translate the semiotic codes of one into another. For example, Tsing's (2015) study of matsutake mushrooms explores how mushroom-pickers in Oregon establish patches of autonomy based in their knowledge of how to find mushrooms; the mushrooms' ability to survive in an industrial waste-ground; the ability of mushroom-buyers to convert these mushrooms into packages appealing to the Japanese market; and, the cultural significance of the matsutake in Japan. As in Gibson-Graham's (2008) ground-breaking work on diverse economies, the emphasis here is on giving attention to plural meaning-making systems and economies of value, rather than reifying one larger (always precarious, partial and wobbly) logic of violence (capitalism) by making it the central object of study.

Still, questions remain: can deconstructive work assert connective imaginaries that destabilise logics of enclosure in more than temporary ways? How do we emphasise and work to prevent the foreclosure of singular worlds, as well as enable political alternatives forged 'in the mix'?

My term terra plena articulates an ontological starting point for connecting these diversely positioned decolonial projects. Terra plena invokes a manner of approach that starts from the premise that the land is not empty but full. Rather than overpopulated, terra plena denotes an earth teeming with communities of species life and knowledge-practices for knowing them. Besides acknowledging the complexity of multispecies communication systems that are always-already in relation, the term foregrounds the prevalence of multispecies thinking in non-Western ontologies. Distinctively, I bring a relational notion of ontology, which understands knowing as a partial and hybrid process, into dialogue with notions of multinaturalism. This hybridity is critical. In a notable example, the claims being made by transnational agrarian movements (TAMs) increasingly depend on innovative forms of dialogue between smallholder and indigenous communities from very different contexts, made possible through new technologies of communication and transnational forms of association. Yet, the political claims forged in the process defend indigenous forms of knowledge and traditional agricultural practices. Recalling an epistemic ethics advocated by the feminist Science and Technology Studies scholar Donna Haraway (2008) – through whom many aspects of ontology, assemblage, and multispecies thinking reached the social sciences – terra plena thinking articulates meaningful ethical principles for engaging already-existing, situated systems of knowledge production from within their material imbrication, rather than attempting to stand outside them.

To develop terra plena thinking, I focus in this chapter on the context of agricultural struggle in Central America, where hybridity is being negotiated in the form of everyday environmental ethics and shared political claims. Edelman (1998) has long emphasised that transnational peasant organising in Central America 'raises significant questions regarding social scientific approaches to transnationalism, collective action, and agrarian change' (p74). For at least the last three decades, emerging movements have combined elements of the class-based interests of 'old' social movements with fresh attention to cultural difference and specificity. This organising was largely prompted by efforts like the consolidation of new decision-making loci above the scale of the national state; the slashing of social services like agricultural extensions; the liberalisation of the grain trade in Central America, which brought grain producers into competition with foreign farmers; and, the growing effects of agrochemical contamination (Edelman, 1980). Responses to material and cultural enclosures at both national and transnational scales have increasingly led to cross-border alliances that connect issues such as food poverty, cultural dispossession, and environmental degradation, rather than treating them in isolation.

To foreground these relational issues, I focus on the permaculture movement in El Salvador, which emerged in the dynamic contexts of civil war, the Green Revolution, and transnational dialogues surrounding agricultural futures. Emphasising permaculture practices highlights the mixedness embedded into agricultural practices as a basis for innovating design solutions, but it does this by bringing principles of scientific method into dialogue with agricultural tradition and elements of indigenous cosmology. Here terra plena also meets Madre Tierra as a trope mobilised increasingly between diversely situated TAMs. In the second part of the chapter I examine how Madre Tierra acquires resonance between agrarian movements, raising issues with the potential 'essentialising' functions such images can play, but showing how its circulation also serves to keep ontological questions open. What the Latin term terra plena aims to enable within post- and de-colonial scholarship is already being expressed within food and land policy through such tropes – but they have important questions to ask one another.

My goal with this chapter is ultimately to signal emergent pedagogical practices that work precisely to resist ontological closure in contexts that are characterised by cultural dispossession, hybridity and interaction. Through the course of the chapter, I propose seven terra plena principles as tools for this collective decolonial work. It is my hope that these principles can help define an ethos for habitually destabilising colonial geographies as part of everyday reading and listening practices.

Situating permaculture

'Permaculture' emerges from the idea of creating 'permanent cultures': interacting ecological, social, and biophysical systems that replenish and aid one another. First named in Australia during the 1970s, permaculture defines a catalogue of tools and principles centred on simulating, or directly utilising, patterns observed in ecosystems and environments within the design of human food production and social systems.[3] Permaculture entered El Salvador via a series of 'brokers', who saw it as a means of regenerating exhausted soils and deforested countryside, but also of addressing social and political issues. While it is possible to think of permaculture as a colonial import to solve colonial problems, the attention to, first, observing the interactions between other animate systems, and second, enabling all learners to both create their own designs and teach others, pushes in other directions. Indeed, permaculture principles are explicitly articulated as alternatives to the cultural imperialism associated with histories of colonialism and international development. For example, part of permaculture's learning process involves engagement with local and ancestral practices of growing (and knowing), with the understanding that these have been progressively sidelined by modern agriculture. Acknowledging indigenous knowledge as a repository of pragmatically useful practices avoids placing indigenous voices in the past, without reifying localism as the only possible solution.

Salvadoran permaculture is distinct here because of its elaboration in the context of popular education practices associated with the growth of peasant resistance prior to El Salvador's long running, and Northern-implicated, civil war (1979–1992). Permaculture also found resonance in El Salvador because it echoed with the concerns of growing farmer-to-farmer (campesino-a-campesino) movements, seeking to revive traditional practices of agriculture as part of a critique of the social and ecological impacts of large-scale modernisation projects. Third, permaculture caught the imagination of people seeking to reconnect with indigenous cultures and heritages virtually obliterated in El Salvador in the early twentieth century. In this section, I contextualise permaculture by giving more detail to the intersection between the political, increasingly transnational movements it articulates. As I do so, I suggest how permaculture complements decolonial academic thought, helping us to think about terra plena in terms of ethical principles that can be practised by anyone.

The fieldwork that informs this argument involved ethnographic observation in the two regions of El Salvador where permaculture has been most active (Cuscatlán and Morazán: see Figure 4.1). Data collection included 36 interviews with small-scale farmers; local, national and representatives of international non-governmental organisations (INGOs); regional and municipal governors, to provide broader perspectives, as well as oral histories with founding members.

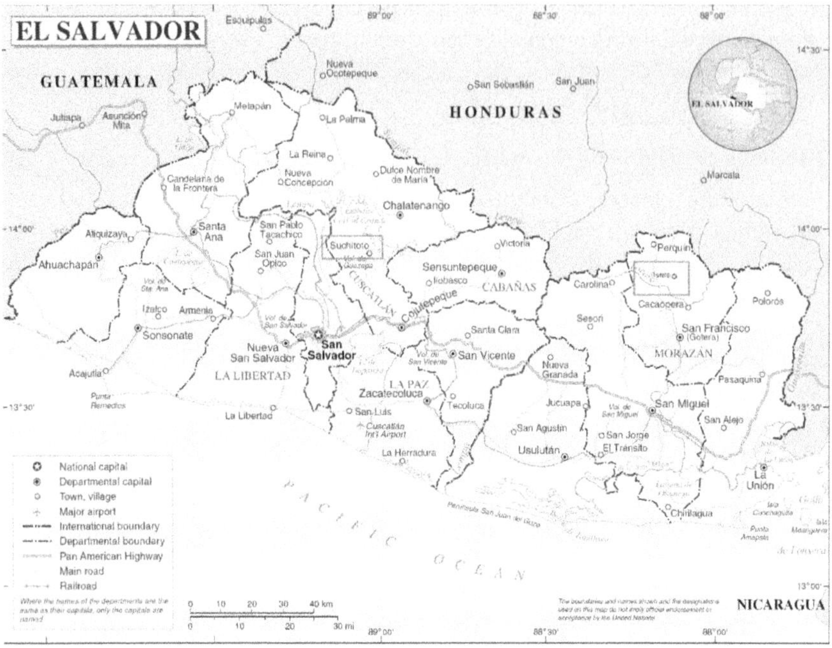

Figure 4.1 Map of research sites in El Salvador; Field sites: Suchitoto and Toroloa (indicated in rectangles above).

I also observed and took part in a permaculture design course in Suchitoto, Cuscatlán that runs across a year for campesinos in the surrounding area, and, after an 18-month involvement, I co-designed a series of eight four-hour participatory workshops with my research contacts at the Instituto de Permacultura de El Salvador (IPES) in both regions, aimed at fostering knowledge exchange on the topic of everyday environmental know-how. After gaining the appropriate consents, audio and video capture of interviews and fieldwork also enabled a second translation of data upon return to the UK, as well as the production of a short film.

Tracing the politics of everyday environmental expertise

Recent attention to, and actions by, TAMs like La Via Campesina [LVC] give context to the indigenous 'tropes' emerging at the intersection between movements, including permaculture. LVC was set up during the early 1990s by agricultural producers from around the world against a backdrop of global agrarian crisis and the withdrawal of support for domestic agricultural sectors across the Global South (Edelman, 2014). Founding members linked the subsidisation of large-scale export crops with the large-scale dispossession of small-scale farmers, as well as endemic forms of pollution (Desmarais, 2007; McMichael, 2012). These activists also rejected definitions of food security consolidated between the 1970s and 1990s for presuming that colonially grounded, structural inequalities can be solved by solutions based in market-based economics and industrial agriculture.

The contradictions embedded within United Nations (UN) institutions were a particular target. Coercive tactics employed by the Bretton Woods Institutions (the World Bank and the International Monetary Foundation) to enforce the liberalisation and deregulation of indebted national economies were seen to stand directly at odds with aspirations mapped out by the Food and Agriculture Organisation (FAO), the UN Development Programme (UNDP) and World Food Programme (WFP) (Ziegler et al., 2011). In 1996 LVC introduced their counter-concept, 'food sovereignty,' to the UN World Food Summit, where definitions of food security were being approved. Food sovereignty was presented as a kind of 'agrarian citizenship' (Wittman, 2009) within whose basis the constitutive roles, and correlative rights, for small-scale farmers were articulated. Security is meaningless, they argued, unless it also involves protecting the autonomy of small-scale farmers to choose how to grow, and to refuse to open the best agricultural land to the vagaries of the global food market. Today LVC boasts a 'peasant internationalism' comprising hundreds of thousands of small-scale producers (Martinez-Torres and Rosset, 2010) and 200 (sub-)national organisations from more than 56 countries (Borras, 2010).[4]

Indeed, the knowledge production processes of movements like LVC have, more latterly, been heralded as a potential model for democratic knowledge production (Pimbert, 2006), because of the distinctive dialogic practice ('dialogo de saberes') that has shaped the movement since its inception (Rosset and Martinez-Torres, 2013). The point is that the agenda's force emerges from an exchange not only between knowledge content, but between forms of knowledge that derive from diverse worldviews, cultural contexts, and experiences of neoliberal globalisation (Borras, 2010). As such, momentum proceeds from a dialogue between absences: shared experiences of dispossession and colonialism, elaborated through shared experiences in agroecological movements and peasant training schools in the Americas, Africa, and Asia. Meanwhile, much of what is distinctive in this ideologically autonomous and pluralist coalition emerges from the negotiation of internal differences: disagreements over the basic unit of politics (such as the family, community or collective), or the appropriate vehicle for agency (workers, families or militants). This differentiation is internalised within and between TAMs, and shapes their agendas and strategies as they interact with international development institutions (Ibid.). The 'counter-hegemonic globalisations' articulated by such movements, alongside other growing coalitions, including the World Social Forum (WSF) and the Brazilian Movimento dos Trabalhadores Rurais Sem Terra (MST) articulate a critical dimension of decolonial alliance-building that I propose as a first principle for terra plena thinking. Namely, such movements reveal the importance of deriving shared claims not from a common starting-point or solution, but from common experiences and questions, combined with shared mechanisms for managing disagreement. Thus:

1 It is the spaces in between the singular dimensions of a specific experience that provide impetus for common claims that do not rely on reducing the range of possible solutions.

In Central America, shared contours of this specific experience were critically tied to the introduction of Green Revolution technologies, which were set to modernise agriculture and significantly intensify production. The well-known Mexican Agricultural Program (MAP) was set in motion after United States Vice-President Henry Wallace, an experienced Iowa corn breeder, expressed astonishment at rates of low yield he witnessed on a tour of Mexican farms in 1940 (Mangelsdorf, 1951). Wallace entered talks with the U.S.-based Rockefeller Foundation, leading to the creation of the semi-autonomous Office of Special Studies in 1943, to carry out research on Mexican strains of wheat and corn (Edelman, 1980). Early projects were of limited success despite huge investments, but the major 'Plan Puebla,' beginning in 1967, was considered successful enough to form a prototype for other regions in Central and Latin America (Ibid.: 33). Such projects were contextualised by a deliberate policy on the part of United States and quasi-U.S. aid organisations of directing their assistance primarily to countries featuring in their global strategy (Ibid.; Paré, 1990). Such biases were also expressed through an anti-communist rhetoric, which pervaded the discourses of the Rockefeller Foundation and associated agricultural research institutes. Thus, national governments were persuaded that failure to introduce 'modern solutions' would lead 'underdeveloped' countries to accept communist promises and systems (Carey, 2009). From the 1960s onwards, projects based on the Mexican model were rolled out in Guatemala, El Salvador, Venezuela, and Brazil, mostly under the auspices of the U.S. Department of Agriculture. Meanwhile, the 'miracle' high-yielding varieties – distributed freely at first, but subsequently purchased anew each year – required higher and higher inputs of fertilisers as soils became polluted and impoverished.

A shared sense of agricultural dispossession in the wake of these agricultural 'revolutions' created a strong impetus for bottom-up organising in subsequent decades, as well as scholarship that revalorises the sidelined agricultural know-how. Long considered 'backward' or peripheral, peasant or campesino farming is regarded the locus of agricultural innovation and activism in contemporary TAMs. In the surrounding literatures we are asked to attend to the 'residualisation' of agricultural knowledges as a long-standing issue of food insecurity: it is not only soils but know-how and rural solidarities that have been eroded (Van der Ploeg, 2014; Desmarais, 2007). Political ecologists have been pushing in this direction since the late 1970s by attending to overlooked sites of knowledge production, and revealing the centrality of 'peripheral' rural transformations to global economic processes (Dahlberg, 1979; Stonich, 1993). This is also the premise of Harriet Friedmann (1982) and Philip McMichael's (2009) work on food regimes since the late 1980s, which exposes the entanglement of rural and urban processes in food production from the 1870s onward. Friedmann's work shows how food aid from the United States to formerly self-sufficient agrarian societies significantly shaped the international food order, leading to the establishment of new (unequal) divisions of labour and urban concentrations of dispossessed people in the 1950s and 1960s. Meanwhile, McMichael (2014), amongst others, uses the food regime terminology to situate food sovereignty historically, and to highlight contradictions in the transitioning global food regime as important conditions for possibility for LVC's emergence.

Here I identify a second key principle for terra plena thinking:

2 Innovation is not the property of Euro-American urban processes, and has long been situated in rural and agricultural sites of experimentation.

Thus, it is critical that contemporary research into emerging TAMs also draws on at least two decades of work siting small-scale farming as a locus of innovation in its own right. For example, the landmark study by ecological anthropologist Robert Netting (1993) presents smallholder agriculture as a distinctive cultural ecosystem that can thrive even producing for growing populations – except where industrialised agriculture is disproportionately incentivised. Netting frames a common repertoire of smallholder techniques that is always being reproduced afresh because of the suitability of specific techniques (such as the terracing of slopes) to the scale of practice. The politics of this approach lies in the emphasis on the aptness of these techniques for future food production, and the refusal to doom small-scale agriculture by framing it in terms of exploitative global economic systems.

More recently, scholars have highlighted that the conduits of knowledge-sharing, through which smallholders can improve practice and share concerns, are a crucial part of this politics. It is not only corporate capitalism that is characterised by innovative networks (Van der Ploeg, 2014). Others underline the way that networks like LVC only emerge because of the unevenness of the ways that capital reshapes peasant agriculture over time (Edelman, 2014). This unevenness is critical, because Netting's model tends to romanticise the patriarchal norms in the smallholding contexts he investigates, rather than linking them with economic relations that might also be transformed. It is important to acknowledge that there are gendered power relations in play in the smallholding economy, and that small-scale farmers are also actively redefining broader economic landscapes.

Either way, the notion of the peasant as 'backward' or 'passive' is replaced by an understanding of small-scale farming as an evolving and innovative mode of relating with surrounding environments. Following Netting's efforts to politicise the prevailing cultural ecology approaches, political ecology has since gained pace as an interdisciplinary field for setting environmental uses and transformations into historical contexts. Alongside political ecology, awareness of everyday environmental expertise as a contested political field has also fostered the emergence of 'agroecology' as a field of agronomy, where small-scale and traditional techniques are systematically aggregated and testes (Cox and Atkins, 1979; Gliessman, 1990). From this vantage, the politics of environmental know-how cannot be separated from post- and decolonial concerns.

Seeing double: Applying agroecological principles

Agroecology has, since its inception, in many ways embodied terra plena thinking into the domain of agriculture. Providing significant context to the emergence of the permaculture movement in El Salvador, agroecology also highlights a third principle for terra plena thinking:

3 Issues of social justice and environmental degradation are crucially intertwined, and need to be approached in in holistic connection.

This is vital, because issues such as food security, gang violence, and soil degradation are often approached through development strategies that reinforce, rather than addressing, their connection with colonial histories. In contrast, the kind of 'double-vision' promoted in agroecology and permaculture works by strengthening the close ties between historically embedded environmental know-how and the thriving of multiple species communities on the other.

Agroecology arrived in Central America as part of responses to Green Revolution technologies. Although it spread initially through the support of INGOs, the movement has always foregrounded local leadership and existing organisations, and has been characterised by 'bottom-up' forms of organising. The first documented agroecology initiative in Central America was a small programme in Guatemala in 1972, which aimed to empower a group of indigenous Maya-kaqchikel campesinos to teach ecological techniques that were embedded in their traditional culture (Holt-Giménez, 2006). Projects centred on sharing long-standing principles for agriculture that were being eroded, such as recycling biomass, minimising nutrient losses, and restoring degraded soils (Altieri, 1995). Bilingual campesinos were also trained as farmer 'extensionists', and used simple instruments – a machete, a tape measure, and an apparato A (a simple apparatus for measuring land gradients) – together with oral traditions and sociodrama to communicate with other farmers. These were so successful that several encounters [encuentros] and exchanges [intercambios] were organised subsequently. First in Guatemala, and later in Nicaragua, small INGOs arranged for intercambios [exchanges] between indigenous farmers and other small-scale farmers, promoting agroecological knowledge exchange in response to increasing interest in interregional cooperation[5] (Holt-Giménez, 2006). This saw the emergence of a Campesino-a-Campesino (CaC) [farmer-to-farmer] model, which grew rapidly during the late 1980s when heavy flooding exposed the difference between traditionally terraced farms, and modern farms, which were stripped of topsoil. INGOs subsequently supported village-level projects through into the early 1990s throughout Central America in conflict-affected areas.

Agroecology and permaculture take on a particular social significance in El Salvador, however, in relation to the 1979–1992 civil war, which centred on issues of access to land and colonially inflected power relations. Before the civil war, liberalisation of the state and a period of intense capital accumulation in the 1920s led to varying degrees of proletarianisation and dispossession throughout the country (Cabarrús, 1983). Meanwhile, the advance of agrarian capitalism had devastated the material basis of indigenous communities and contributed to a widespread rejection of indigenous markers from this point onward, such as language and dress (Gould and Laria, 2008). Particularly poignant in collective memory is La Matanza: a massacre of more than ten thousand people in 1932 that marked increasingly brutal tactics to suppress growing social unrest, as well as the emergence of an oligarchical alliance between 14 powerful, land-owning families sought to liquidate blacks and Indians whilst establishing coffee as the country's principle export. Meanwhile, between the late 1940s until the 1980s the number of campesinos forced to rent plots [microfundias] to make a living rose considerably, with rents increasingly demanded in advance. Elections were tightly controlled, and in some places sterilisation was practiced as an imposed form of birth control (Roseberry, 1991). The long struggle, which ended in a land retribution programme mandated by UN Peace Accords,[6] led to democratisation of the national government, but brought issues of race and historical marginalisation firmly to the centre of environmental concerns.

Agroecology had some presence in El Salvador before the war but was sidelined under the pressures of the drawn out political conflict. The CaC movement only truly took off in El Salvador in the late 1990s via Nicaragua and Guatemala as leftist coalitions were breaking apart under the pressure of organising without the unifying mission of guerrilla warfare. Rising concerns in Europe with gender issues, discrimination against indigenous peoples, and environmental degradation were reflected in NGO project involvements, and later in bilateral and multilateral funding (Pearce, 1988) – although many of such projects were short-termist in focus and quickly disintegrated. The CaC movement was distinct for its 'bottom-up' models of community development and conflict resolution, and several meetings were organised for Salvadoran campesinos, who brought back practical ideas to their own regions, including Cuscatlán and Morazán. This led to the development of first an Eastern Commission, and later a Western Commission for the CaC movement in El Salvador, with the objective of rehabilitating useful traditional agricultural practices. To do this, CaC pedagogies borrow from the lexicon of scientific methodology, and foregrounding training in experimental investigation as a basis for exchange between campesinos. Each farmer, or community of farmers, is encouraged to test the capacity of proposed techniques by constructing twin plots [parcelas gemelas] and measuring the differences in production. Each farmer is also encouraged to become a promotore [promoter] for principles they had found effective. Agroecology, from this perspective, marks a conjuncture between scientific principles for establishing translatable authority, and notions of tradition that resist notions of universal translatability.

Permaculture itself was first encountered in Guatemala by CaC campesinos in the late 1990s, although it became established only in 2000, when Juan Rojas, a Salvadoran man exiled to Australia during the civil war, returned to become a 'permaculture missionary,' sponsored by churches sympathetic to the revolutionary movement. Karen Inwood, a community development worker from the United Kingdom, met Rojas at a permaculture course at a Scottish eco-village in 1999, and subsequently spent 12 years in El Salvador supporting the development of incipient permaculture networks, run today by campesino men and women. The new networks differ from existing agroecological movements in several key ways: high numbers of women and young people are in leadership roles; they have strong discourses of health that connect growing and social activities; and, the networks frame sustainable practices in relation to inequalities in access to food and land.

It is not by accident that permaculture and agroecology are most highly concentrated in formerly guerrilla-dominated regions, as the popular education practices that informed the revolutionary uprising also provided the foundation for agroecological farmer-to-farmer learning after the end of the war. Many of those involved in permaculture played important roles in guerrilla movements, although some also fought on the opposite side. Meanwhile, the twelve-and-a-half-year conflict meant a complete, or partial, interruption of school education for many young people, such that permaculture also plays an important role in providing adult education. Suchitoto, a colonial town in the Cuscatlán region, was an important guerrilla hub during the civil war and today hosts the largest permaculture demonstration site where training takes place. Morazán was also an important area for guerrilla activities and suffered some of the worst damages. The regions surrounding these two towns are the most active in terms of permaculture, although there are smaller networks in San Salvador, Chalatenango and La Libertad.

Permaculture, thus, draws strongly on a repertoire and ethos of popular education that informs agroecological practices, and encourages the use of simple instruments that can be mastered and taught by anyone. Permaculture pedagogies, or principles for teaching and learning, centre on the 'design course': a curated series of training sessions in which students become teachers and produce designs for their own settings. In El Salvador, this normally runs for three days per month over a year, to suit subsistence farming work patterns. While participating in the course at Suchitoto, I observed that 33 of 36 participants were small-scale campesino farmers, while three were students from the capital interested in making a film about permaculture. Roughly half were under 30 and considered jóvenes [youth]; 16 were women. The emphasis in the course is on ecological observation and learning to experiment by trying out new ideas, and testing the efficacy of old ones. These are transmitted through reflection on the 12 permaculture 'design principles' that connect the practices as they travel globally (see Figure 4.2). The idea is to create designs for agricultural and social purposes based on close observations of the interactions between diverse species communities, micro-climates, and already-existing functional uses. For example, the eleventh principle, 'use edges and value the marginal,' encourages designers to recognise how hedges and boundaries may support the proliferation of pollinators, but it also encourages recognition of how 'in-between' social spaces can form valuable interfaces between distinct cultures. The core values of 'earth care', 'fair share', and 'people care' express that such practices mean to support vibrant ecological and social relationships, as well as a more vibrant soil. Course activities include making field-maps to show water resources, species populations, gradients, and micro-climates; sharing agroecological techniques; and, exploring local social and environmental history through videos, dialogue and improvised sociodramas. When running for the first time in a new area, courses also explore capacities to establish a local permaculture association and plant nursery, as well as local farmer's markets to sell products. Such associations aim to become autonomous within

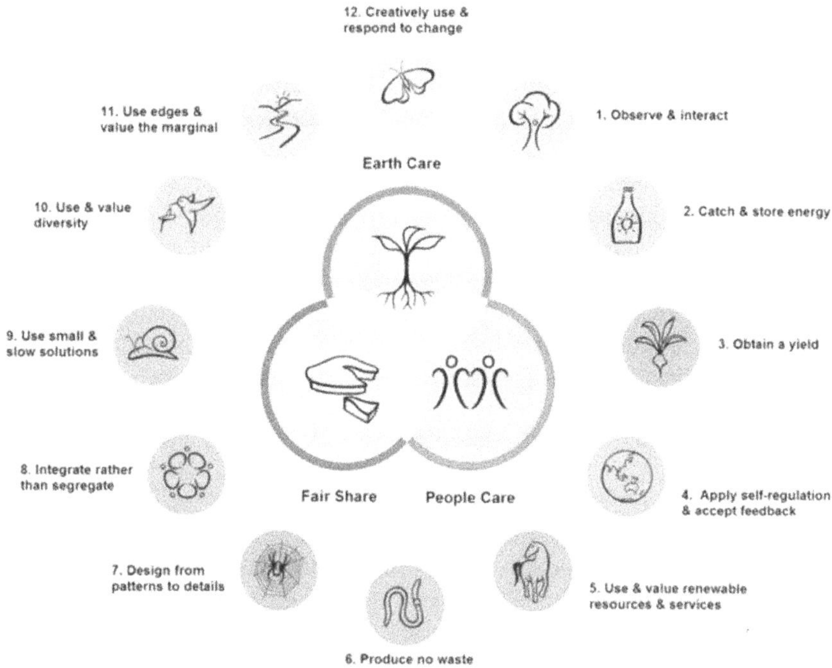

Figure 4.2 David Holmgren's 12 permaculture design principles.

Source: permacultureprinciples.com. [Accessed 22.05.2017].

three years, running their own design courses. This process is similar in locations around the world, although the way that permaculture has become established as part of decolonial environmental justice practices is highly specific.

Tradition and experimentation as decolonial ethos

Understanding the pragmatism of agricultural tradition

At the intersection between popular education principles and principles of scientific method, permaculture principles treat more-than-human environments, encountered through the body, as a kind of universal foundation for knowledge. Whilst this universalisation may seem to reintroduce a kind of ontological essentialism – by which I mean, a fixing of the diverse meanings of 'nature' into the Euro-American, romantic sense – what makes permaculture distinct is the way these principles are brought into dialogue with long-standing agricultural traditions, as well as particular material sites. From a terra plena perspective permaculture models decolonial double-vision into the micro-contexts of everyday life, allowing the production of designs for food production that combine environmental sustainability with a historical sense of environmental justice. There is a sense in which justice translates into environmental 'healing', as it has more to do with the restoration of soil wealth and cultural capacity than in seeking retribution for specific wrongs. This brings us to a fourth principle for terra plena:

4 Tradition can hold a bearing on the future, as well as the past, but only when it is apprehended in terms of capacities and agencies, rather than essential truths.

Tradition has a complicated place within terra plena, and indeed, within a decolonial politics of knowledge. It can refer to quenching forms of authority that discourage radical change, as well as long-standing practices eroded by globalising and industrial forms of production. Traditional and indigenous practices are highly valued in permaculture because they have been developed in perpetual dialogue with specific climactic and soil conditions, and evolving seed varieties. However, not all traditional practices are effective, and what is considered 'indigenous' often relies more on imaginative reconstruction than a continuity of practice.

In El Salvador, a small indigenous minority of the Nahua-Pipil population still live in the southwestern region of El Salvador, and there are smaller populations of Lenca and Cacaopera in the Eastern regions. However, most permaculturists are not themselves 'indigenous' in this sense. Rather they draw on their mixed ancestry, identifying indigenous practices as those long-practiced in the region, and deriving an indigenous 'ethos' that draws as much from discourses circulating among TAMs in the region as from particular indigenous traditions. In Central America, the Mayan indigenous past, embodied in well-preserved ruins throughout the Maya Forest, as well as in contemporary communities and literature, holds particular resonance among transnational environmental movements. This is partly because of the richly symbolic visual representations of ecological and planetary relationality in the Maya 'cosmovision' that have been preserved until this day (de valle Escalante, 2011)

In reviving so-called lost agricultural traditions that embody these forms of relationality, indigeneity becomes associated with connectedness to the land. For example, the uses of plants remembered from indigenous communities are considered sacred knowledge in permaculture, to be preserved as an integral part of agricultural futures. Thus, in one design course session, students shared herbal remedies learned from grandparents or at home. At the session's close, Angélica Gonzalez, a regular permaculture volunteer in her 40s, remarked on what was striking for her:

> we've forgotten how to see what living things can do. We see food in the shop and we think of the taste, or we look at a seed and we imagine a tree. But once our ancestors [nuestras indígenas] looked and saw: this one, good for the stomach. That one, good for bad lungs. This one, good for my stew!
>
> (Field notes, April 19th 2014)

Angélica highlights a joyful sense of reconnection with the agency of her environment through traditional know-how – although participants also explored in this session how to differentiate between practices forgotten because others work better, and those that actually work. What is critical here, though, is that Angélica associates the indigenous 'mode' of relating with a set of capacities that change how she sees, rather than with a fixed view of the world she must reproduce. This emphasis on capacity allows for the creation of solutions that can also be tested and modified. It also makes tradition future-oriented: rather than what we used to do, tradition means what we continue to do now.

Thinking about tradition in these terms helps us to notice the operation of a practical ethos within the everyday practices of permaculture and agroecology, and functions to determine which practices to conserve; to adopt; to adapt, or to discard. While exchange takes place with grandparents, with indigenous people from nearby, and with agroecologists in other regions, what we will continue to do rests on two factors: an evaluation of the capacity of a practice to promote balance and interaction between species-systems, based on ecological observation; and, an evaluation of the capacity of a practice to nourish human and nonhuman wellbeing, based in a 'worlding' of the site in dialogue with local traditions.

This interplay of experimental testing with an ethics of care is strikingly illustrated in Angélica's contribution to a later participatory workshop wherein groups created timelines showing changing food resources in their local communities:

> There was deforestation before the civil war, due to the large population. In Palau Grande there were 600 families [there are now about 60] living from where we live right up to the Pacayas river . . . from the time of the 1932 uprisings. Now there is more vegetation there than before the war. [. . .] But the large forests have gone. [. . .] The traditional practices were like the tree-roots holding the land together, all the way down, and they've almost disappeared. We can't go backwards . . . but we have to make solutions that go all the way down like that – like the tree-roots.
>
> (Angélica as quoted in fieldnotes, May 5th 2014)

There is a certain pragmatism to tradition reimagined in this powerful metaphor of the forest landscape, populated by living eco-systems and infrastructures of intricate know-how. Angélica uses the sense of social history derived from her permaculture practice to resist the common assumption of a linear narrative in which deforestation is universally worse now than it was before: she points out there was more of an issue when the area was more populated. At the same time, she highlights the thinning of another 'forest' – that of traditional environmental know-how. From a terra plena perspective, what is important here is the way Angélica invokes a future characterised by restored root-systems. They might not be the same as before, but they have the capacity to hold things together. In this sense, principles of careful observation are applied to evaluate long-standing agricultural practices as part of rebuilding a repertoire for redesigning food systems in an integrated way. If they do not work, they will not be revived. If they only work for one part of the system, they will be discarded. Of course, there is disagreement, especially where cultural memory is concerned; however, the emphasis is on what works, and for whom, with ecological observation as the principle means for evaluation.

Authorising personal experience

Whilst there may be some reduction of historical 'indigeneity' here, what we are observing is precisely an ethos for translation that operates by refusing ontological closure rather than trying to dictate one pattern for organising all other systems. An example is manifest in the ongoing reinvention of the recipe for organic compost, known amongst Salvadorian permaculturists as 'bocashi.' There are several words used for compost on a Salvadoran farm: 'abono' describes decomposed vegetable matter, while 'fertilizante' denotes chemical fertilisers. 'Bocashi', on the other hand, is active, fermented organic matter. Surprisingly, this term entered the permaculture lexicon via Japan, where agroecologists learned to mix 'Efficient Microorganisms' (EM) – cultures of naturally occurring microorganisms – into compost matter to increase the microbial diversity of soils and plants. Today, bocashi is produced by agroecological farmers throughout Central America according to local recipes, and is considered 'traditional' in that it enhances existing local methods for creating fermented compost; it demonstrably improves soil quality; and it supports the autonomy of campesinos because it costs nothing. Oscar Lopez, Suchitoto's Community Environmental Officer, who was enrolled on a Permaculture Design Course, explains how this contributes to the kind of integrated environmental recovery plan that Angélica invokes:

> In introducing homemade organic fertiliser to people we are also creating an economy that allows people to save money and get away from the hybrid seeds. Because people are dependent on those hybrid seeds; they are the standard programme available. But this is a better kind of food security. It makes people more independent, and they can see how it works for themselves.
>
> (Interview with Oscar Lopez, 20th April 2014)

Here Oscar emphasises a key dimension to food security reimagined through decolonial double-vision. 'Healing' ecological landscapes involves restoring soil and biodiversity, but also recreating rich infrastructures of know-how as part of meaningfully establishing the autonomy of small-scale farmers to determine what and how to grow.

Given this focus on knowing the more-than-human world via pragmatic principles and traditions, a significant part of permaculture training is a training of the senses. Design students learn to feel the soil and notice how it heats and then cools through the process of making active compost. They also learn how to detect when there is 'life' or 'no life' in the soil through their fingertips, and what healthy soil smells like. This reflects the biological principle at work in permaculture, which is, writes Puig de la Bellacasa (2010) fundamentally based in ethos and doings. That is, permaculture design does not just revolve around human selves and actions, but principles of collective interdependency, which reflect a different conception of 'bios' and 'biopolitics' than those usually employed in the social sciences. This bios is expanded to include soil life, plant life, and communities of microbes as well as the human body (Ibid.) – while the politics is also expanded to address questions of whose knowledge counts, and why. Permaculture practices are thus essentially carried by habit – by embodied repetitions which cultivate ways of seeing, knowing and doing, and change the future in terms of the way that particular capacities become associated together.

This emphasis on designing in the context of interrelated eco-systems also asks participants to explore what agricultural technologies, social spaces, and forms of production promote vibrant interrelationships between humans and other species-systems. In the process, what counts as health comes under much discussion. In articulating the end-goal of the design process the notion of health [salud] is also associated with the process of healing [sanar] food systems from within. This notion of health/healing as the integrated goal of permaculture was highlighted in two comments made in the final workshop in Suchitoto, where participants selected health as a key theme for future participatory inquiry. When I asked one group what they meant by health, Evelio Alos, a permaculture volunteer in his early 30s, who often helped organise others, remarked:

> Health is everything: a good body, a good mind, a good corn plant, a good soil, a good place to rest, a good society. You cannot have a healthy human without all these kinds of health. [. . .] Look at the diagrams we made about seed preservation [Indicates the five-part diagram of traditional seed-saving his group made previously]. The corn is healthy because inside it is the whole process that made it. And so, it makes the campesino healthy. If we know better what makes us healthy we can really make change in our communities.
> (Fieldnotes, May 8th, 2014)

The point of Evelio's incisive reflection for terra plena thinking is that you cannot meet nutritional needs without also engaging the broader social context. The point is that food is its context. This idea was reinforced by Nelson Garcia, a young man in his twenties, during the same session. Each person selected an object to express something learnt during the knowledge exchange process. Nelson presented a pumpkin seedling he had planted after an informal seed exchange:

> A plant is synonymous with hope. It's synonymous with life. And why is there life in this plant? Because of the relationship it has with the earth. [. . .] You know the plant is natural because it's in the earth, but so it is with food. And we say that health is for humans, but food can be healthy too, and so can plants. So, I think that the right to food is in nature, because we are all in nature, and what permaculture means is keeping it all connected.
> (Fieldnotes, May 8th, 2014)

Nelson makes a sophisticated point here: what we call 'culture' and 'nature' are socially constructed ideas that in fact have no easy separation. Meanwhile, food security, often treated as an issue of human poverty, is vitally connected to environmental processes, environmental history, and environmental politics. Nelson's sense of hope is very much grounded in holding a sense of environmental justice/healing connected to the living systems that make up particular contexts, at the same time as linking the human right to food with the right to be part of the solution.

Connecting the double vision of environmental healing/justice with the broader knowledge politics at stake also gives us a fifth key principle for terra plena thinking:

5 Rather than import ready-made solutions, support pedagogical practices that authorise personal capacities and experience as a basis for making specific claims.

In both agroecology and permaculture, campesinos are experts of their own experience, and protagonists for social and environmental restoration, while learning to experiment as an essential part of training. This is critical because it enables dimensions of individual experience become part of the invention of a fresh language for specific power relations. Such is the basis of the long history of pedagogical practices that have spread through Latin America and beyond in the wake of Freire's (1972) pioneering methodology for using literacy as a basis to teach groups to 'read' and 'name' oppressive power relations for themselves.

I was particularly struck by the empowering nature of this pedagogical training when, one day, I found Lucy Flamenco, a regular volunteer in her forties, at work at the Suchitoto demonstration site and busy with trays of seeds. I asked Lucy what she was engaged in and I was caught by the comment of Cesar Ramírez, a young volunteer who had just attended the workshop:

Lucy:	I'm making up an almácigo [a seed tray for germinating seedlings] of fresh tomatoes as an experiment, to see how many seeds sprout, and how long it takes . . .
Cesar [Laughing]:	Naomi, it's like what we are doing when we look back at the traditional practices, no? Watching the seeds planted by our ancestors . . . waiting to see which ones can still grow again . . .
Lucy:	But you still have to plant the seeds to see! (Fieldnotes, 24th April 2014)

Cesar's words in relation to Lucy's experiment emphasise the forms of translation at stake. On the one hand, permaculture trains individuals to create and test new hypotheses based on observation, while on the other, groups of permaculturists create fresh narratives integrating environmental pasts with the present. In terms of terra plena, these forms of expertise position campesino practitioners, along with a revived sense of tradition, at the forefront of international food politics.

Madre Tierra: Between singular and specific voicing difference

In unpacking terra plena thinking I have differentiated between 'singular', or non-reducible, dimensions of experience, which emerge in relation to distinctive ontological frameworks, and 'specific' political claims, responding to shared aspects of historical (and geographical) experience. The problem this produces both for TAMs and for terra plena scholarship is how to make specific political claims without reducing singular experience. In this last section I explore how tropes derived from indigenous cosmologies, especially Madre Tierra [Mother Earth], are employed to this end, especially in contexts characterised by hybrid interactions. In the permaculture movement and in other TAMs, the figure of Madre Tierra captures the relational dimension of the reimagined Maya cosmology, connecting all life in a living, breathing balance of interactions. This makes close observation of eco-systems the fabric through which individuals translate personal, excessive dimensions of experience into shared political claims. Whilst terra plena helps identify some of the risks involved, the polyphonic conversations at play around Madre Tierra also help define the sixth principle of a decolonial spatial imagination:

6 Rather than reify the aesthetic 'tropes' being used to voice shared claims, look for the moments of disagreement through which these circulate.

Madre Tierra in the Salvadoran permaculture movement is the concept that the earth is one living system of which we are a part, which gives us life. The term is used to refer to the living vibrancy of the material world of plants, people and soil; to the sense of 'spirit' that these share; and to express the refusal of life to obey or even heed human laws. Rather than reflecting a distinct cosmology, this notion signals a plurality of ways of knowing the world, as well as a plurality of living eco-systems, progressively sidelined by Euro-American knowledge systems. However, the planetary complex of eco-systems also forms a basic condition of sharing that cuts across the boundaries of nation-state belonging, rooting environmental ethics in an embodied ethos of attentive observation and mutual care (Puig de la Bellacasa, 2010). This allows for plural cosmological orientations even within particular movements. Amongst permaculturists one can honour Madre Tierra whilst also upholding Catholic, evangelical, Muslim, or atheist beliefs. Meanwhile, both the subject and object of environmental healing, the articulation of Madre Tierra is used functionally to show the importance of keeping solutions to social issues connected to their material contexts.

In permaculture, Madre Tierra has a direct impact on the way design practices are conceived as part of this process. Rather than carrying their own ontological content, design practices are comprised of ethical and pragmatic principles that allow collectives to (re)design environments in dialogue with the more-than-human world, via the ontological forms of knowing and traditions that are practised locally. The resultant designs are ultimately improvisations based on sustained observation, identification of resources to hand, and reflection on the intended uses of spaces viewed from a multiplicity of perspectives. This pushes against a conservation ethics that defines carc in terms of the protection of future resources, placing emphasis on multiple horizons of environmental healing/justice – as one of the two members of the Nuevo Amanacer Health Association of Torola emphasised:

> Morazán is one of the areas that was most devastated by the civil war. Since then, the worst problems we've faced have been housing, education, health

> and food. [. . .] The problem for us is that organisations come in and they just want to deal with one of these things. They want to help, but they don't want to think about how these things became connected. [. . .] Permaculture is having an impact because it connects these things together. Plus, it gives people the tools to understand what food is made of and what makes it better for bodies and why. And also what makes it better for Madre Tierra [. . .] So we are looking for holistic solutions that mean that people can also take back control over their lives.
>
> (Interview with Nuevo Amanacer, March 18th 2014)

Two things stand out in this insightful analysis of food insecurity in the region. First, we learn that local people want solutions that integrate the various dimensions of poverty, without losing sight of the historical and environmental context. Meanwhile, Madre Tierra, mentioned in Nuevo Amanacer's interview, forms a key way that a holistic ethos of health is imagined that can repair these complicated disconnections. Without rooting design in a particular cosmological narrative, Madre Tierra encourages a 'perspectival' modality of problem-solving that invites reflection on the problem from various angles, including the observed functions and capacities of different species-systems. What this allows, second, is a redefinition of the problems and solutions at stake. Instead of a problem of food insecurity to be solved by international actors, the problem becomes a monocultural agricultural system premised on a 'mononatural' notion of environmental knowledge.

Like food sovereignty, Madre Tierra carries something of a strategic essentialism – an ontological gap filled in by a figure of a question mark – without itself being an essentialist concept. It is a narrative-vehicle into which diverse experiences and cultural commitments can be translated, without requiring that they compromise their 'autochthonous' singularities (Rosset and Martinez-Torres, 2013). This articulation also finds resonance in the Quechua imagining of Pachamama, the name of an Andean 'Mother Earth' deity who forms the ethical mainstay in the political and environmental campaigns of indigenous Bolivian president Evo Morales. Pachamama appeared on the global stage in the Bolivian Peoples' Conference on Climate Change and the Rights of Mother Earth in April 2010, which brought together people from more than 140 countries as a direct response to the UN COP15 Copenhagen Climate Change Conference in December 2009, which was perceived as a popular failure. This appearance constituted an international dialogue over what it would mean to make Mother Nature a subject of legal rights (De Angelis, 2011), as well as offering an important insight into the significance of Mother Earth imagery for contemporary TAMs. The peoples' alternative to COP15 featured wide representations from Latin American indigenous groups with a focus on specific struggles, and adopted a dialogical format leading to the formulation of a shared manifesto. Within the format, Madre Tierra functioned as a vehicle of aesthetic resonance that invoked political testimony from both non-human and non-white/Western-European interlocutors (Gudynas, 2010). Given the range of contexts invoked in these terms, there was a tendency for the strong script in play to reinforce a somewhat utilitarian definition of the environmental resources at risk, as well as to push dissenting voices to the margins (Ibid.). However, in practice, the invitation to share from individual experience into this broader field of resonance keeps the ontological question (re)open(ing).

Through my analysis of permaculture, I suggest that this ontological (re)open(ing) is most prevalent when the accompanying practices keep ecological observation and intercultural dialogue at the heart of everyday ethics. Of course, such practices do not always prevent essentialising such experiences. For example, Peace (2008) offers a fascinating anthropological account of the biannual 'Terra Madre' meeting of the Slow Food movement in Turin, Italy, which brings together several thousand members (Peace, 2008). This example reveals the importance of the trope equally to TAMs in the Global North, where agrarian politics are entangled with, but different from, the claims arising in former European colonies (Edelman, 2005). However, analysing the secular rituals and discursive events through which a sense of global community arises at the meeting, Peace finds a tendency to romanticise indigenous cultures and peasants, and a pronounced fetishism built into the discourse surrounding Terra Madre/Madre Tierra, not least in the use of the term 'community,' which constituted a particular kind of rural imaginary based on face-to-face association, coherent identity, and moral obligation (Ibid.: 38). Participants consistently emphasised that small-scale production characteristic of 'the natural economy' could renew and sustain 'the traditional culture,' 'the culture of the past,' and 'the culture which made earlier generations what they were' as far as foodstuffs were concerned (Ibid.: 36), without acknowledging cultural diversity (or multinaturalism). Yet Peace claims that the theatrical performances associated with Terra Madre constituted a political field in which the prospects for cultural resistance 'from below' were given serious consideration, allowing for the depth contextualisation of specific struggles and cases. Peace concludes her analysis optimistically, with the proviso that fetishising the dignity of the small-scale producer within the figment of 'a natural economy' is a clear danger of mobilising such a powerfully resonant trope. Naturalising elements of smallholder agriculture can reduce the potential for ontological disagreement, resulting in the failure to recognise the indignities and inequalities imposed on numerous others within the narrow bounds of the rural community.

This is critical as it shows that it is possible to perform terra nullius even when speaking in the name of Madre Tierra. As when, in the forests of British Columbia, environmentalists spoke 'in the name' of the forests and their inhabitants and silenced both, overemphasising the specific platform of sharing can end up annulling difference between cosmologies, as well as diverse ecological contexts. The issue here is the extension of a model in such a way as to obscure the difference between worlds – which tends to be accompanied by the practice of speaking for others, rather than hearing others speak (Alcoff, 1991; Spivak and Harasym, 1990). This is tricky because, as Boff (2012) points out, Madre Tierra finds additional resonance in the way she captures new insights into the relational and informational nature of every being derived from quantum physics and evolutionary history. It is easy to slip between images of ecological complexity and a 'true nature' of how smallholder agriculture is and should be. However, what is most interesting in Boff's account is his emphasis on the emergence of Madre Tierra in relation to historical moments where different senses of perspective became possible to think. In particular, he highlights the 'overview effect;' when astronauts visited the moon for the first time, images circulated of the earth 'looking back on itself,' leading to fresh environmental campaigns premised in a sense of the earth as a coherent and fragile planet. For Boff, Madre Tierra is a name for this planetary-based environmental thinking. What is central here is the fresh capacity that Madre Tierra enables in terms of thinking the future. Rather than naturalising the image as an 'always true' originary narrative, it can be understood as a translative device in dialogue with plural situated practices.

Healing damaging dichotomies

How, then, do we think Madre Tierra through terra plena? I have suggested that we focus on the way that singular and specific dimensions interrelate in particular instances, and, rather than valorising the trope as such, we look to see what (and how) it enables. This argument is reinforced when examining the broader literary field through which the trope has been converted into a vehicle for political testimony. Indeed, imagery and poetry related to a hybrid Madre Tierra were in circulation raising issues of gender, colonialism, and ethnicity decades before the emergence of TAMs on the global stage. For example, in the Central American context, images of the 'Madre' have long been used to evoke to questions of resistance and healing in the context of colonial history. Specifically, Madre Tierra is situated in relation to what Lara (2008), a radical Chicana-feminist scholar, calls the virgin/whore dichotomy. In the Americas, and in Mexico in particular, this dichotomy is represented by La Virgen de Guadalupe, an indigenous or mestiza spiritually pure Maria figure, and La Malinche, the indigenous mistress of, and allegedly treacherous translator for, conqueror Hernan Cortes.[7] This binary, together with the negative effects of the dichotomy in the development of female subjectivity, is a foundational theme in Chicana feminist thought and Latin/Central American literature. Healing this dichotomy, Lara argues, entails, firstly, deconstructing the ways it regulates spiritual and sexual practices; and secondly, reconstructing a 'decolonial imaginary' by remembering the goddesses and indigenous conocimientos [knowledges] rejected from Guadelupe's cultural genealogy. For example, another name for Guadelupe was Tonantzin, another 'mother earth,' who, in Aztec folklore, was also a human mother. Neither entirely pure nor impure, Tonantzin is associated with Mesoamerican indigenous images of agricultural and human fertility as well as the earthly energy of fiercely creative destruction. Through their re-memberings, Lara argues, Tonantzin-Guadalupe becomes a decolonial figure capable of healing the virtuous virgen/puta [virgin/prostitute] split that Western patriarchy perpetuates.[8]

While this resonance will not be present for all those who employ the trope of Madre Tierra, it is critical to recognise that in literary traditions the hybridity of the term already performs its own deconstructive work on images of motherhood, femininity, and sexuality, rather than essentialising one normative version. Secondly, given its wide aesthetic resonance, the decolonial dimension of Madre Tierra is not automatic but invitational. As in the popular education practices of agroecology and permaculture, the trope calls on observation, collective reflection and active deconstruction through attention to experience and context. Because of this process, and not prior to it, specific political claims are articulated as a response to the (colonial) repression of singular experience.

This nuance is particularly clear in the work of Maya-kaqchikel writer Francisco Morales Santos, who published his first major works just as members of his indigenous community were taking part in the first Guatemalan agroecology exchanges. A central theme in Morales Santos' work is the Mayan literary trope of a journey into tradition, as part of the healing and empowering process necessary to confront actually existing violence. The journey to the origins is a prevalent theme in contemporary indigenous literature, otherwise known as Abya Yala, and invokes a situation where an individual has to go back to the ancestors to recover memories, or other discursive elements like cultural or religious values, in order to transcend the colonial experience in the present. In his poem 'Madre, nosotros también somos historia' [Mother, We, Too, Are History] (1988), the poet develops this theme, using the journey to the origins to express a critique of modernity and its hegemonic narratives, particularly official historiography and capitalism. This is del Valle Escalante's (2011) point as he concludes an analysis of the Madre in the poem, suggesting that the elicitation of the Mayan cosmovision in the poem empowers and validates the knowledge and prominence of subaltern indigenous peoples from rural areas, while also developing the preconditions for a process of mental and political de-alienation on the other. What is important here is not that the protagonists return to their 'pure' origins in order to divest themselves of violence, but that a cyclical notion of time is invoked as part of an alternative ontological foundation, as a means to heal the present. This adds a further dimension to Madre Tierra, and, in turn, a final principle to terra plena:

7 The decolonial project concerns not only the 'worlding' of nonhuman relationships but the worlding of time.

Terra plena confers a vision of an earth full of other nonhuman agents and ways to know them, but also with other modalities to know memory, future, and connection with others.

This sense of alternate temporality is critical to the way Madre Tierra elicits political testimony on the global stage. The work of Morales Santos, alongside his contemporaries – including the Maya-kaqchikel writer Luis de Lión, who similarly sites decolonial force in the moment that people begin 'walking backwards' toward their ancestors – reached popular visibility nationally and internationally after the 1990s, particularly with the reception of the autobiography of Guatemalan indigenous activist Rigoberta Menchú (Burgos, 1985). The wide circulation of the autobiography and later associated narratives marked a surge of interest in Maya literature as well as the testimonial genre. While I lack the space to develop the interconnection between the arrival of the Maya cosmovision on the scene of universal human rights and the place of figures like Madre Tierra in TAMs, this dialogue between articulations of 'universal' human experience, specific (colonial) political conditions, and singular life narratives sees certain forms of political articulation gaining a much wider audience. Specifically, the testimonial format allowed the shared coordinates of experiences of colonisation create waves of resonance in the Global North in this moment, and fresh platforms for political speech. Within such narratives, hybrid tropes like Madre Tierra form an invitation into solidarity that opens not only to indigenous peoples, but to all those who have been subjected to capitalist and state violence. However, the very transformation of the aesthetic field such that this resonance carries with the weight of a call to action, creates new risks: the possibility of constituting a new form of terra nullius through the assertion of a fresh 'ground' for politics that papers over disagreement. The challenge for terra plena thinking is to neither decry nor uncritically affirm the appearance of figures like Madre Tierra, but instead to foster practices of reading, listening, and attention, that enable the perception of her appearance as a declaration of universal difference, rather than universal sameness.

Box 4.1: Seven design principles of 'terra plena'

Seven Terra Plena principles, or, an ethos for decolonial design

1. It is the spaces in between the singular dimensions of a specific experience that provide impetus for common claims that do not rely on reducing the range of possible solutions.
2. Innovation is not the property of Euro-American urban processes, and has long been situated in rural and agricultural sites of experimentation.
3. Issues of social justice and environmental degradation are crucially intertwined, and need to be approached in in holistic connection.
4. Tradition can hold a bearing on the future, as well as the past, but only when it is apprehended in terms of capacities and agencies, rather than essential truths.
5. Rather than import ready-made solutions, support pedagogical practices that authorise personal capacities and experience as a basis for making specific claims.
6. rather than reify the aesthetic 'tropes' being used to voice shared claims, look for the moments of disagreement through which these circulate.
7. The decolonial project concerns not only the 'worlding' of nonhuman relationships but the worlding of time.

Conclusion

Borrowing from Marilyn Strathern, Marisol de la Cadena (2015: xxv) calls communication across such networks of communication a 'partially-connected' conversation. Different ways of 'worlding' the world do not, in fact, create fundamentally separate worlds. While the images and sign-systems we use to make sense of the earth are plural, and to an extent incommensurable, communication is still possible between worlds, because worlds share (some) common materialities – including, but not limited to, hunger, death, pleasure, violence, grief and belief. Ontologies are constructed from very different foundations, but they are always partial, in-construction and hybrid, as they are (re)shaped through dialogue with the patterned communication systems of other social and species systems. On the other hand, I have suggested that awareness of the gaps and fissures between meanings is a vital part of a decolonising process that fosters environmental healing/justice. I have named the logic of this process terra plena, setting it in disagreement with the terra nullius claims through which cultural and material forms of colonisation acquired legitimacy.

Through the course of this chapter, I have identified seven principles for enacting scholarship according to this logic, and in solidarity with the claims and spatial imaginations already being developed by intersectional rural movements and actors. These principles are summarised in Box 4.1. Terra plena, as a kind of ethos for decolonial design, affirms processes of dialogical sharing that make room for difference as part of developing common claims. It locates innovative and future-oriented potential in geographical sites considered 'marginal' and 'backward'. It refuses to allow issues of social justice and environmental degradation to be treated in isolation. And, finally, it insists on the bearing that tradition may have on the future, when approached in terms of capacities. This emphasis on capacity is carried through to its commitment to connect and acknowledge pedagogical practices that authorise personal capabilities and experiences as a basis for making shared claims. Similarly, it identifies visual 'tropes' used to mediate between experience and claims without naturalising them, and it acknowledges the rich potential of the alternate forms of inhabitation and temporality embodied by diverse ontological patternings and the spaces of interaction between them.

Through my empirical analysis I have suggested that permaculture offers to small-scale farmers – but also to scholars and policy-makers – an embodiment of terra plena principles that reveals the central place of ecological observation and intercultural dialogue to what we may call decolonial praxis. By establishing principles for experimentation and design, permaculture recreates 'tradition' as the vehicle for future experimentation, rather than hypothesising a model than can be applied to all settings. Through tropes like Madre Tierra, this tradition is refreshed and recreated as part of a solution to interconnected issues including food insecurity, environmental injustice, and a broader colonial politics of knowledge. While Madre Tierra seems to propose a hybrid blend of indigenous cosmovisions as a fresh ground for politics, I have suggested that the broader aesthetic field constitutes a trope characterised by an ontological empty space, or question mark, which makes room for conversation between diverse traditions. This 'empty space' is what makes these principles so valuable for learning within other spheres, as it creates an invitational opportunity for knowledge production across multiple scales that exerts friction in relation to imposed universals.

Terra plena opens a postcolonial project to what de la Cadena, after Schmitt, terms the 'pluriverse.' Signalling political philosopher Jacques Rancière's notion that politics is not made up of power relations, but relations among worlds, de la Cadena's pluriverse is a utopian, but practical, project in which partially connected heterogeneous 'socio-natural worlds' negotiate their ontological disagreements politically – that is, via conflict. There are two steps to the enactment of the pluriverse in this account: first, to recognise that the world is more than one socionatural formation; and second, to interconnect such plurality without making the diverse worlds commensurable (de la Cadena, 2010: 361). The appearance of Madre Tierra in social protests, like de la Cadena's (2010: 336) 'earth beings,' promise a moment of 'rupture' of modern politics and an 'emergent indigeneity' – which is not a new mode of being indigenous, but an 'insurgence of indigenous forces and practices with the capacity to significantly disrupt prevalent political formations.' What unifies such movements is not one ontology, as in cases where human rights are expounded as a 'fit-all' solution, but a commitment to diversity: to biodiversity, but, before this, to ontological diversity, without which biodiversity cannot be achieved.

Notes

1 This notion of ontological violence builds on Gayatri Spivak's extensive work on 'epistemic violence', which refers to the active obstruction and undermining of non-Western approaches to knowledge (see Spivak and Harasym, 1990). The emphasis on ontology here makes the point that not only categories of knowledge but worlds of knowing are at stake.

2 For example, a sense of 'nationhood' is produced in relation to specific senses of nationality and nationalism, rather than being simply associated with a scale that sits above local and regional belonging.

3 It leans heavily on the principles of tree ecology set out by Joseph Russell Smith (1929) in his book Tree Crops: A Permanent Agriculture, as well as developing concepts of agroecology (the application of ecology to the design and management of sustainable agroecosystems) and agroforestry.

4 Within the Salvadoran permaculture movement, food justice [justicia alimentaria] is used to describe a political commitment to reshaping power and knowledge dynamics within the global food system. LVC are regarded as allies in food justice struggles, although the movement is not part of the network, and food sovereignty [soberanía alimentaria] is used more to refer to the capacities of nations or regions to cultivate the full range of food crops needed for nutrition. On the other hand, Salvadoran permaculture shares LVC's insistence on the rights of smallholding campesinos to select what and how to grow. This a resolutely confrontational notion of food justice that likewise firmly rejects the idea that the "food poor" are passive victims in need of developmental solutions. When I use the term food justice here, I articulate this political sense latent within both concepts.

5 In Nicaragua the main clearing ground was the UNAG (Nicaraguan Uni6n Nacional de Agriculturores y Ganaderos) founded in 1983 by smallholders, cooperatives and medium-d landowners who felted underrepresented in Sandinista dominated rural workers' unions (Edelman, 1998: 58). The UNAG received visitors from abroad and assumed a central role in the CaC program.

6 El Salvador's 1992–1997 Land Transfer Program, known as El Programa de Transferencia de Tierras (PTT), was mandated within the Chapultepec Peace Accords concluded on January 16, 1992. The Program laid out principles for legalizing tenancy in occupied conflict zones, although its ambiguity led to considerable tensions in the following decade.

7 Guadelupe was crowned the Patron Saint of New Spain and the Queen of Mexico in the mid-eighteenth century, while Malinche became known as 'la Chingada' after the Mexican Revolution; the violated, 'fucked' mother of the first mestizo population, and ancestor of the superstitious, demonised native woman.

8 As part of the legacy of decolonial imagining initiated by Chicana artists, such as Ester Hernandez in the 1960s and Yolanda Lopez in the 1970s, the Chicana-feminist scholar Gloria Anzaldúa was among the first writers to theorise about Tonantzin Guadalupe from

a Chicana feminist perspective. In Borderlands/La Frontera: The New Mestiza (1987), Tonantzin-Guadalupe appears as a symbol of the spiritual borderlands connecting Chicana/os to their Indian ancestry. She inhabits the interstices and the spaces between different worlds, including those of the spirit and the flesh, influenced by the des/conocimientos [(counter-)knowledges] of indigenous, European, and mestizo [mixed race], cultures.

References

Alcoff, L. 1991. 'The Problem of Speaking for Others', *Cultural Critique*, vol. 20, pp. 5–32.

Altieri, M. 1995. *Agroecology: The Science of Sustainable Agriculture*. Boulder: Westview Press.

Barrera-Bassols, N. and Zinck, J. 2003. 'Ethnopedology: A Worldwide View on the Soil Knowledge of Local People', *Geoderma*, vol. 111, no. 3, pp. 171–195.

Blaser, M. 2014. 'Ontology and Indigeneity: On the Political Ontology of Heterogeneous Assemblages', *Cultural Geographies*, vol. 21, no. 1, pp. 49–58.

Boff, L. 2012. 'La Madre Tierra, sujeto de dignidad y de derechos', *América Latina en Movimiento*, vol. 479 [online]. www.democraciaycooperacion.net/IMG/pdf/La_Madre_Tierra_sujeto_de_dignidad_y_de_derechos.pdf [Accessed 01.04.2016].

Borras, S.M. 2010. 'The Politics of Transnational Agrarian Movements', *Development and Change*, vol. 41, no. 5, pp. 771–803.

Braun, B. 2002. *The Intemperate Rainforest: Nature, Culture, and Power on Canada's West Coast*. Minneapolis: University of Minnesota Press.

Burgos, E. 1985. *Me llamo Rigoberta Menchu y asi me Nacid la Conciencia*. Mexico City and Madrid: Siglo Veintiuno.

Cabarrús, C. 1983. *Génesis de una Revolución: Análisis del Surgimiento y Desarrollo de la Organización Campesina en El Salvador*. Mexico: Centro de Investigaciones y Estudios Superiores en Antropología Social.

Carey, D. 2009. 'Guatemala's Green Revolution: Synthetic Fertilizer, Public Health, and Economic Autonomy in the Mayan Highland', *Agricultural History*, vol. 83, no. 3, pp. 283–322.

Cox, G. and Atkins, M. 1979. *Agricultural Ecology: An Analysis of World Food Production Systems*. San Francisco: WH Freeman.

Dahlberg, K. 1979. *Beyond the Green Revolution: The Ecology and Politics of Global Agricultural Development*. New York and London: Plenum Press.

De Angelis, M. 2011. 'Climate Change, Mother Earth and the Commons: Reflections on El Cumbre', *Development*, vol. 54, no. 2, pp. 183–189.

de Castro, E. 2004. 'Exchanging Perspectives: The Transformation of Objects into Subjects in Amerindian Ontologies', *Common Knowledge*, vol. 10, no. 3, pp. 463–484.

de la Cadena, M. 2010. 'Indigenous Cosmopolitics in the Andes: Conceptual Reflections Beyond "Politics" ', *Cultural Anthropology*, vol. 25, no. 2, pp. 334–370.

de la Cadena, M. 2015. *Earth Beings: Ecologies of Practices Across Andean Worlds*. Durham and London: Duke University Press.

del Valle Escalante, E. 2011. 'El viaje a los orígenes y la poética "decolonial" maya en Madre, nosotros también somos historia de Francisco Morales Santos', *Revista de Crítica Literaria Latinoamericana*, vol. 37, no. 74, pp. 351–372.

Descola, P. 2005. *Par-delà Nature et Culture*. Paris: Gallimard.

Desmarais, A. 2007. *La Vía Campesina: Globalization and the Power of Peasants*. Halifax: Fernwood Publications.

Edelman, M. 1980. 'Agricultural Modernization in Smallholding Areas of Mexico: A Case Study in the Sierra Norte de Puebla', *Latin American Perspectives*, vol. 7, no. 4, pp. 29–49.

Edelman, M. 1998. 'Transnational Peasant Politics in Central America', *Latin American Research Review*, vol. 33, no. 3, pp. 49–86.
Edelman, M. 2005. 'Bringing the Moral Economy Back in . . . to the Study of 21st-Century Transnational Peasant Movements', *American Anthropologist*, vol. 107, no. 3, pp. 331–345.
Edelman, M. 2014. 'The Next Stage of the Food Sovereignty Debate', *Dialogues in Human Geography*, vol. 4, no. 2, pp. 182–184.
Freire, P. 1972. *Pedagogy of the Oppressed*. London: Penguin.
Friedmann, H. 1982. 'The Political Economy of Food: The Rise and Fall of the Postwar International Food Order', *American Journal of Sociology*, vol. 88, pp. 248–286.
Gemein, M. 2016. ' "Seeds Must Be Among the Greatest Travelers of All": Native American literatures Planting the Seeds for a Cosmopolitical Environmental Justice Discourse', *Interdisciplinary Studies in Literature and Environment*, vol. 23, no. 3, pp. 485–505.
Gibson-Graham, J.K. 2008. 'Diverse Economies: Performative Practices for Other Worlds', *Progress in Human Geography*, vol. 32, no. 5, pp. 613–632.
Gliessman, S. 1990. *Agroecology: Researching the Ecological Basis for Sustainable Agriculture*. New York: Springer.
Gould, J. and Lauria-Santiago, A. 2008. *To Rise in Darkness: Revolution, Repression, and Memory in El Salvador, 1920–1932*. Durham: Duke University Press.
Gudynas, R. 2010. 'La Pachamama: ética ambiental y desarrollo', *Le Monde Diplomatique*, no. 27 (junio-julio), La Paz (Bolivia) [online]. http://herencia.org.bo/webdocs/pdfs/GudynasPachamamEticaLeMondeBolJun10.pdf [Accessed 25.10.2016].
Haraway, D. 2008. *When Species Meet*. Minneapolis: University of Minnesota Press.
Harvey, D. 2006. 'Neo-Liberalism as Creative Destruction', *Geografiska Annaler B*, vol. 88, no. 2, pp. 145–158.
Hayden, C. 2003. *When Nature Goes Public: The Making and Unmaking of Bioprospecting in Mexico*. Princeton: Princeton University Press.
Holt-Giménez, E. 2006. *Campesino a Campesino: Voices From Latin America's Farmer-to-Farmer Movement for Sustainable Agriculture*. Oakland: OUP.
Kohn, E. 2013. *How Forests Think: Toward an Anthropology Beyond the Human*. Berkley, CA: University of California Press.
Lara, I. 2008. 'Goddess of the Americas in the Decolonial Imaginary: Beyond the Virtuous Virgen/Pagan Puta Dichotomy', *Feminist Studies*, vol. 34, no. 1/2, pp. 99–127.
Li, M. 2014. *Land's End: Capitalist Relations on an Indigenous Frontier*. Durham: Duke University Press.
Makki, F. 2014. 'Development by Dispossession: Terra Nullius and the Social-Ecology of New Enclosures in Ethiopia', *Rural Sociology*, vol. 79, no. 1, pp. 79–103.
Mangelsdorf, P. 1951. 'Hybrid Corn: Its Genetic Basis and Its Significance in Human Affairs', In: L. Dunn, ed., *Genetics in the 20th Century: Essays on the Progress of Genetics During Its First Century*. New York: MacMillan Press, pp. 555–572.
Martinez-Torres, M. and Rosset, P. 2010. 'La Vía Campesina: The Birth and Evolution of a Transnational Social Movement', *The Journal of Peasant Studies*, vol. 37, no. 1, pp. 149–175.
McMichael, P. 2009. 'A Food Regime Genealogy', *The Journal of Peasant Studies*, vol. 36, no. 1, pp. 139–169.
McMichael, P. 2012. 'The Land Grab and Corporate Food Regime Restructuring', *Journal of Peasant Studies*, vol. 39, pp. 3–4, 681–701.
McMichael, P. 2014. 'Historicizing Food Sovereignty', *Journal of Peasant Studies*, vol. 41, no. 6, pp. 933–957.
Morales Santos, F. 1988. *Madre, Nosotros También Somos Historia*. Guatemala: Editorial Universitaria.

Morton, T. 2007. *Ecology Without Nature: Rethinking Environmental Aesthetics*. Cambridge, MA: Harvard University Press.

Netting, R. 1993. *Smallholders, Householders: Farm Families and the Ecology of Intensive, Sustainable Agriculture*. Stanford: Stanford University Press.

Paré, L. 1990. 'The Challenges of Rural Democratisation in Mexico', *The Journal of Development Studies*, vol. 26, no. 4, pp. 79–96.

Peace, A. 2008. 'Terra Madre 2006: Political Theater and Ritual Rhetoric in the Slow Food Movement', *Gastronomica*, vol. 8, no. 2, pp. 31–39.

Pearce, J. 1988. 'From Civil War to "Civil Society": Has the End of the Cold War Brought Peace to Central America?' *International Affairs*, vol. 74, no. 3, pp. 587–615.

Pimbert, M. 2006. *Transforming Knowledge and Ways of Knowing for Food Sovereignty*. London: IIED Press.

Puig de la Bellacasa, M. 2010. 'Ethical Doings in Naturecultures', *Ethics, Place and Environment*, vol. 13, no. 2, pp. 151–169.

Rist, S. and Dahdouh-Guebas, F. 2006. 'Ethnosciences – A Step Towards the Integration of Scientific and Indigenous Forms of Knowledge in the Management of Natural Resources for the Future', *Environment, Development and Sustainability*, vol. 8, no. 4, pp. 467–493.

Roseberry, W. 1991. 'La falta de brazos', *Theory and Society*, vol. 20, no. 3, pp. 351–382.

Rosset, P. and Martínez-Torres, M. 2013. 'Rural Social Movements and diálogo de Saberes: Territories, Food Sovereignty and Agroecology', paper presented at Food Sovereignty: A Critical Dialogue International Conference, Yale University.

Smith, J. R. 1929. *Tree Crops: A Permanent Agriculture*. New York: Harcourt, Brace & Co.

Spivak, G. and Harasym, S. 1990. *The Post-Colonial Critic: Interviews, Strategies, Dialogues*. London: Psychology Press.

Stonich, S. 1993. *"I am Destroying the Land!": The Political Ecology of Poverty and Environmental Destruction in Honduras*. Boulder, San Francisco and Oxford: Westview Press.

Tsing, A. 2005. *Friction: An Ethnography of Global Connection*. Princeton: Princeton University Press.

Tsing, A. 2015. *The Mushroom at the End of the World: On the Possibility of Life in Capitalist Ruins*. Princeton: Princeton University Press.

Van der Ploeg, J.D. 2014. 'Peasant-Driven Agricultural Growth and Food Sovereignty', *Journal of Peasant Studies*, vol. 41, no. 6, pp. 1–32.

Wittman, H. 2009. 'Reworking the Metabolic Rift: La Vía Campesina, Agrarian Citizenship, and Food Sovereignty', *The Journal of Peasant Studies*, vol. 36, no. 4, pp. 805–826.

Ziegler, J., Golay, C., Mahon, C. and Way, S-A. 2011. *The Fight for the Right to Food: Lessons Learned*. Geneva: Graduate Institute of International and Development Studies & Palgrave Macmillan.

5 Refracting colonialism in Canada

Fish tales, text, and insistent public grief

Zoe Todd

In 2015, the 'Truth and Reconciliation Commission of Canada on the Indian Residential School System' released its 'Executive Summary' on its five years of hearings about the impacts of the Residential School system across Canada (the 'TRC Report'). Pundits across the country issued their thoughts, reducing over a century of Residential School history to sound bites. Declarations were made, some racist editorials urging Indigenous peoples to 'get over it' saw the light of day, and eventually the news cycle ran itself out.

Five months later, as I write this chapter, the 94 Calls to Action issued in the TRC Executive Summary were barely acknowledged by any of the party leaders in Canada's 42nd Federal Election (Kinew et al., 2015). But the Canadian government remains ostensibly committed to the goals of 'reconciliation' outlined in the TRC Report. In fact, incoming Minister of Indigenous Affairs and Northern Development, Carolyn Bennett, told news reporters at her swearing-in that she had been informally advised to see herself as the 'Minister of Reconciliation' (Barrera, 2015).

Many Indigenous thinkers in the country remain sceptical of promises of reconciliation between Indigenous peoples/nations and the Canadian State (see: Coulthard, 2014; Simpson, 2015), and with good reason. As literary scholar Thomas King (2013) reminds us in his book The Inconvenient Indian, there is truly very little difference between the past and the present when it comes to colonial approaches to engaging with Indigenous peoples in North America (King, 2013: 159–192). In fact, he goes so far as to cheekily urge readers to conduct a thought exercise in order to 'forget history', as so many settler-colonial pundits urge Indigenous peoples to do when they proclaim Indigenous peoples must 'get over it'. However, in conducting the thought experiment, King concludes: "You see my problem. The history I offered to forget, the past I offered to burn, turns out to be our present. It may well be our future" (King, 2013: 192). Past, present, and future meld together; the Canadian and American state approaches to acknowledging Indigenous sovereignty have changed very little, in real terms. Both nation-states still occupy unceded territories, and Treaties between Indigenous nations and the State continue to be violated (Asch, 2014: 100–115; King, 2013: 244–269). Scepticism is, therefore, healthy, and arguably, necessary. As King (2013) shows us, ideas like postcolonial and colonial, past and present, mean almost the same thing when a government's legitimacy derives from: a) ongoing occupation of stolen land; b) questionable interpretations of Treaty negotiations (Asch, 2014: 100–115); and, c) sweeping structural violence against Indigenous peoples under the rug – witness the shelving of the TRC's recommendations and the similar shelving of the five volume, four thousand page Report of the Royal Commission on Aboriginal Peoples tabled in 1996.

Such is the ebb and flow of colonial recognition of Indigenous peoples in Canada. It surges in moments of public remorse or defensive denial of Canada's colonial history. And then it ebbs back into collective non-memory for those who are not viscerally impacted by the day-to-day legacies of Canada's genocidal history.

At least, that is how it seems on the surface. Beneath the surface lurk creative, insistent and dynamic modes of resistance to ongoing colonial relations and storytelling in Canada. And these modes of resistance involve not only human, but also more-than-human, entities. A central focus of, and theoretical lens in my work is the principle of 'ethical relationality' that Papaschase Cree scholar Dwayne Donald (2009) proposes in his own writing. Through this principle of relationality, I seek to understand the impulses and imperatives that weave Indigenous and non-Indigenous peoples into (often fraught) shared experiences in settler-colonial Canada – and I seek to understand how tending to this relationality can build robust and meaningful transformation of current settler-colonial violence. Donald (2009: 6) defines this ethical relationality as:

> an ecological understanding of human relationality that does not deny difference, but rather seeks to more deeply understand how our different histories and experiences position us in relation to each other. This form of relationality is ethical because it does not overlook or invisibilize the particular historical, cultural, and social contexts from which a particular person understands and experiences living in the world. It puts these considerations at the forefront of engagements across frontiers of difference.

In mobilising an ecology of 'ethical relationality', this chapter examines, first, what it means to assert an insistent, public grief as an Otipemsiw/Red River Métis woman and scholar in the face of ongoing colonial violence in Canada. Second, I explore the role of the more-than-human, specifically fish, in bearing witness to colonialism. I examine how grief extends beyond human experiences of colonial suffering, and I demonstrate how fish-as-non-human persons are present actors and agents in the experiences, and mourning, of Indigenous peoples in a country deeply impacted by the British and French colonial legacies. I do so by detailing three important moments in my own work as an Indigenous woman in the academy: a) my work with nehiyaw iskwew scholar Erica Violet Lee and Inuk lawyer Joseph Paul Murdoch-Flowers on a public reading project we entitled #ReadTheTRCReport; b) my doctoral research on human-fish relations, Indigenous legal orders and colonial sites of engagement in northern Canada; and c) my invocation of the decolonial trickster fish the 'Ness namew' who carries stories of colonial grief and suffering in Canada's ruined waterways back to the United Kingdom. These three examples demonstrate how an insistent, public mourning of the impacts of colonial violence – between human and more-than-human nations and peoples that occupy and inhabit what is today known as Canada – serves as a mode of what I term 'refraction' in the so-called postcolonial moment in North America.

Humans and fish, together, can and do resist colonial narratives, and turn these discourses on their head to craft self-determined stories that transcend categories or limits of impact imposed by the State. Grief is shared across human/non-human boundaries because Indigenous legal orders, ontologies and governance incorporate the more-than-human as integral actors in self-determination. This shared grief deeply informs the futures we can and will build here in North America as we imagine a world informed by what Papaschase Cree scholar Dwayne Donald (2009) deems our 'ethical relationality' with one another.

#ReadTheTRCReport

I start here with an examination of the #ReadTheTRCReport Project as a form of insistent public grieving. The Truth and Reconciliation Commission of Canada was established under section 'N' of the Indian Residential Schools Settlement (TRC undated). Residential schools were an officially sanctioned program of schooling established by the Canadian government in the 1870s (TRC undated b). British colonial anxieties about Indigenous self-determination and Indigenous thinking in what is now Canada manifested in a desire to completely assimilate not only Indigenous people, but to also subsume and erase their epistemologies, laws, practices, and languages. In 1883, Canada's First Prime Minister, Glasgow-born John A. MacDonald, argued the rationale for Residential Schools in such terms (TRC 2015: 2):

> When the school is on the reserve the child lives with its parents, who are savages; he is surrounded by savages, and though he may learn to read and write his habits, and training and mode of thought are Indian. He is simply a savage who can read and write. It has been strongly pressed on myself, as the head of the Department, that Indian children should be withdrawn as much as possible from the parental influence, and the only way to do that would be to put them in central training industrial schools where they will acquire the habits and modes of thought of white men.

As MacDonald demonstrates in the above quote, the goal of the Residential School system was to rend the relationships between children and their families, and to control and obliterate the transmission of Indigenous language, laws, and stories from one generation to the next. The goal was utter annihilation of Indigenous thinking and epistemologies in order to install Euro-Western language, thinking, laws, and ideologies. Over the course of the residential school program, over 150,000 Indigenous children were forced to attend the 130 residential schools run by the Anglican, Catholic, United and Presbyterian Churches across the country (CBC, 2008; TRC undated b). Violence, starvation, and abuse were hallmarks of the institutions. Death rates at residential schools were comparable to those of Canadian soldiers in action in World War Two. Daniel Schwartz with the Canadian Broadcasting Corporation reports that the odds for children of dying in Residential Schools was 1 in 25; the odds of dying for Canadians serving in World War Two was 1 in 26 (Schwartz, 2015).

Survivors of Residential schools across the country pushed for legal recognition of the physical, emotional, and sexual violence they experienced at the schools. In 2006, a class action lawsuit launched by Residential School survivors "was effectively settled by Dec. 15, 2006, when documents were released that said the deal had been approved by seven courts: in Alberta, British Columbia, Manitoba, Ontario, Quebec, Saskatchewan and the Yukon" (CBC, 2008). The settlement ordered the formation of a Truth and Reconciliation Commission, which outlined its work in terms of healing through "truth telling" and "holistic and comprehensive response to the Indian Residential School legacy" (TRC undated). Over six years, the TRC gathered testimony from people across Canada, hearing from over 6,000 witnesses (TRC, 2015: v).

On 02 June 2015, the TRC released its Executive Summary of the testimony of survivors as well as former teachers, janitors, principals, church officials, and non-Indigenous community members wishing to bear witness to the impacts of these schools on Indigenous peoples. The Executive Summary was the first report to be released by the TRC, and it was followed by the December 2015 release of six full volumes prepared by the research staff of the Truth and Reconciliation Commission from all the materials gathered during the TRC's mandate. The Executive Summary, in its most recent draft, is a full 528 pages long. It is a public, open-access document, available for free on the Truth and Reconciliation Commission of Canada website.

As soon as the Executive Summary report was released, editorials appeared in national newspapers, some in favour, some refuting the findings. For example, Rodney Clifton and Dan Rubenstein (2015) wrote a piece in the National Post entitled "Debunking the half-truths and exaggerations in the Truth and Reconciliation Report"; meanwhile, Mark Kennedy published an editorial in the National Post on the same day entitled "Atoning for 'cultural genocide': Truth and Reconciliation Commission lays out its blueprint". Perhaps the most contentious issue for many was that the report clearly and unambiguously states that the Residential School system is an example of cultural genocide (TRC, 2015: 1). While non-Indigenous pundits like Clifton and Rubenstein (2015) balk at the invocation of genocide with respect to the Residential school system, some Indigenous commentators like Pam Palmater (2015) argue that labelling it 'cultural' genocide diminishes the meaning, arguing instead that Residential Schools were an example of genocide, full stop.

As the media rushed to publish opinions on the TRC Executive Summary and its findings, Métis (Indigenous) educator and writer Chelsea Vowel took notice of the national opinion editorials that were published on the first day of the report's release. She asked a salient question: how could people be forming meaningful opinions of the report on the first day, given that the report is hundreds of pages long? She set out to read the entire report, tweeting her reflections as she read through the report. She then wrote a nuanced reflection on the report and published it to her nationally popular blog 'Âpihtawikosisân' on 08 June 2015

(Vowel, 2015). In this post, she calls upon individuals to read the report in full before stepping on to their soapbox to make pronouncements about it.

Thinking about how to step up and meet Chelsea Vowel's challenge – to read the entire TRC Executive Summary Report so that I could make informed opinions about its findings – I mused on Twitter on the quiet Sunday morning of 07 June about how to accomplish this daunting task while also trying to finish my doctoral work, gear up for a cross-Canada move, and start a new tenure-track job. First, I suggested a podcast project. Why not gather individuals to read sections of the report and turn it into an audiobook? My friend and colleague Erica Violet Lee, in turn, suggested something even more affective and powerful: to encourage individuals to film themselves reading sections of the report and upload these videos to a collective playlist, and to turn the report into a living document for people all over the world to access in an oral and visual format.

I was deeply struck by this suggestion. It felt like a call; something really important; something I needed to do. It was a Sunday morning, quiet, the June prairie sun was streaming into my living room. Within an hour, I had set up my computer in the living room, tested out a short intro script and recorded myself reading the first eight pages of the TRC Executive Summary. I uploaded this clip to YouTube and sent a note to Erica to let her know I had done so. She was traveling in California, but she sat down and recorded herself reading the next section of the Introduction. Seeing the tweets flying back and forth between Erica and I on Twitter, my friend Joseph Paul Murdoch-Flowers contacted us and asked if he could also contribute. He recorded himself reading the last section of the Introduction while sitting out on the land near his home in Iqaluit and uploaded it as soon as he could.

People were immediately moved by the videos we uploaded. Offers to assist us began to roll in. We began to coordinate between California, Edmonton (Alberta, in the Canadian prairies) and Iqaluit (Nunavut, in the Arctic). By the end of the September, 140 people across Canada – and a few people from the UK and further abroad – had filmed and uploaded 140 videos encompassing the entire report (excluding appendices and references). The project was covered by national press and was even featured on Buzzfeed's website (Strapageil, 2015).

Insistent grief

In taking up Erica Violet Lee's call for people to read the report on film, I was not prepared for how emotional it would be both to read the report in real time, and for an audience to later access. The act of reading privately is a quiet affair: eyes skim over pages, seconds pass in a flash. However, the act of reading out loud shifts the engagement beyond just the self and the text to include a third actor: the audience. In my video reading of the first part of the Introduction to the report, I stumble over words, and I cry when I read John A. MacDonald's pronouncements about children being savages on page two of the report. This moment of insistent grief – a feeling of horror and anger at the words of the 'founder' of the State of Canada and his violent views of Indigenous peoples (which includes my own ancestors) – is now alive forever on #ReadTheTRCReport playlist.

Many of the videos uploaded to the project are similarly affective: feelings of sadness, pain, guilt, shame, and anger that various passages in the report elicit are laid bare on the faces of those reading them. While I initially approached the #ReadTheTRCReport project as a textual one, aiming to work with Erica and Joseph to increase the accessibility of the findings to people who cannot sit down and read the whole report, it quickly became an emotional grieving process for me. The process of receiving videos from volunteers and viewing these pieces before uploading them to the playlist was emotional. Seeing the vulnerability and grief on the faces of individuals who offered to read incredibly heavy passages – including passages that deal with the sexual and physical abuse that children experienced in Residential Schools – filled me with a mixture of despair and gratitude to volunteers for communally sharing their own grief and anger regarding Canada's acts of cultural genocide. Seeing Indigenous friends bravely offer to read and upload sections, knowing how heavy and difficult it is to viscerally re-encounter the pain of colonial violence, even if it is 'only' in writing, filled me with equal measures hope and sadness. Hope that our work might have some small impact in dismantling denial of the colonial moment as a genocidal one, but also sadness that we are still forced to tell our stories insistently over and over and over to convince people that what the French, British and Canadian governments did to Indigenous peoples in North America was in fact utterly violent and inexcusable.

In the autumn of 2015, we were invited by artists Amy Kazymerchyk, cheyanne turions, and Alexander Muir, who form the collective No Reading After the Internet, to mount an exhibit about #ReadTheTRCReport as part of their contribution to the show Reading Exercises at the Leonard and Bina Ellen Gallery at Concordia University in Montreal, Canada. The show sought to engage the relationship between people, place, and text that are invoked and enlivened by the process of reading (Ellen Gallery, 2015),

> The exhibition Reading Exercises juxtaposes recent works that, each in their own way, speak to a renewed relationship to reading, viewed as a space for productive experimentation and critical investigation. As the title suggests, reading is addressed and deployed in this exhibition specifically as an exercise – physical, mental, cognitive, pedagogical, epistemological, political, ethical, etc. Whether considered in the sense of training (requiring repetition and work, as in long-term effort), or associated with the social, cultural and political conditions of individual as well as collective practice, the expression "reading exercises" aims to convey the complexity of this activity that cannot be confined to the mere notion of a profound literary experience cut off from the outside world.

Curator Katrie Chagnon (2015) mounted six iPads in the centre of the Ellen Gallery, on a custom-made table, and invited patrons to sit and watch the #ReadTheTRCReport videos and peruse printed materials about the TRC. The Ellen Gallery also organised a reading marathon, inviting members of the public to come and record themselves reading sections of the French version of the Executive Summary in order to generate a French playlist for public consumption.

This show positioned our project alongside other oeuvres that explicitly draw out the visceral and embodied processes of reading within the social and political worlds we inhabit. The No Reading After the Internet collective builds their work around the phenomenon we stumbled upon accidentally in our own project: that the experience of reading aloud brings people, stories and places together in ways that are complex, beautiful, and affective.

Through the experience of the #ReadTheTRCReport project itself, and through the experience of being included in the Reading Exercises show at the Ellen Gallery, I learned some powerful truths about reading. What could have simply remained a private act – reading the report in solitude, without witnesses – became itself a way of marking, through body language and palpable emotions – collective, insistent, visible, and resistant grief. You may ask why this matters. First, what is so amazing, to me, about this collective, insistent, grief that people tapped into through the #ReadTheTRCReport is that I was not afforded the space to express insistent grief during my time living in the UK as a doctoral student. Claiming a space to cry and emote without apology, as a Métis woman with my own entanglements with intergenerational trauma from British colonialism in North America, while reading about deliberate, violent, calculated acts by the Canadian-British actors against small Indigenous children was, and is, liberating for me after my years of living in the United Kingdom. In Britain, I came upon the oft-repeated notion that the violence of colonialism is in the past. Phrases like "that was our ancestors' fault, not mine" or "independence already happened, you can't blame us for things today" (wordier versions of 'get over it') were levied any time I had the temerity to bring up colonialism in Canada as an ongoing reality that ties British and Canadian narratives very closely together. I vividly remember the time my Gwich'in colleague Crystal Fraser visited me in Scotland in the autumn of 2013, and we discussed colonialism on the bus in Dundee as we rode home from the train station. A woman in a long duffel coat with a 'Better Together' pin on her lapel began to snort – snort! – at us as we spoke. She was far too polite to tell us to shut up, but the intent was clear: "do not bring your insistent anger or grief back to us: we don't want it".

Second, the act of reading the report out loud imbues the text with emotion and weight that is impossible to ignore for many who view the videos. It captures, in some small way, the visceral feeling I felt upon listening to survivors share their stories in the TRC hearings I attended in Paulatuuq, Northwest Territories, Canada, during my doctoral research. There is, of course, the risk that some people may view the act of reading as reparation enough for the hundreds of years of colonial violence that shape North America. This is not my aim in working on #ReadTheTRCReport. I hope that the affective response of the non-Indigenous readers and viewers of the project instead encourages participants and audiences to view the process of aural-visual engagement with reports of colonial violence as beginning a collective, decolonising healing process. The performative, intra-active, affective, and circulatory unfolding of this spontaneous project is one dynamic step amongst many in a critical politics of decolonial resistance. But, more importantly, as I have argued elsewhere (Todd, 2016b), it also enacts reciprocity, love, accountability, and care as equally, if not more important, and powerful political tools required to face uncertain futures and the continuity of precarious worlds as we know them.

For me, the #ReadTheTRCReport is also an act of reclaiming Indigenous ontologies and epistemologies and political aims from those non-Indigenous scholars who claim to use Indigenous thinking and voice in the service of decolonisation. I want non-Indigenous scholars who view the videos in the #ReadTheTRCReport playlist to squirm and feel shame and think hard about their use of Indigenous epistemologies and lofty claims of re-instating Indigenous ontologies within academic and educational systems that not very long ago were used as models for the Residential School system to completely obliterate and deny Indigenous thinking, language, stories and laws. I want, and seek, and yearn for a reckoning of the 'epistemic violence' that Kwakwaka'wakw (of the Kwagiulth Band) scholar Sarah Hunt plainly demonstrates non-Indigenous scholars are guilty of when they reduce embodied, lived Indigenous ontologies to mere 'trinkets' for academic pontification (Hunt, 2014: 30). As Erica Violet Lee (2015) points out in a recent piece on Indigenizing the academy,

> The centering of Indigenous knowledges in universities is important, and it must be done right. If the foundations of the settler colonial state are not challenged, the incorporation of so-called "Indigenous content" into classrooms is a method of continuous recolonization; furthering claims of ownership and authority over Indigenous knowledges and Indigenous lands.

It is important for non-Indigenous scholars who claim and appropriate Indigenous epistemologies and ontologies in their work to also engage directly, generously, and unambiguously with the ongoing (de)colonial stories that we tell as dynamic and diverse Indigenous peoples. It is also necessary for postcolonial scholarship to engage the structural inequalities that erase and dismiss the work of Indigenous scholars and thinkers in the academy (Hunt, 2014; Lee, 2015; Todd, 2016a; Sundberg, 2014; Watts, 2013). If non-Indigenous scholars seek to employ Indigenous thinking and philosophies in their work on colonial/postcolonial realities, then they must also contend with the visceral, embodied, and living experiences of Indigenous peoples. And that means being willing to engage with the affect and insistent grief of women, like me, who share hallways, seminars and workshops with non-Indigenous peers.

Fish as witness

The thing about grief, colonialism, and reconciliation is that much of the focus in Canada still tends towards the human (Todd, 2014). The Canadian State, after all, does not view animals as non-human persons in the way that many Indigenous legal orders frame animals as sentient beings with whom humans share territories, stories, and reciprocal duties.

In my doctoral dissertation work, I examine human-fish relations in the arctic hamlet of Paulatuuq as an 'under-theorised site of colonial engagement'[1] which are part of a local Inuvialuit legal order. Cree legal scholar Val Napoleon and legal scholar Hadley Friedland (2014: 227) remind us that: "all Indigenous groups had self-complete, non-state systems of social ordering that were successful enough for them to continue as societies for tens of thousands of years". These legal orders encompass "law that is embedded in social, political, economic, and spiritual institutions" (Napoleon, 2007: 2). In other words, Indigenous peoples in North America had, and continue to have, legal traditions which are bound up with the specific spatial-temporal histories of diverse and dynamic societies, which weave together stories, kinship, language, cosmologies, as well as relationships with the more-than-human, in everyday practices that guide how to relate well with the world.

Articulated as non-human persons that circulate through local stories, local thinking, and local praxis, fish in Paulatuuq are asserted as social components of a local Indigenous legal order. In Paulatuuq, people refract the colonial laws and principles imposed upon local lives by negotiating across simultaneous and often contradictory 'sameness and difference'. In so doing, they apply aspects of local Inuvialuit legal orders that govern how to respect fish while also contending with the Canadian State's colonial legal systems that seek to control the legal-governance relationships between humans, animals, lands and waters. Within this context, reciprocal duties between humans and fish are enacted through careful attention to the well-being of fish. For example, the hamlet has been recognised for its ground-breaking community-based monitoring work on the Hornaday River char (Ayles et al., 2007), and the community actively closed a government-imposed commercial char fishery on the Hornaday in 1986 in order to further the well-being of humans and fish in the community (Todd, 2014).

Paulatuuq was the first community within which I experienced directly the work of the Truth and Reconciliation Commission. Elsewhere (Todd, in press) I describe the experience of public mourning and grief through the public hearings of the TRC held in Paulatuuq in April 2012 during my doctoral work. Working in this context, during the active operations of the Truth and Reconciliation Commission and during a time when community members impacted by Residential School were pursuing claims with the government under the terms of Indian Residential Schools Settlement, meant that many of my interviews on fishing in the community also elicited reflections on the impact of the Residential School system on northern communities like Paulatuuq.

Once I finished my fieldwork in Paulatuuq, I moved on to archival research. My aim was to include an analysis of colonial understandings of human-fish relations in the region in the 1920s-1960s, during a period recognised as one of 'northern Industrial transformation' (Piper, 2009) in sub/arctic Canada. As I researched the archives of the Oblates of Mary Immaculate in June 2013, searching for any and all documents relating to fish in Paulatuuq in the early-to-mid twentieth century, I held the stories of Residential School survivors from the community at the top of my mind. It is here that I began to encounter the stories of human-fish relations as a 'micro-site' across which Inuvialuit asserted their legal orders vis-à-vis missionary, state and commercial agents. Fish were an integral part of local life, and when missionaries came to the Paulatuuq area, they came to rely heavily on the fishing skills that Inuvialuit taught them.

There are other less obvious ways[2] that fish were entangled in colonial relations in northern Canada, and this is as active witnesses and participants in the colonial encounter in their own right. As I read the 'Paulatuk Codex Historicus'[3] from 1936, I came upon a section where the Oblates of Mary Immaculate discuss, rather matter-of-factly, sending children to Aklavik by boat to attend the Residential School there (Paulatuk Codex Historicus, 1936: 53). The Oblates' schooner, the 'Our Lady of Lourdes', was used to take the son of a local Paulatuuq widow, Uniqaun, to the town of Aklavik, where he was forced to attend the Catholic Indian Residential School run by the Oblates. Subsequent interviews and conversations[4] in the community revealed that Uniqaun's only son never returned from the school, dying there. As I read this entry in the Paulatuk Codex Historicus, and after reading through sections of the Oblates' codices where the Fathers and Brothers of the mission record how they came to rely so heavily on local men to teach them how to fish for survival, I could not help but imagine the way the fish bore witness to the children as they were stolen from their parents. How the fish bore witness to the Oblates, police, Hudson's Bay Company men, doctors, nurses and others moving into the region and coming to rely on local fish and waters for sustenance. And how the fish would have noted the absence of many local children at fish camps that spring and for decades of springs thereafter.

One August afternoon in 2012, I met with a community member in Paulatuuq, Jane,[5] to conduct an interview for my research on human-fish relations in Paulatuuq. She described what it was like to come home for the first time in seven years after being sent to the Indian Residential School in Inuvik as a child. The first thing her dad did, as soon as she got back home, was take her to fish camp. Drawing on memories from before being sent away to school, she reminisced over what it was like being at fish camp as a child as we continued to chat:

> My grandparents, my parents, and my uncle and his wife and all the kids . . . every summer we'd have a camp over at Billy's Creek. Just a couple of kilometres down this way [motions towards area on map]. I grew up summertime swimming in the creek. Early in the morning my grandfather would be up and he'd be up pretty early waiting for the boys to get up so they could help him check his net so know he could see the fish on the net because that's where the sunrises in the East so that's how he could see the fish on the fishing net. So that's how I remember growing up, being around Billy's Creek, being around fishing, all the of the grandchildren of [my granddad] and his boys, his boys too would be busy checking the net and my grandmother and my nan, they'd be cleaning fish and making dry fish. And the kids younger, teenagers would be washing all the fish on the ocean, washing them, there'd always be somebody hanging the fish. We just grew up that way.

What strikes me about our conversation back in 2012 is that her father re-introduced her to community life, to the land, and to fish as soon as he could. Her return to community life was marked, deeply, by the importance of fish camp in shaping local kinship relations and local stories, memories and relationships to the more-than-human.

I have elsewhere argued that any notion of reconciliation in Canada can and must incorporate the more-than-human (Todd, 2014). This move to the more-than-human decentres concepts of grief as solely a human response to the upheavals and violence of colonialism in Canada, for it manifests Donald's ecology of 'ethical relationality' in the insistent, material resonances of how grief becomes discernible. Fish can and do experience the impacts of colonialism, and indeed through my research, Inuvialuit interlocutors taught me that animals are political-legal agents in their own right. Fish bear witness to the destruction of waterways, which are Indigenous legal orders that incorporate fish as kin and reciprocal economic and political agents. Fish bear witness to, and even resist and shape, the changes that colonialism wrought and still wreaks throughout Canada.

Ness Namew

If Britons, who continue to live with the material culture and wealth taken from Indigenous homelands around the globe, respond to even the mildest of public conversations about colonialism on the bus with snorting, or with aggressive rattling of their newspapers to try to silence uncomfortable conversations about British Imperialism on the train, is there a way to return the Insistent Grief of This Place (North America) in another (non-human) form? If fish can and do bear witness to the colonial suffering wrought in the territories they share with diverse Indigenous nations in North America, and I believe they do, then perhaps fish can better instigate and provoke British awareness of colonial violence than any fleshy, insistent, and inconvenient Indigenous body like mine can.

In January 2015, I wrote a piece entitled "Decolonial Dreams: Unsettling the Academy through *namewak*" for a collection published by Green Lantern Press in Chicago. It is my first attempt to describe and honour the Decolonial Trickster Fish I encountered and dreamed of while living in the UK. The Ness *namêw* is a lovelorn, heartbroken prairie lake namêw (sturgeon) who finds her way from the ruined waterways of my home territory – ravaged by overfishing during the colonial fur trade; polluted and drained by oil and gas activity in the infamous Tar Sands in my home province of Alberta today – to the highland lochs of Scotland. She is the decolonial Trickster fish who teases and haunts European interlocutors by lurking deep within Loch Ness, carrying stories of suffering and rage back to the British Empire. To give you a taste of her movements, I include an excerpt from the piece here (Todd, 2015: 116–117):

> She swims in Scottish waters without apology. She tells her stories through her movements. Through her simultaneous presence and absence, she weaves a narrative that ties Scotland to the distant lands and waterways its people dispossessed for England. Sometimes we have to tell stories that make people uncomfortable. Sometimes we have to move inscrutably through waters, relentlessly teasing and provoking. Sometimes we carry stories of suffering and dispossession back with us. People will react with horror, fear, indifference or curiosity.

> . . . To me, Ness namêw is a decolonial story-teller haunting the Scottish psyche until it can see how intimately it is woven into our own stories of dispossession and loss as Indigenous peoples. And from there we can dismantle the hierarchies that separate our experiences – hierarchies and beliefs that unfairly situate ultimate authority to know, describe and tell within non-Indigenous bodies – and bring our stories and voices as Indigenous peoples, so quickly erased in British academia, to the fore. To tell a decolonial dream: scaly and slippery and present. And to build something more accountable, reciprocal and loving in place of the structures and narratives that currently exist.

In 'Decolonial Dreams', I explain how I learned of the Loch Ness monster's possible true identity (or, perhaps, one of her many monster pluralities) during a tourist trip to Inverness. It is here on the shores of Loch Ness that I learned that the Loch Ness monster is possibly a large Baltic sturgeon who wanders accidentally up the River Ness and circling the Loch in search of love before leaving again (Todd, 2015: 109–110; National Geographic undated). Regardless of her definitive form, I take a great deal of joy in the Ness namew's relentless provocation of the Scottish psyche. TV specials are devoted to unlocking her mysteries and puzzling through her identity (National Geographic undated). Kitschy plush toys and doctored postcards sold across Scotland hint at her shenanigans. Imagining her movements from the ruined waterways of Treaty Six territory and her bending of space and time to bring urgent stories back to Europe buoys me as I wrangle with my own complicity in colonial realities in Canada.

In my writing and telling of her story, the Ness Namew elicits an affective response: bewilderment, lust, passion, curiosity, fear, rejection. Her movements are deliberate and inscrutable. I inscribe in her flesh a story of insistent grief but also hope. In traveling the River Ness, and in circling Loch Ness in her own ways, she is carrying the grief of my own homelands back to the Scottish people, to lay this suffering bare to them within one of their own cherished waterways. But she also carries stories of hope, stories of persistence and movement beyond colonial violence. She even carries stories illegible to us, stories that whisper of worlds and possibilities perhaps beyond our comprehension at the moment, but which matter all the same.

Refraction

I borrow a phenomenon from physics to help explain the simultaneous and often contradictory negotiations across 'sameness and difference' that Indigenous peoples mobilise every day as they engage with the colonial State in Canada. This phenomenon is refraction, which is the bending of light through a glass material and its related phenomenon, dispersion – which is the scattering of the wavelengths that make up white light. To draw out the analogy: just as prisms work to bend and disperse light, Indigenous legal orders, stories and movements work to bend and scatter colonial thinking to reveal its constitutive components. The insistent grief and affective stories being told by human and more-than-human entities about colonial realities in Canada today work to refract colonial understandings of Canada and its histories. While colonial actors may have intentions and expectations regarding how processes like reconciliation or redress should unfold, the affective responses of Indigenous peoples, and more-than-human kin, refract and disperse these intentions. I also apply this metaphor of refraction to describe the intellectual labour that I observe Indigenous peoples applying in their own assertions of Indigenous laws, stories, language and presence in Canada. Refraction, therefore, in the context of my work with the #ReadThe TRCReport, through my study of fish pluralities in Paulatuuq, and through the movements of the trickster dream fish the Ness Namew, is a metaphor for the intellectual and affective labour that Indigenous people (and the more-than-human!) perform every day to bend, scatter, challenge and transform the ideologies and ethics that the State imposes regarding human-animal, human-environmental and legal-governance norms. Through nuanced engagement with colonial laws and norms, Indigenous peoples bend and re-shape the State's hard realities regarding relationships to land, water, fish, stories, histories, memories, pasts and futures in order to assert our own insistent grief, insistent needs, and insistent dreams.

In this sense, refraction can play a role in complicating and redirecting settler-colonial understandings or mobilisations of reconciliation discourses in Canada. However, this refraction must also be understood within the webs of power relations, dispossession, and continued denial of Indigenous sovereignty that operate in Canadian settler-colonial realities (Simpson, 2014). Both Simpson (2014, 2015) and Coulthard (2014) bring our attention back to land and dispossession, reminding their readers that discourses of reconciliation in Canada occur on and in land stolen by the Canadian State, land through which Canada legitimises itself. Refraction operates to bend and scatter the State's intentions and narratives about itself, but resistance through refraction can only be applied so far. When and where the State is intent on denying Indigenous agency, Indigenous legal orders and Indigenous self-determination – and where Canada insists on not addressing dispossession – it has made it very clear that it will do so.

With this in mind, Canada's broad, collective understandings of reconciliation have arguably been co-opted into the 'politics of recognition' of the Canadian State (Coulthard, 2014; Simpson, 2015) which seek to delimit how, when, where, why and what reconciliation and Indigenous sovereignty shall be, within the confines of State logics. However, as I have learned from projects like #ReadTheTRCReport and from my research in Paulatuuq, there are ways to complicate, disrupt and redirect the State's understandings of what 'reconciliation' looks like in Canada towards logics, stories and experiences determined by Indigenous peoples. There are ways to refract the State's aims and negotiate alternatives within and alongside Indigenous philosophies and legal orders. Reconciliation will necessarily be more than operas and ballets[6] – it will be, and must be, informed and challenged and re-narrated through the fleshy, affective responses and stories of trickster fish, fish-as-witnesses, and those stories insistently told by still-present Indigenous peoples on their terms. Refraction of reconciliation and recognition through Indigenous legal orders, art, cosmologies, philosophies, histories, geographies and resistances are necessary in order to subvert the impulse of the State and its agents to tell the story of colonialism, and reconciliation, solely on its terms. Refraction, therefore, is one tool with which we can foster both a) Dwayne Donald's 'ethical relationality' and b) an unambiguous acknowledgement, and addressing, of Indigenous dispossession.

Conclusion

To contend with the violence of the colonial moment, those affected seek out ways to tell stories. These stories root us in place, and root us in kinship and care networks that distribute the grief and pain of collective experience across many bodies, many spaces, many tongues, hands and hearts. These stories also allow us to work across the contradictory and complicated relationality between coloniser and colonised, invoking the 'ethical relationality' that Donald (2009) brings to light in his work. Through the work of the #ReadTheTRCReport, I became aware of the affective power of reading my grief and rage into the world, and I also became aware of the power of watching others transform text into words and video. Through my work in Paulatuuq, I learned that the ethical relationality of fish bears witness to the insistent grief of peoples deliberately dispossessed from their lands and waters by colonial governments. And in my dreams and stories of the Ness Namêw, I imagine a fishy trickster carrying the insistent grief of humans and more-than-human alike back to the heart of colonial Empire, where erasure of Indigenous stories still dominates public imaginaries. States can and do try to co-opt these stories towards their own ends (Coulthard, 2014; Simpson, 2015) and we must be careful about how these narratives become enmeshed in the ongoing power-relations of the settler-colonial State. However, paradoxically, the stories that Empire and its governments and agents try to tell about itself can be – and are – refracted through the insistent, fleshy, present bodies, voices and movements of humans and more-than-human entities. These refractions of the colonial and postcolonial offer us possibilities that expand our understanding of what it is to live well, together, across space and time.

Notes

1 I refer here to anthropologist Ann Fienup-Riordan's (2000: 57) notion of 'active sites of engagement' in her work in Yup'ik communities in Alaska.
2 Less obvious to colonial agents, anyway.
3 The 'Paulatuk Codex Historicus' is an archival record of the daily activities of the Oblates at their Mission, first in Letty Harbour and later in Paulatuuq.
4 Upon explaining to local educator and historian Rosemary Kirby that Paulatuuq elder Annie Illasiak had given me the Inuvialuktun name Uniqaun (in honour of Uniqaun), Rosemary told me the story of Uniqaun's son's attempt to escape the Aklavik school. He was found dead in the forest, his escape unsuccessful.
5 An alias is used in order to not impact her claims in legal proceedings regarding her Residential School experiences.
6 At the time of writing, the National Art Centre in Ottawa, Canada, is showing a ballet entitled 'Coming Home Star', which dwells on the experiences of Residential School survivors.

References

Asch, M. 2014.*On Being Here to Stay: Treaties and Aboriginal Rights in Canada*. Toronto: University of Toronto Press.

Ayles, B., Bell, R. and Hoyt, A. 2007. 'Adaptive Fisheries Co-Management in the Western Canadian Arctic', In: F. Berkes, D. Armitage and N. Doubleday, eds., *Adaptive Co-Management [Electronic Resource]: Collaboration, Learning, and Multi-Level Governance*. Vancouver: UBC Press, pp. 126–151.

Barrera, J. 2015. 'New Indigenous Affairs Minister Speaks Reconciliation With Sage in Her Boots, Loaned Eagle Feather in Hand'. http://aptn.ca/news/2015/11/05/new-indigenous-affairs-minister-speaks-reconciliation-with-sage-in-her-boots-loaned-eagle-feather-in-hand/

Canadian Broadcasting Corporation (CBC). 2008. 'A History of Residential Schools in Canada'. www.cbc.ca/news/canada/a-history-of-residential-schools-in-canada-1.702280 [Accessed 08.02.2016].

Chagnon, K. 2015. 'Reading Exercise, Exhibit Catalogue'. http://ellengallery.concordia.ca/wp-content/uploads/2015/11/LBEAG_Exercicesdelecture_ENG.pdf [Accessed 08.02.2016].

Clifton, R. and Rubenstein, D. 2015. 'Debunking the Half-Truths and Exaggerations in the Truth and Reconciliation Report', *National Post*. http://news.nationalpost.com/full-comment/clifton-rubenstein-debunking-the-half-truths-and-exaggerations-in-the-truth-and-reconciliation-report

Coulthard, G. 2014. *Red Skin, White Masks: Rejecting the Colonial Politics of Recognition*. Minneapolis: University of Minnesota Press.

Donald, D. 2009. 'Forts, Curriculum, and Indigenous Metissage: Imagining Decolonization of Aboriginal-Canadian Relations in Educational Contexts', *First Nations Perspectives*, vol. 2, no. 1, pp. 1–24.

Ellen Gallery. 2015. 'Reading Exercises', Exhibition website. http://ellengallery.concordia.ca/?exposition=exercices-de-lecture&lang=en [Accessed 08.02.2016].

Fienup-Riordan, A. 2000. 'An Anthropologist Reassess Her Methods', In: A. Fienup-Riordan, W. Tyson, P. John, M. Meade and J. Active, eds., *Hunting Tradition in a Changing World: Yup'ik Lives in Alaska Today*. New Brunswick: Rutgers University Press, pp. 29–57.

Hunt, S. 2014. 'Ontologies of Indigeneity: The Politics of Embodying a Concept', *Cultural Geographies*, vol. 21, no. 1, pp. 27–32.

Kinew et al. 2015. 'Wab Kinew et al: Make Reconciliation as an Election Issue'. http://news.nationalpost.com/full-comment/wab-kinew-et-al-make-reconciliation-an-election-issue [Accessed 08.02.2016].

King, T. 2013. *The Inconvenient Indian: A Curious Account of Native People in North America*. Canada: Doubleday.

Lee, E.V. 2015. ' "Indigenizing the Academy" Without Indigenous People: Who Can Teach Our Stories?' *Moontime Warrior blog*. http://moontimewarrior.com/2015/11/09/who-can-teach-indigenous-philosophy/ [Accessed 08.02.2016].

Napoleon, V. 2007. 'Thinking About Indigenous Legal Orders', Research Paper for the National Centre for First Nations Governance. http://fngovernance.org/ncfng_research/val_napoleon.pdf [Accessed 05.06.2014].

Napoleon, V. and Friedland, H. 2014. 'Indigenous Legal Traditions: Roots to Renaissance', In: M. Dubber and T. Hörnle, eds., *The Oxford Handbook of Criminal Law*. Oxford: Oxford University Press, pp. 225–247.

National Geographic. 'The Truth Behind: The Loch Ness Sturgeon?' http://channel.nationalgeographic.com/the-truth-behind/videos/the-loch-ness-sturgeon/ [Accessed 07.02.2016].

Palmater, P. 2015. 'What Happened in Residential Schools Was Genocide: But What Really Matters Is Justice', *Rabble Blogs*. http://rabble.ca/blogs/bloggers/pamela-palmater/2015/06/what-happened-residential-schools-was-genocide-what-matters-j

Paulatuk Codex Historicus. 1936. Archives de Hurlevent. Diocese of Mackenzie-Fort Smith Collection: Paulatuk, N.W.T. Vol. 1.

Piper, L. 2009. *The Industrial Transformation of Subarctic Canada*. Vancouver: UBC Press.

Schwartz, D. 2015. 'Truth and Reconciliation Commission: By the Numbers', Canadian Broadcasting Corporation (CBC). www.cbc.ca/news/aboriginal/truth-and-reconciliation-commission-by-the-numbers-1.3096185

Simpson, A. 2014. *Mohawk Interruptus: Political Life Across the Borders of Settler States*. Durham: Duke University Press.

Simpson, A. 2015. 'Reconciliation and Its Discontents. Talk, February 19, 2015, World of Matter: Extractive Ecologies and Unceded Terrains', Montreal. http://lowbidinal.tumblr.com/post/111902612002/here-is-audra-simpsons-talk-from-world-of-matter

Straptigeil, L. 2015. 'The Most Powerful Way to Experience the TRC Report'. www.buzzfeed.com/laurenstrapagiel/this-might-be-the-most-powerful-way-to-take-in-the-truth-and?utm_term=.yn5gXD02Yo#.tlW2yAKxX5 [Accessed 07.02.2016].

Sundberg, J. 2014. 'Decolonizing Posthumanist Geographies', *Cultural Geographies*, vol. 21, no. 1, pp. 33–47.

Todd, Z. 2014. 'Fish Pluralities: Human-Animal Relations and Sites of Engagement in Paulatuuq, Arctic Canada', *Etudes/Inuit/Studies*, vol. 38, no. 1–2, pp. 217–238.

Todd, Z. 2015. 'Decolonial Dreams: Unsettling the Academy Through *namewak*', In: C. Picard, ed., *The New (New) Corpse*. Chicago: Green Lantern Press, pp. 104–117.

Todd, Z. 2016a. 'An Indigenous Feminist's Take on the Ontological Turn: "Ontology" Is Just Another Word for Colonialism', *Journal of Historical Sociology*, vol. 29, no. 1, pp. 4–22.

Todd, Z. 2016b. '"Relationships" Theorizing the Contemporary', *Cultural Anthropology*. www.culanth.org/fieldsights/799-relationships

Todd, Z. In press. 'Métis Storytelling Across Time and Space: Situating the Personal and Academic Self Between Homelands', In: J. Christensen, L. Szabo-Jones, C. Cox and A. Boisselle, eds., *Activating The Heart: Storytelling, Knowledge Sharing and Relationship*. Wilfred Laurier Press.

Truth and Reconciliation Commission of Canada (TRC). 2015. 'Honouring the Truth, Reconciling for the Future: Summary of the Final Report of the Truth and Reconciliation Commission of Canada'. www.trc.ca/websites/trcinstitution/File/2015/Honouring_the_Truth_Reconciling_for_the_Future_July_23_2015.pdf

'Truth and Reconciliation Commission'. (TRC) Undated a, Mandate. www.trc.ca/websites/trcinstitution/index.php?p=7

'Truth and Reconciliation Commission'. (TRC). Undated b, Residential Schools. www.trc.ca/websites/trcinstitution/index.php?p=4

Vowel, C. 2015. 'Reaction to the TRC: Not All Opinions Are Equal or Valid', *Âpihtawikosisân. blog*. http://apihtawikosisan.com/2015/06/reaction-to-the-trc-not-all-opinions-are-equal-or-valid/

Watts, V. 2013. 'Indigenous Place-Thought and Agency Amongst Humans and Non-Humans (First Woman and Sky Woman go on a European Tour!)', *Decolonization, Indigeneity, Education and Society*, vol. 2, no. 1, pp. 20–34.

6 Unsettling the urban geographies of settler-colonial cities

Aporetic encounters with the spatiotemporal dynamics of modern logic

Delacey Tedesco

Introduction

Within postcolonial urban studies, it has been recognized for some time that transition narratives of modern urbanization often reproduce political constructions and colonizing practices of Western modernity (e.g. Robinson, 2006). Extensive work has since been employed to "blast open the geographies of theory" (Roy, 2009: 820). My challenge here, however, is not simply to relocate the site of urban theorizing outside the West but to dislocate the geographical and conceptual boundaries that produce, within the Western city, an erasure of contemporary settler-colonial relations. Such boundary constructions operate through sedimented configurations of spacetime, ontology, epistemology, and phenomenologies of modernity. Work on decoloniality (Mignolo, 2007, 2009) has emphasized that configurations of the modern are sustained by, and serve to mask, comprehensive logics of coloniality. Efforts to address directly the challenges of the pervasive boundaries of modern urbanization have emphasized a transition to contemporary modes of global urbanization. At the same time, modernist boundaries of colonization have been challenged by numerous 'turns': an "ontological" turn in social and cultural geography (Blaser, 2014, Sundberg, 2014); an "epistemological" turn that guides decoloniality (Mignolo, 2007, 2009); and, an "Indigenous" turn from Western to Indigenous onto-epistemologies (Hunt, 2014, Cameron et al., 2014), each of which both engages and troubles emerging work in geography broadly, and postcolonial studies specifically. These 'turns' enable alternative accounts of space and time, ontology, epistemology, and phenomenology, particularly accounts that enable a plurality of simultaneous configurations. Yet these accounts are frequently accompanied by critical attempts to demonstrate how such 'turns' have a tendency to return to distinctively modern forms of ontology and epistemology (Blaser, 2014; Sundberg, 2014). This pattern of turns away from modern configurations, which subsequently return to modern configurations, suggests a troubling spatiotemporal dynamic that reproduces the conceptual and generative boundaries of the modernization/urbanization/colonization nexus, even in efforts to unsettle boundaries. The following chapter explores how this happens as a function of the political logic of limits.

The chapter focuses on a recent downtown revitalization project in the Canadian settler-colonial city of Kelowna, BC. In Kelowna, a multivalent emphasis on "authentic place-making" enables the investigation of two related problematics: the spatiotemporal dynamics embedded in claims to open possible reconfigurations of settler-colonial urban community; and, patterns of closure that these turns to openness also seem to generate. I present an analysis of the logic of the aporia as a means of engaging with these patterns of turns and returns, but also suggest that an aporetic analysis troubles strategies of 'turns' being figured in epistemological and ontological terms. Aporetic analyses do not offer alternative configurations of spacetime, ontology, epistemology, and phenomenology. Rather, they direct attention to the ways that inherent insecurity in these terms generates and undermines transition narratives (Derrida, 1993; Tedesco, 2012). I use aporetic boundaries and the figure of the hiatus to engage the complicated place-making practices in Kelowna and other settler-colonial urban contexts. Yet, as a theoretical resource drawn from Western institutional, linguistic, and political traditions, the aporetic analysis potentially reaches its own limits when faced with decolonial efforts in this place.

While in conversation with dense, dispersed webs of academic and practical efforts, the work offered here is contextualized within Kelowna as a site of Western settlement, development, and increasingly intensive urbanization on the unceded land of the Syilx, or Okanagan Nation.[1] Kelowna's settler-colonial origins continue to configure the city according to the dominant transition narratives of modern urbanization and the imaginaries of authentic community idealized by these narratives. The city is now trying, if in limited ways, to reconfigure itself in relation both to global processes of urban transformation and to Indigenous efforts to advance postcolonial or decolonial possibilities in settler-colonial urban contexts. I focus on the recently completed 'Bernard Avenue Revitalization Project', a functional and aesthetic reconstruction of Kelowna's downtown main street that mirrors the patterns of turns and returns that pervade webs of academic and practical decolonization efforts. While the project is embedded within competitive pressures assumed through global processes of urban transformation, pervasive claims to "authentic" place also repeat dominant spatiotemporal logics of modernity and coloniality: linear transition narratives (rural to urban, nature to culture, tradition to modern), and dominant configurations of the singularity of place within universal space and linear time.

The Bernard Avenue Revitalization project uses "place-making" strategies to build a more "authentic" sense of place by incorporating Syilx cultural imagery and language, an element that arose from decolonization efforts locally. The emphasis of Kelowna's revitalization project on representations of Syilx language and imagery bears an uneasy resemblance to the way that "[p]rocesses of colonialism in North America involved representational strategies that transformed Indigenous peoples and their lands conceptually and materially, in order to facilitate their displacement and render them less than human" (Hunt, 2014: 29; see also Edmonds, 2010: 10). In Kelowna, such pervasive representations have, over time, erased or minimized Syilx presence, worldview, and cultural practices. Historic orchard company logos, which celebrated the British Empire and displaced

Syilx labour in the orchards, are now emplaced as decorative wraps on electrical boxes around the city. Nxaʔxʔitkʷ,[2] the Syilx guardian of Okanagan Lake and its waterways, becomes represented as Ogopogo, a secretive water creature in the figure of the Loch Ness Monster. And, Syilx life is relegated to an imaginary of the past, as in the information panels at the Pioneer Pavilion (Tedesco, in press). Within the Revitalization project, gestures such as the inclusion of Syilx language on paving stones are potentially dwarfed by the repetition of dominant visual accounts of colonial settlement, such as prominent archival images of pioneer life and settler city spaces. Finally, the unsettling possibilities of Syilx art and politics were constrained by the City of Kelowna's recent policy that absorbed and subordinated its independent public art committee into the multiple economies of its comprehensive urban planning and design processes.

Despite these pressures towards forms of closure, containment, and erasure, the revitalization project presents some tentative, unstable openings, most notably in the form of the Gateway Banners, public artworks designed and produced by members of the Syilx Westbank First Nation (WFN). Insofar as "a number of Indigenous scholars have pointed to stories, art, and metaphor as important transmitters of Indigenous knowledge" (Hunt, 2014: 27), the Gateway Banners offer an important aesthetic site for engaging Syilx identifications of their own Okanagan Indigeneity (Armstrong, 2010). This chapter acknowledges, but cannot answer, the powerful question: "what [does] it [mean] for Indigenous knowledge to be moved from spaces of lived Indigenous governance and culture, such as a potlatch ceremony, to a conference session" (Hunt, 2014: 31), or, for that matter, into an urban planning project led by a settler-colonial municipal government. However, this chapter takes as foundational the notion that "particular examples of artistic endeavour intervene in our world and prompt us to think and feel differently and perhaps more cogently about it" (Danchev and Lisle, 2009: 777). The intervention enacted by the Gateway Banners unsettles the urban revitalization project's claims to authenticity through representations of a singular, shared history. More importantly, the banners propose not simply a transition from one configuration of place to another, but an opening towards multi-spatial and multi-temporal configurations of place-beyond-place (Massey, 2006), and what might be called place-within-place.

Ultimately, this chapter stages an aporetic encounter with the Bernard Avenue Revitalization Project. It observes, obliquely, the turns and returns, borders and boundaries, spaces and limits that emerge when multiple configurations of the possibilities of settler-colonial urban community are articulated within the same urban planning project. Finally, it argues that the Gateway Banners offer prompts to think and feel differently, more cogently, more openly, about the repetitive dynamics of openings and closures that have marked the emplacements and displacements of settler-colonial urbanization in Kelowna.

Modernization, urbanization, colonization, and its limits

Kelowna is about 400 kilometres northeast of Vancouver, British Columbia, at the mid-point of the 110-kilometre-long Okanagan Lake, which creates a north-south, semi-arid valley between forested mountains. At 120,000 people, Kelowna is now the biggest city in the Okanagan Valley. But less than 200 years ago, this land was settled only by the Syilx people, now also called the Okanagan Nation, who lived on and cared for the land from time immemorial. The Syilx people ranged from the north end of the lake to the south end, and along the Okanagan River to where it drains into the Columbia River. With the decision to recognize the 49th parallel as the dividing boundary between sovereign states, the Okanagan people became a transboundary nation, divided into two with seven tribes in the Okanagan Valley north of the border, and an eighth tribe known as the Okanogan Confederacy in Washington State.

The first permanent European settlement in Kelowna occurred in 1859, when Father Pandosy left Washington State and established a mission on a creek that drains into Okanagan Lake. When Kelowna was officially incorporated, in 1905, as a City under provincial legislation, the historic town site was centred on the lakeshore, several kilometres from the original mission site,[3] at a narrowed bend in the lake that the Okanagan people used as a canoe crossing point. As with most of British Columbia, Kelowna became a site of predominantly British immigration, though the orchards of Kelowna, which were seen as providing a life of culture and leisure suitable to upper-middle class families, attracted a narrow segment of British immigrant, particularly younger sons of wealthy families (Bennett, 1998). Kelowna thus transformed from a French community to a strongly British one, fashioning an idyllic middle-class orcharding imaginary: a city that would remained tied to rural life, and whose growth might therefore avoid the perceived social diseases of the metropolises of the UK and the United States (Bennett, 1998: 75). The imaginary of a shimmering white Kelowna (Aguiar and Martin, 2011), emerging from and purifying the processes of modern urbanization and industrialization, set a pattern of complex, interwoven, translocal communities and conflicted developments that proceeded to shape the materialities of Kelowna.

The population of Kelowna has doubled in the past 30 years, and is set to double again in the next 30 years, leading, today, to built forms and economic networks that are more now recognizably urban. On the west side of the bridge that enables the crossing of Okanagan Lake, contemporary urban development and population growth has followed a similar trajectory. This development has occurred on land that was only recently incorporated as the municipality of West Kelowna, and also on land designated as Westbank First Nation (WFN) reserve land, including Tsinstikeptum Reserves 9 and 10, which border on Okanagan Lake.[4] The WFN is unique within the eight Okanagan Nations, and within Canada more generally, because they are party to a self-government agreement with the federal Government (Vogl, 2013). The trajectory of urbanization on WFN land has followed the morphology of urban development in Kelowna and West Kelowna: strip malls and box stores strung along Highway 97, which runs north-south through the valley; gated communities and scenic golf courses are often combined into single developments (Cooper, 2013).

This dominant account of the history of Kelowna satisfies the modern narrative of urbanization as a series of connected transitions: rural to urban, nature to culture, tradition to modern. The dominant linear transition narrative of urbanization holds that while forms of social, economic, and political organization change, and even change our relationships to the landscape, "the land [itself] is geographically fixed" (Hessing, 2010/2011: 133) within the dominant configuration of modern spatio-temporality. Yet the transition narrative of urbanization also claims cognate boundaries, such as savage and civilized, primitive and advanced, as crucial elements within the intensive and extensive processes of modernization and colonization, particularly through the "stadial syntax," otherwise known as the four stages of development assumed by modern economic theory: hunter, pastoralist, agriculture, and commerce (Edmonds, 2010: 6–8). These transition narratives have replicated and enacted the assumptions underlying modern colonization, forming a modernist/urbanist project (Meagher, 2012) through which urbanization has been perceived and practiced as a key part of state-building colonial projects.

Such linear accounts never capture the complexity of urbanization in places outside the modern west – if they do even in 'the West'. Rather, they exemplify the sovereign ambitions and colonial/imperial practices of modern urbanization and urban theory (King, 1990, Isin, 2002; Robinson, 2006). Postcolonial studies have drawn attention to forms of urbanization that might identify non-Western processes, including distinctive forms of participation and exclusion, in cities of the Global South (Roy, 2009; Simone, 2011; McFarlane, 2011), particularly postcolonial cities, understood as places that have emerged from direct colonial rule (Yeoh, 2001).

Until recently, serious consideration of settler-colonial contexts and patterns of urbanization has been limited, producing a gap in geographic, social, and political analyses of how urbanization participated in modern colonial projects in these places (Edmonds, 2010; Freeman, 2010). Even within settler-colonial contexts, there has been a marked absence of recognition for the place of Indigenous peoples in urban settings (Edmonds, p. 4–5). These gaps are not incidental, but speak to the logic of modernity/coloniality that underlies that transition narratives of urbanization. This logic effectively colonizes diverse productions space, place, and time (Mignolo, 2007: 470) into a modern, Euclidean configuration that is universal and linear, in which each entity occupies a unique place and can be positioned at distinct points on the temporal line. It configures ontology as singular identities, often in binary, within universal space, and thus not only produces the particular "ontological split between nature and culture *as if it is universal*" (Sundberg, 2014: 35; emphasis in original), but also produces the spatiotemporal assumptions that enable this split. It configures epistemology as the universal transferability of knowledge, constituting spatial and hierarchical relations between knowing subject and known object, while producing "the illusion of the zero point epistemology" (Mignolo, 2009: 160). Finally, it configures phenomenology as a singularity of experience, each independent subject rationally processing its sensual encounters with the world. Together, these function as "a meta-narrative rooted in Enlightenment thinking and globalized through colonial discursive practices" (Sundberg, 2014: 35), including those associated with urbanization. Not only has the configuration of the modernization/colonization logic been entangled with the modern, Western account of urbanization. The techniques by which modernity erases coloniality – "[t]he crooked rhetoric that naturalizes 'modernity' as a universal global process and point of arrival hides its darker side, the constant reproduction of 'coloniality'" (Mignolo, 2007: 450) – are the same techniques that make it particularly difficult to see the 'Western' city in settler-colonial contexts as a colonial site.

The opening account of Kelowna's settlement, incorporation as a city, and subsequent urbanization emplaces it within this logic of modernity/coloniality. The more it became possible to represent Kelowna as a site of modern urban progress, repeating the binary constructions that enable this narrative of transition, the more were violent colonial displacements excluded from view. Kelowna is exemplary of the dominant narrative of settler-colonial cities that "depicts a transition from 'Edenic nature to improved settlement; from common to private entitlement'" (Edmonds, 2010: 8; quoting Blomley, 2005). Yet Kelowna was literally and figuratively incorporated to embody the spatialized relations of political authority that enabled the colonization of Canada. The peculiar model of incorporation here, one that established "the city as a corporation without citizens" (Isin, 1992), was developed to turn settler communities into tools of colonial governance, and this municipal structure was institutionalized within the federalized government to ensure that urban political authority was understood as a delegation from the (provincial) state. The dismissal of Syilx relationships to land, in favour of concepts of ownership over land, was made possible in part by a modern political geography of place within a singular extension of space, with the result being the restriction of the Okanagan people to reserves. The purpose of this political structure was to put both people and places in place, locatable with singular clarity.

These transition narratives of modern urbanization appear less and less able to account for the uncertainties of contemporary global urbanization, now widely considered the shared condition of contemporary life (Amin and Thrift, 2002, McFarlane, 2011). The world is increasingly figured as a global city, urbanization as a singular, if differentiated, planetary process (Brenner, 2013), and place as distantiated (Amin and Thrift, 2002) or as place-beyond-place (Massey, 2006). Kelowna can be located within this transition from localized to globalized urbanization: from its "origins" as a rural orcharding community to the 'world-class mid-sized' city of agri-tourism, vineyards, golf courses, and technical development it pictures itself as now; from its local, semi-isolated foundations to its inclusion in global networks and mobilities, including migrant temporary agricultural and construction workers from Mexico and Jamaica (Aguiar et al., 2010/2011; Aguiar et al., 2005);[5] and from its position within the colonial state-building project of Canada to its location within the urban transformations of neoliberal economic restructuring (Aguiar and Marten, 2011). These transformations unsettle presumed relationships between local and global and generate profound uncertainty in the foundational modern relationship between space and place (Massey, 2006).

Even as modern transition narratives appear to lose their analytic purchase in the face of the complexities of global urbanization, and even as transition narratives of global urbanization project a clarity of vision that cannot always be sustained when contextualized, what remains are a series of transition metanarratives, whereby contemporary global urbanization represents a fundamental transformation to our abilities to locate urbanization, to say what urbanization is, to define and analyze relevant phenomena, and to understand our experiences of and encounters with the urban and with each other within urban spaces and times. In other words, the narratives of transitions in the spatiotemporality of contemporary urbanization create a moiré pattern with narratives of transitions in the ontologies, epistemologies, and phenomenologies of contemporary urbanization: not parallel, but rather, layered patterns of uncertain intensification and dispersal created by the transformational processes of urbanization itself.

Yet the problem with these transition metanarratives of urbanization is that they suggest a troubling repetition of the linear, progressive spatiotemporal logic that sustains modern transition narratives. These transition metanarratives in accounts of contemporary urbanization are paralleled by attempts within contemporary postcolonial and decolonial work to unsettle the dominant binary boundaries of Western modern/colonial logic. Such efforts suggest that the boundaries that generate so many of the effects of the modernization/urbanization/colonization nexus are not simply the binary boundaries of rural/urban, nature/culture, tradition/modern, or even savage/civilized, but the non-binary boundaries that divide the messy complexity of worlds into spacetime, ontology, epistemology, and phenomenology as parameters that can be distinct, reliable, and comprehensive.

Efforts to disrupt the modernization/colonization/urbanization matrix through approaches such as decoloniality, postcoloniality, or posthumanism require "serious engagement with Indigenous epistemologies, ontologies, and methodologies" (Sundberg, 2014, p. 40). Yet the transition metanarratives of urbanization suggest that a crucial part of the postcolonial/decolonial challenge is a serious engagement with the spatiotemporal dynamics of turn and return, progress and repetition, which repeatedly pull attempts at new accounts of spacetime, ontology, epistemology, and phenomenology back to familiar configurations. These patterns are evident in iconic texts of contemporary urban geography (Amin and Thrift, 2002, Massey, 2006; see also Tedesco, 2012), and they are noted in substantive and wide-ranging works trying to develop a postcolonial or decolonial approach to settler-colonial urban contexts and effects. Diverse works note patterns of binaries and inversions, borders and margins, at work in settler-colonial city building (Edmonds, 2010: 11), and patterns of anxiety and security, hope and fear, escape and return as characteristic of settler-colonial cities and their extension into suburban developments (Veracini, 2012: 344, 346, 350). Also noted are patterns of efforts that seek to unsettle the logic of modernity/coloniality and the attendant configurations of ontology and epistemology, only to continually re-embed features of the logic they seek to challenge. Blaser (2014: 52), for example, observes that the theoretical efforts of posthumanism are hampered by "cases in which modernist ontological assumptions are enacted (and thus reinforced) even when they are being denounced." Sundberg (2014: 35) similarly notes that the posthumanism literature "repeatedly references such dualisms as if they are universal foundations of thought, which only serves to perpetuate their presumed universality." This pattern of return is most explicitly recognized, and cautioned against, in the warning by Cameron et al. (2014: 21), drawn from their reading of Blaser (2014) that "to make a move of engaging different ontologies without thoroughly interrogating the ontological frame from within which analyses proceeds . . . is to risk reasserting modern ontological claims to knowledge."

These repetitions are examples of a limit experience: they represent attempts to move beyond the limited and limiting configurations of the modernization/colonization/urbanization logic, only to reproduce these limits, though perhaps with some new or revised configuration. The dynamics of this limit encounter, simultaneously challenged and reasserted, are a function not simply of Western configurations of knowledge (Mignolo, 2009: 177) but of the modern logic of boundaries themselves, which, in putting anything other than modern outside the boundary of acceptability, ensures that the limits are continually set to be breached. While it seems clear, as articulated in terms of ontology, specifically, that "investigations into Western ontological possibilities are bounded in ways that limit their ability to fully account for Indigenous worldviews" (Hunt, 2014: 27), it also seems necessary to continue to understand the function of these boundary practices, lest they be reasserted through the acceptance of boundaries between spacetime, ontology, epistemology, and phenomenology as the appropriate or necessary boundaries that configure plenitude into allowable parameters of place, being, knowledge, and experience.

Aporia and hiatus

The figure of the aporia has been used in postcolonial and decolonial projects to analyze the production and reproduction of binaries, boundaries, and the limit experience. Writing about Indigeneity and political identity in the United States, Piatote (2013: 47) focuses on "the meaning of aporia as logical impasse, reflecting a key set of paradoxes," in this case, the "impossible knot" that produced a necessary but incompatible relationship between Indian and Agent. For Piatote (2013: 47), the aporetic boundary is represented as a slash, the "and/or" within a binary that "marks the incommensurate nature of both categories." She notes, particularly, the challenges of opening and closing that this boundary-as-limit creates, arguing that "[o]ne of the ways the Indian/Agent aporia functioned was to foreclose other ways of thinking about political agency. Getting outside of the aporia was its own puzzle" (Piatote, 2013: 58–59). This sense of an aporetic boundary as creating a limit that must be exceeded or reconfigured, but that is impossible to progress beyond, is repeated in Lund and Wainwright's (2008) analysis of what they term the aporia of postcolonial geography. Here, concerned with the reproduction of modern, Western, colonial configurations of race and space within a literary work articulating new possibilities of Indigenous Mayan identity in place within Guatemala, the authors caution that

> [j]ust as postcolonial historiography exposes the limits of rethinking the colonial present, so too must postcolonial geography contend with its aporia, which resides in the irreducible challenges that result when attempts to

> liberate the world from geographies of domination play out in and across spaces fashioned in the crucible of colonialism.
>
> (Lund and Wainwright, 2008: 142)

In struggling to understand how colonial logic attempts to overdetermine the possibilities emerging from places that have experienced colonization, the authors point again to the aporia as linked to a pattern whereby open-ended possibilities of transformation produce or reproduce limits and closed spaces (Lund and Wainwright, 2008: 154).

While evocative in their descriptions of the patterns of turn and return that seem to haunt efforts to develop contemporary postcolonial or decolonial analyses of urbanization in settler-colonial contexts, these works offer no real analysis of the aporia itself, no theoretical framework that articulates how and why it generates the patterns of effects that it does, and no reflection on the limits of applying this figure, as associated as it now is with Western post-structuralist thought, into a decolonial setting. As articulated by Derrida (1993), the aporia is the result of a line drawn, a line that both unites and divides the entities it generates as it is drawn. This characterization is consistent with Piatote's (2013: 47) claim that the aporetic "slash mark simultaneously connects the two subjects and reveals the chasm between them" (Piatote, 2013: 47). Yet Derrida proceeds to argue both that the aporetic boundary does not exist as such, but only through its patterns of effects, and that aporetic boundary, as a line that is not a line, is threatened with instability from the moment it is drawn. By implication, an aporetic boundary generates an unstable dynamic: each boundary drawn creates an ideal of unity, yet creates the conditions of instability that require continual acts to resecure the boundary. Foucault (2002: 371) calls this relationship of possibility, necessity, and impossibility "the hiatus between the 'and.'" Where an aporia is popularly figured as a void, gap, or black hole, and by Derrida or Piatote as a line, the figure of the hiatus suggests, instead, an opening, a temporary space, a temporal pause, prior to the act to redefine the boundary, redetermine the limit, and regenerate the dynamic of repetition and return.

The logic of aporetic boundaries, with this specific spatiotemporal dynamic, positions entities not as opposing poles that one might transition between, but as unstable productions of particular boundary acts. The binaries of urbanization (rural/urban, nature/culture, tradition/modern) can be understood as the unstable results of aporetic boundary practices (Tedesco, 2012), and the imbrication of the modern urban project with the modern colonial project suggests that aporetic boundaries are at play in the dynamics of repetition and return that have already been noted. Mignolo's (2007: 455) claim that "[d]elinking in my argument presupposes border thinking or border epistemology in the precise sense that the Western foundation of modernity and of knowledge is on the one hand unavoidable and on the other highly limited and dangerous" encapsulates perfectly how the aporetic boundary appears to reappear in decolonial settings: simultaneously creating, requiring, and inhibiting a border practice of thinking that is necessary, possible (in the sense that it is already existent), but always threatened and always confronting its own conditions of impossibility. Crucially, aporetic boundary practices affect not only the binary boundaries usually presumed, but the boundaries of possible, necessary, or impossible configurations of spacetime, ontology, epistemology, or phenomenology individually, and together as a configuration that states the parameters of vitality or plenitude, and its limits.

The spatiotemporal dynamic of the aporia, with its central and decentralizing figure of the hiatus, further clarifies both how transition narratives take hold, and why they never seem to arrive at the expected end point. Mignolo (2007: 476) suggests that "[t]he problem with the idea of 'transition' is that, once the new appears, the old vanishes out of the present, which is precisely the problem with the rhetoric of modernity for those who are not lucky enough to be in the space where time and history move forward." An aporetic mode of analysis, instead, suggests that the problem with transition narratives is not that the old is exceeded and forgotten, but that the figure of the 'old' is continually reproduced, in spatial terms as now out of place, in temporal terms as in the past, but in logical terms as always present, brought to life again with every repetition of the transition narrative. Thus the logic of aporetic boundaries continually reproduces what is supposed to have been surpassed.

Thus, "we are not dealing with a shifting ontology" (Blaser, 2014: 51) – or spatiotemporal, epistemological, or phenomenological configuration – where shifting presumes a transition from one (the old) to another (the new). However, Blaser's (2014: 51) proposed alternative – the possibility of "dealing with multiple ontologies, a multiplicity of worlds animated in different ways" – continues to be subjected to the processes of repetition and closure that have marked the dominant response to aporetic insecurity. It thus becomes necessary to trace "the work of discourse in creating and sustaining boundaries around what is considered real and, by extension, what is unable to be seen as real (or to be seen at all)" (Hunt 2014: 29), work which proceeds through materializations of these boundaries in built communities, in this case, in settler-colonial urban settings.

In this complicated context, Kelowna becomes a valuable, if counter-intuitive, site of investigation. It is a place that easily satisfies the transition narratives of modern urbanization as social, cultural, economic, and political development, and just as easily satisfies the narrative of a transition from modern to contemporary global urbanization. Against these dominant transition narratives, Kelowna is the site of crucial counter-narratives of exclusions, colonizations, inequalities, and uncertainties. The challenge, restated in more concrete terms, is unsettle the dominant configurations of the content of and boundaries between spatiotemporality, ontology, epistemology, and phenomenology, without being caught within the spatiotemporal dynamic of progress and return, opening and closure, singularity and authenticity, that tend to reinscribe possible reconfigurations within the dominant narratives once again. The range of boundaries that enable the transition narratives of urbanization are, in Kelowna as elsewhere, more uncertain and more contested than the transition narratives themselves allow, and efforts to definitively reconfigure boundaries (whether of urbanization or of spatiotemporality, ontology, epistemology, and phenomenology) risk reproducing the strategies of spatiotemporal authorization that continue to the colonial erasure of Indigenous peoples in this place. Attempts to decolonize contemporary global urbanization, both as complex phenomena and as narrative accounts of these phenomena, must unsettle authoritative claims to the actual and desired path of urban development in this place, as much as any other. By observing how the boundaries operate – how they are drawn, how they are stabilized, and under what conditions they are destabilized – it might be possible to learn to unsettle these narratives and the settler-colonial cities they sustain.

Aporetic boundaries in context

Bernard Avenue revitalization

Bernard Avenue is considered the historic main street of downtown Kelowna. Rapid urban growth in Kelowna, and its pattern of dispersed, sprawling development, has generated constant concerns about the vitality of the downtown core, which has come to look not so much "historic" as just old and tired. Despite glossy contemporary high-rise projects outside the downtown core, the downtown district itself is far from satisfying the image of a "world-class" city, the frequently repeated aspiration of the City of Kelowna. At its westernmost end, Bernard Avenue meets the lakeshore and City Park, and it extends east out of downtown into the residential communities beyond. The intersection of Bernard Avenue and Richter Street functions as the eastern entry, or "gateway," to downtown Kelowna. Walter Gray, Mayor of the City of Kelowna during the project, identifies the central importance of Bernard Avenue to the City, noting that "[f]or me personally having grown up in Kelowna when the city was only 4,700 people and eight square miles, Bernard Avenue was Kelowna." The Bernard Avenue Revitalization project addresses these concerns by promising, in Mayor Gray's words, "an opportunity to re-define what Kelowna is about" (Moore, 2012: n.p.). This attempt to emplace a redefined urban experience in Kelowna responds to the spatiotemporal boundary insecurities produced by contemporary global urbanization and local demands for decolonization. The project demonstrates familiar patterns: the collapse of complexity and uncertainty into the desire to fix the boundaries of the space and time of community; the insecurity of these determinations and the resulting spatiotemporal dynamic of the aporetic logic; the desire for authenticity and identity in the community and the desire for an authority that can keep this stable across space and over time. Precisely because of its repetitions and closures, the Revitalization project is the site of a limit experience. It therefore opens the possibility of locating an aporetic hiatus in process, and the possibility of pausing to encounter, rather than secure, the limit experiences it offers.

This Revitalization project deals with the insecurity of contemporary urbanization by managing space, speed of movement, and activities and relations in the downtown core. It clearly delineates the space of downtown from the outside, through elements such as the "gateway" and distinctive light posts and sidewalk pavings, while concurrently ensuring continued flows of people and vehicles through downtown. Bernard Avenue becomes spatially constructed as a hub in a diverse network, an off-centre centre in the horizontal and vertical extension of the city. Similarly, the Revitalization project defines the time of downtown as now, new, and thus appropriately urban, while ensuring that the natural and cultural "heritage" of the region grounds the downtown transformation. Further, it clearly delineates how people, objects, and vehicles are supposed to occupy and navigate the space and time of Bernard Avenue: it constructs and directs the "travel/traffic zone" of the road; the "furnishings" zone where benches, light posts, and plantings are located; the "passage" zone for easy movement; and the "frontage" zone that segues into the commercial activity of businesses (Kelowna, 2011: 24). The Revitalization project works not simply to reconfigure the material space of downtown Kelowna, but to secure particular spatiotemporal configurations within this location.

But the revitalization of Bernard Avenue also attempts to define a place-based identity for Kelowna and to secure this identity in space and over time. The design standards and public statements by the City explicitly participate in the contemporary urban planning emphasis on development as "place-making," efforts undertaken to improve the capacity for residents to "live, work, and play" in the community. This articulation of place-making is bound up in the perceived patterns of global urbanization, which are supposed to make it necessary for cities, particularly mid-sized cities (Bell and Jayne, 2006) to amplify the specificity and authenticity of their local identity to be competitive at the global level. The design standards draw the connection between its aestheticized built environment and the goals of place-making by arguing that

> a variety of permanent and ephemeral elements contribute to giving places a distinct and memorable identity. . . . Bernard Avenue's identity is rooted in its history and the experiences people share here. . . . above all else, experiences are what make a place memorable.
>
> (Kelowna, 2011: 9)

In order to support this vision of a shared experience of downtown, the redesign emphasizes an increase in spaces for the "public" to congregate. Ultimately, this place-making project represents an effort to materialize a particular experience of Kelowna's historic identity – Kelowna as the idealized British orcharding community – as universally shared by constructing new spaces, and generating new forms, practices, and experiences that can fill this space with appropriate, "authentic" contemporary relations among residents, and between residents and the place they inhabit.

If revitalization needs to satisfy the requirement that artistic expression in public spaces root sensibilities of place, the challenge the City faces is to create an ostensibly "authentic" built environment that confirms Kelowna's tentative urban status, while embedding a uniquely local, natural aesthetic that references dominant conceptions of Kelowna's rural history. Thus the comprehensive design document (Kelowna, 2011) that guides construction and ornamentation states that "[t]he design theme that emerged from the community consultation process was 'landscape inspired' and the execution is contemporary rather than historical in form" (Moore, 2013: n.p.). Bernard Avenue is divided into three zones, each meant to "evok[e] a distinct sense of place and the natural beauty of the area:" an aquatic zone by the lake; an agriculture and cultural history zone in the middle; and a hillsides and grasslands zone by Richter Street (Zielinski, 2013: n.p.). Further, "[s]treet features where chosen to be in keeping with the natural elements and history of the Okanagan, with benches and pavers made of natural stone, and light poles that, like the Downtown library, draw their inspiration from the heavy timbers used in the Kettle Valley Railway trestles" (Moore, 2013: n.p.). The inclusion of 'the rural' within the Bernard Avenue Revitalization project is achieved primarily through these spatialized and temporalized references to a 'natural' history, including agricultural history, of Kelowna.

Mayor of the time, Walter Gray, claims that the revitalization project is an act that redefines the core of Kelowna. He states that through this reconstruction "we are starting to perform a heart transplant here on Bernard" (Moore, 2012: n.p.), and that in the finished product we "will see Kelowna's main street become the centre piece of downtown – the heart of the city" (Moore, 2012: n.p.). Revitalization, then, becomes a figurative attempt to reinvigorate the familiar form of the human; the decline and death of the city centre mirrors the finitude that haunts modern conceptions of the human. Contemporary urban planning is envisioned as offering a route to resecure the spatiotemporality of modern, Western subjectivity: if the individual is insecure through death, transmuting the 'body' into the urban built environment offers the hope of a limitless community. By focusing on the heart, the ability to embody life is configured as centralized.

The redesign standards materialize a centralized attempt to redefine who gets to participate in public spaces, and thus in the "community" of Kelowna.[6] Pat McCormick, city planner for the project, claims that not only are the themes of the revitalization "derived from our local natural and cultural heritage," but they act as "tributes to the First Nation people of this area" (Waters, 2013: n.p.). In the sidewalks, "sporadically placed polished basalt pavers" that feature engravings of local animals and plants (Waters, 2013: n.p.) provide the names in English, Latin, and Nsyilxcen, the language of the Syilx people. The Gateway to downtown provides an even more prominent recognition, with banners produced by WFN artists Janine Lott and Jordan Coble, in partnership with the City of Kelowna, supported by metal posts that say 'Welcome' in English and 'Limlint' ('thank you') in Nsyilxcen (KCN, 2014: n.p.). According to the artist, "[t]he theme of the banners is 'the land is our culture' and celebrates the Okanagan/Syilx people and their connection to the land" (KCN, 2014: n.p.). WFN chief Robert Louis called "the inclusion of First Nations art on the new-look Bernard Avenue in the heart of Kelowna 'very significant'" (KCN, 2014: n.p.). Indeed, whereas the presence of the Syilx people has historically been negated, voided, these efforts at inclusion may suggest complicated attempts to reconfigure relations between the settler community and the Syilx. However, as largely decontextualized uses of Syilx cultural imagery, these inclusions are also consistent with neoliberal place-making and neocolonial urban development. Further, as a City project, these inclusions foster a practical, and mutually beneficial, relationship between the City and WFN, and attempt to solidify Kelowna's claims to authentic identity through a positive relationship, not only with place as an external, objective, singular notion, but with the Syilx people.

Therefore, the Revitalization project continues to mobilize the entangled modern colonial logic of spatiotemporality in its efforts to emplace its limited vision of authentic, shared community: "[t]he design of the street will support . . . the hierarchy of spaces and site elements that provide for authentic experiential programming" (Kelowna, 2011: 9). The glancing reference to 'hierarchies' of spaces refracts the work of this aestheticized and spatiotemporalized intervention into the built environment of Kelowna, making visible the relationship between authenticity and the authority to determine, define, and delineate appropriate or stable boundaries and spatiotemporalized relations. There is an implicit claim to improve the condition of the community by developing bonds of affinity with each other and with shared public places, while continuing to restrict this community to those who identify with this vision of a 'shared' history. The spatiotemporal strategies of the Bernard Avenue Revitalization seek to determine, specify, and authenticate a complicated form of 'place:' one that satisfies the modern, Western, colonial transition narrative of a linear progression from rural heritage to urban future; one that can participate in the spatial and temporal reconfigurations identified with the intensive and extensive transformations of global urbanization; one that can acknowledge its position on Syilx land and its relations with the Syilx community, in the form of WFN. However, for the most part, it is manifestly unwilling and incapable of acknowledging Kelowna as a continued settler-colonial community founded by a violent emplacement on Syilx land and through the violent displacement of Syilx people. Instead of undertaking this form of 'redefinition' of Kelowna as a settler-colonial city, the revitalization project repeats the patterns of modern spatiotemporal and political determinations, specifications, and authentications, patterns that display the aporetic boundaries operative in constituting the problem.

Gateway banners: The land is our culture

A more focused consideration of the Gateway Banners (see Figure 6.1), by Westbank First Nation artist Janine Lott and WFN apprentice artist Jordon Coble, can help to clarify the limits of Kelowna's repeated spatiotemporal construction of political authority and political security, as well as suggest the spatiotemporal openings that this Syilx articulation of community in place offers. For Lott and Coble, the banners exceed the limited frames cast by the City of Kelowna and the Revitalization project. These banners were designed to represent the continued vitality of the Syilx people, despite the myriad violences and ongoing effects of colonization.[7]

The banners incorporate both traditional cultural imagery, particularly the four food chiefs that have guided Syilx relations with the land and pictographs found on Syilx territory, and non-traditional elements that represent contemporary Kelowna. One banner includes imagery of food chief Siyaʔ (Chief Saskatoon Berry), shown as a bush "constructed of hundreds of individual shoots," which represents the "strength that results from being in one place for extended generations . . . as the Syilx have resided on this land for thousands of years."[8] Another banner visualizes Kelowna as seen, across the lake, from the traditional territories

Figure 6.1 "The land is our culture," Gateway Banners. Artist Janine Lott and apprentice Jordan Coble. Bernard Avenue, Kelowna.

Photo by author.

of the Westbank First Nation. Two women look across the lake to a city-scape, identifiable as Kelowna by the presence of the iconic "Sails" statue and fountain that sits at the western end of Bernard Avenue, at the gateway to City Park and the lakeshore. Instead of the bridge, there is a canoe with two paddlers in the lake. While a reference to the traditional crossing place used by the Syilx, it also refracts singular transitions narratives into multiple spatialities and temporalities being cooperative, sending ripples through the lake to both shores. It is not clear which direction the canoe is traveling, and it has not yet completed its passage, which is therefore marked as a pause.

These representations unsettle claims that Western settlement and urbanization define the configuration of politics that determines place-making in this place. Instead, the banners offer a diffraction of the challenges of engaging with the contemporary space-times, forms, categories, and experiences of Kelowna. By asserting an unceded connection between land, language, and people, these banners perform a different form of land claim and suggest other modes of political vitality: they assert a different configuration of the politics of place in space and time.

Thus operative gaps continue to emerge from the efforts of the Bernard Revitalization Project to rebuild the downtown core of Kelowna as the centre of its political community. These gaps at times become literal, as in Mayor Walter Gray being quoted in a local news article "call[ing] the creation of the banners, and their location, another example of how inclusive he feels [blank] has become" (KCN, 2014: n.p.). The gap, the missing word, is present in the published text, and while presumably it is to be filled with the name Kelowna, it reconstructs Kelowna as the site of a void, a hiatus, that can possibly be encountered without panic.

Conclusion – The limits of the aporetic hiatus in the contemporary settler-colonial city

The central problem this chapter engages is the pattern of transition narratives, of turns and returns, which can be reiterated as a problem of borders and boundaries, which can be reiterated as a problem of the logic of spaces of possibility and the encounter with their limits. The analysis of aporetic boundary practices offers an opportunity to engage the figure of the hiatus, as the pause prior to redeterminations, the pause that opens possibilities for multi-temporal and multi-spatial emplacements and displacements. These openings are not necessarily spaces of explicit decolonization: while they hold in abeyance, at least temporarily, the dominant boundary determinations of the modern/colonial/urban logic, these "[l]iminal spaces of opposition are always at risk of appropriation by dominant hegemonies" (Hunt, 2014: 31).

There is a further risk, however, that resides in not taking seriously "the implications of consulting scholarship rooted in Eurocentric thinking in order to address methodological challenges stemming from Eurocentric meta-narratives and dualist ontologies" (Sundberg, 2014: 35). An aporetic mode of analysis does not evade this risk, and as such, it is undertaken in the settler-colonial context of Kelowna tentatively, and with concerns. Its contribution, hesitantly offered, is threefold. First, it offers a theoretically founded caution against the reproduction transition narratives, especially those framed as transitions from Western to Indigenous modes of knowing and being, as the spatiotemporalization of the transition narrative has the tendency to reproduce the aporetic dynamic, where closure reappears as often as openings. Second, insofar as it offers a method of analysis of the patterns of reproduction of boundaries without offering a claim for alternative content for these boundaries, it hopefully contributes to project of opening to what is not known, rather than attempting a mode of "[e]ngagement with Indigeneity [that] involves the establishment of ontological limits around what knowledge is and is not legible – the establishment of boundaries of meaning, the creation of categories, and making them real through their use in geographic knowledge production" (Hunt, 2014: 29). Finally, third, it offers both a support for and a caution over the importance of "the local." As aporetic boundaries do not exist as such, but only through their patterns of affects and effects, the only possible way to engage them, and unsettle the dominant pattern of closures and returns, is to do so in an embodied, emplaced, highly specific context, as this work on Kelowna has sought to demonstrate. However, whether "shifting the geography of reason" (Mignolo, 2009: 172) or exploding "the geographies of theory" (Roy, 2009), the aporetic analysis raises cautions about any presumed unity connecting decoloniality with singularity of place, a presumption that makes it spectacularly difficult, for instance, to include the translocal concerns of migrant temporary workers or global refugees into a decolonial analysis of a place like Kelowna. Here, again, the multi-spatial and multi-temporal places generated through the aporetic hiatus offer figures that may hold valuable openings for decolonial practices, enactments, and embodiments.

Notes

1 For information on the Okanagan/Syilx people, see: www.okanaganfirstpeoples.ca/; www.syilx.org/. Accessed March 17, 2014.
2 Pronounced *N'ha – ah – itk.*
3 The site of the Pandosy Mission, and the broader community that developed around it, only became part of the city boundaries when it was annexed in 1973.
4 A third reserve, #8, is located on the banks of Mission Creek, while two others are located in the hills east of Kelowna.
5 See also RAMA, an anti-racist, anti-colonial migrant justice group in the Okanagan Valley: RAMAokanagan.org.
6 In particular, efforts were made to ensure Bernard Avenue would be more "accessible" for people with visual disabilities.
7 Presentation by Jordan Coble, AlterKnowledge Discussion Event, Alternator Centre for Contemporary Art. November 2014
8 Westbank First Nation (www.wfn.ca/siya.htm).

References

Aguiar, L. and Marten, T. 2011. 'Shimmering White Kelowna and the Examination of Painless White Privilege in the Hinterland of British Columbia', In: A. Baldwin, L. Cameron and A. Kobayashi, eds., *Rethinking the Great White North: Race, Nature, and the Historical Geographies of Whiteness in Canada*. Vancouver: UBC Press, pp. 127–144.

Aguiar, L., McKinnon, A. and Sookraj, D. 2010/2011. 'Repertoires of Racism: Reactions to Jamaicans in the Okanagan Valley', *BC Studies*, vol. 168 (Winter), pp. 65–79.

Aguiar, L., Tomic, P. and Trumper, R. 2005. 'Work Hard, Play Hard: Selling Kelowna, BC, as Year-Round Playground', *The Canadian Geographer/Le Géographe Canadien*, vol. 49, no. 2, pp. 123–139.

Amin, A. and Thrift, N. 2002. *Cities: Reimaging the Urban*. Cambridge: Polity Press.

Armstrong, J. 2010. 'Constructing Indigeneity: Syilx Okanagan Oraliture and tmixwcentrism', Unpublished Dissertation, Greifswald, Germany, Ernst-Moritz-Arndt Universität.

Bell, D. and Jayne, M. 2006. *Small Cities: Urban Experience Beyond the Metropolis*. London: Routledge.

Bennett, J.P. 1998. 'Apple of the Empire: Landscape and Imperial Identity in Turn-of-the-Century British Columbia', *Journal of the Canadian Historical Association/Revue de la Société historique du Canada*, vol. 9, no. 1, pp. 63–92.

Blaser, M. 2014. 'Ontology and Indigeneity: On the Political Ontology of Heterogeneous Assemblages', *Cultural Geography*, vol. 21, no. 1, pp. 49–58.

Blomely, N. 2005. *Unsettling the City: Urban Land and the Politics of Property*. New York: Routledge.

Brenner, N. 2013. 'Theses on Urbanization', *Public Culture*, vol. 25, no. 1, pp. 85–114.

Cameron, E., de Leeuw, S. and Desbiens, C. 2014. 'Indigeneity and Ontology', *Cultural Geography*, vol. 21, no. 1, pp. 19–26.

Cooper, S. 2013. 'Where Did It Go Wrong', *The Province*, 24 February.

Danchev, A. and Lisle, D. 2009. 'Introduction: Art, Politics, Purpose', *Review of International Studies*, vol. 35, no. 4, pp. 775–779.

Derrida, J. 1993. *Aporias: Dying – Awaiting (One Another at) the "Limits of Truth"*. Trans. T. Dutoit. Stanford: Stanford University Press.

Edmonds, P. 2010. 'Unpacking Settler Colonialism's Urban Strategies: Indigenous Peoples in Victoria, British Columbia, and the Transition to a Settler-Colonial City', *Urban History Review/Revue d'histoire urbaine*, vol. 38, no. 2, pp. 4–20.

Foucault, M. 2002. *The Order of Things: An Archeology of the Human Sciences*. London: Routledge Classics.

Freeman, V. 2010. '"Toronto Has No History!" Indigeneity, Settler Colonialism, and Historical Memory in Canada's Largest City', *Urban History Review/Revue d'histoire urbaine*, vol. 38, no. 2, pp. 21–35.

Hessing, M. 2010/2011. 'After the Harvest: Towards a Sustainable Okanagan?' *BC Studies*, vol. 68, pp. 121–134.

Hunt, S. 2014. 'Ontologies of Indigeneity: The Politics of Embodying a Concept', *Cultural Geographies*, vol. 21, no. 1, pp. 27–32.

Isin, E. 1992. *Cities Without Citizens: The Modernity of the City as a Corporation*. Montréal: Black Rose Books.

Isin, Engin F. 2002. *Being Political: Genealogies of Citizenship*. Minneapolis, MN: University of Minnesota Press.

KCN – Kelowna Capital News. 2014. 'First Nation's Art Adorns "Gateway" to Downtown Kelowna', *Kelowna Capital News*, 26 November. www.kelownacapnews.com/news/233548721.html [Accessed 23.04.2014].

Kelowna. 2011. 'Bernard Avenue Revitalization: Design Standards Manual', 23 August. Submitted by Space2Place design.

King, A.D. 1990. *Urbanism, Colonialism, and the World Economy: Cultural and Spatial Foundations of the World Urban System*. London: Routledge.

Lund, J. and Wainwright, J. 2008. 'Miguel Ángel Asturias and the Aporia of Postcolonial Geography', *Interventions*, vol. 10, no. 2, pp. 141–157.

Massey, D. 2006. *For Space*. Los Angeles: Sage.

McFarlane, C. 2011. 'Assemblage and Critical Urbanism', *City*, vol. 15, no. 2, pp. 204–224.

Meagher, S. 2012. 'Unsettling Critical Urban Theory', *City*, vol. 16, no. 4, pp. 476–480.

Mignolo, W. 2007. 'Delinking', *Cultural Studies*, vol. 21, no. 2, pp. 449–514.

Mignolo, W. 2009. 'Epistemic Disobedience, Independent Thought and Decolonial Freedom', *Theory, Culture and Society*, vol. 26, no. 7–8, pp. 159–181.

Moore, W. 2012. 'Bernard: The "Heart" of the Revitalization', *Castanet*. www.castanet.net/news/Bernard-Ave-Revitalization/80114/Bernard-The-heart-of-the-revitalization [Accessed 10.10.2015].

Moore, W. 2013. 'New Gateway Points to Downtown', *Castanet*, 26 November. www.castenet.net/news/Bernard-Ave-Revitalization/103508/New-gateway-points-to-downtown [Accessed 23.04.2014].

Piatote, B. 2013. 'The Indian/Agent Aporia', *The American Indian Quarterly*, vol. 37, no. 3, pp. 45–62.

Robinson, J. 2006. *Ordinary Cities: Between Modernity and Development*. London: Routledge.

Roy, A. 2009. 'The Twenty-First-Century Metropolis: New Geographies of Theory', *Regional Studies*, vol. 43, no. 6, pp. 819–830.

Simone, A. 2011. 'The Surfacing of Urban Life: A Response to Colin McFarlane and Neil Brenner, David Madden and David Wachsmuth', *City*, vol. 15, no. 3–4, pp. 355–364.

Sundberg, J. 2014. 'Decolonizing Posthumanist Geographies', *Cultural Geographies*, vol. 21, no. 1, pp. 33–47.
Tedesco, D. 2012. 'The Urbanization of Politics: Relational Ontologies or Aporetic Practices?' *Alternatives: Global, Local, Political*, vol. 37, no. 4, pp. 331–347.
Tedesco, D. In press. 'Emplacing and displacing whiteness in Kelowna: Aporetic urbanization and the limits of modern politics.' In: L. M. Aguiar and D. Keyes, eds., *Hinterland of Whiteness: White Fantasies in the Okanagan Valley, British Columbia*. Vancouver: University of British Columbia Press.
Veracini, L. 2012. 'Suburbia, Settler Colonialism and the World Turned Inside Out', *Housing, Theory and Society*, vol. 29, no. 4, pp. 339–357.
Vogl, T. 2013. 'Self-Government at Westbank First Nation: Model or Anomaly?' *Public Policy and Governance Review*, vol. 4, no. 2, pp. 105–119.
Waters, A. 2013. 'Kelowna's New Look Bernard Avenue to Have "Themes"', *Kelowna Capital News*, 9 May. www.kelownacapnews.com/news/206850311.html [Accessed 23.04.2014].
Yeoh, B.S.A. 2001. 'Postcolonial Cities', *Progress in Human Geography*, vol. 25, no. 3, pp. 456–468.
Zielinkski, J. 2013. 'Artful Enhancements on Bernard', *Castanet*, 22 July. www.castanet.net/news/Bernard-Ave-Revitalization/95422/Artful-enhancements-on-Bernard [Accessed 10.10.2015].

7 "Well, City Boy Rangoon, it's time to stitch up the evening"

Material, meaning, and Man in the (post)colonial city

Lisa Tilley

Well,
City Boy Rangoon,
it's time to stitch up the evening with beads
Or
the night will drink you up.
(from 'Dressmaker Rangoon' by
Maung Chaw Nwe, 2013,
as translated by Kenneth Wong)

For the Burmese poet Maung Chaw Nwe, Rangoon[1] is a dressmaker. Each morning, the city 'uncovers his sewing machine/ to take a stab at yesterday's order'. The dressmaker pedals the day away making adjustments to material, thinking and rethinking the forms produced as he proceeds to 'cut and stitch', 'trim and bead', and 'line the edges with laces'. The poet's city is both tailor and garment, maker and made, folded together into one plural Rangoon, a city engaged in its own cultivation. Maung Chaw Nwe's urbanity is otherwise innervated with hyper-agential things, such that Rangoon becomes a place where flowers have emotions and rain is 'full of deception.' People appear automated by comparison as they carry their feelings around like luggage: 'At daybreak people arrived/ with their livelihoods and miseries/ cradled on their hips/ or stuffed in head-mounted baskets.' In this playful illustration, the poet's humans are made by Rangoon's urban life as much as they contribute to its making.

But not all of Maung Chaw Nwe's humans are made in the same way. Some are 'splendid [in] outfits with fat belly folds'; they dine on 'roast duck and lamb stew.' Others are harried and depleted, like the frantic porters who run around in 'Japanese slippers with torn backs.' So while some of the poet's humans fit more closely the descriptions of the 'City Boy' as investor, the 'overconsumer' figure of Man[2] (see Wynter, 2003), others, the majority, are both abandoned by, and central to, *homo oeconomicus*, that colonial over-bearer who clothes himself in their 'ragification.' These others are figured outside of the over-representational tropes of the modern, the progressive, the iconic human, but their abandonment is the predicate of the structures that enshrine Man's imperative: 'Struggling to get on/struggling to get ahead/.' Yet, abandonment is also generative of enduring potentialities for living otherwise.

Indeed, how are 'Man' and its matters cultivated in urban life? And how are resistant claims by the Other to be already-human materialised in the city? Through a reading of the human/material of Rangoon, this chapter calls for post-human engagement with the figure of the human as it is outlined in the work of Frantz Fanon, Sylvia Wynter, and cognate scholarship in the sociogenic mode. The basic assertion is that unless we understand how the human has been over-represented in the figure of Man and marginalised as Other-than-Man, then we are far from an understanding of how matter acts in the world's becoming. With reference to urban life in Rangoon, the chapter considers how matter is already formed in a sedimentary way along the Man/Other-than-Man historical split in the human. Drawing on textual sources, as well as fieldwork interviews with urban conservationists, I argue that the co-cultivation of human-material inequalities continues to be shaped according to the figure of Man and its exclusions. Presented here is a reading of the unfolding of meaning and matter in Rangoon from the city's historical moments of foundational colonial planning to the transformative spatial present. Before this, a brief outline of how posthumanism is mobilised by the chapter is followed by an overview of sociogenic thought and its relevance for material approaches.

Unsettling posthuman thought with historico-racial matter

Posthumanism is staged as a further step beyond representationalism where the latter is conceived as the foundational distinction between what is represented and the inherent fallibility within means of representation. Further, posthuman thinking seeks to correct the treatment of language and culture as overdetermining in the making of the world by enacting an analytical shift towards materiality. Work in this vein, like Karen Barad's agential realism, has drawn in the active and mutable qualities of matter, where matter and "humans" are elements in Barad's "intra-active becoming" of the world and where intra-action implies a folding of subject and object into a combined agent engaged in active mattering. Discursive practices become material, enacting tangible qualities in the physical world, and matter is realised in turn as a "congealing of agency" (Barad, 2003: 827).

Posthumanism proposes, then, an "onto-epistem-ology – the study of practices of knowing in being" (Barad, 2003: 829) as a radical break in scholarly thinking. This break, as argued by Zoe Todd (2014), is not new. Multiple genealogies of Indigenous intellectual thought and practice lead to an amalgamation of knowing and being, a folding of subject and object, the enmeshing of the human and the more-than-human, and to an understanding of the more-than-human as agential in the world's becoming. Even so, Felski (2011: 582) speaks of "Latour's idea of the nonhuman actor" as though the nature/culture divide is a natural and universal epistemological feature to be overcome by Latour, rather than a European distortion in the first place. A lack of engagement by many posthuman scholars with Indigenous thought, therefore remains a valid and largely unanswered point of criticism.[3]

A further concern, however, needs also to be advanced. The material reality of the human is already charged with a historico-racial schema, which is often not acknowledged when 'human' relationalities are invoked in posthuman thought. In Barad's terms: "'Humans' do not simply assemble different apparatuses for satisfying particular knowledge projects but are themselves specific local parts of the world's ongoing reconfiguring" (2003: 829). But who are these 'humans' in this ongoing reconfiguring of the world? And how are they differentially involved in the material substantiations of historico-racial schema?

Chakrabarty (2012) has challenged such flattened images of the human and called for engagement with contradictory conceptualisations of humanness. According to his assertion, the rethinking of the human in the context of anthropogenic climate change challenges us to consider the "pure non-ontological agency" of the human as a geophysical force (2012: 13). This pushes related thinking beyond the Enlightenment figure of the human, which is presented as flat and Universalist, as well as beyond the postcolonial figure of the human as a bearer of inscribed difference. Early anticolonial and subaltern studies work centred on the potential for claiming subjecthood (and thus human status) through revolutionary activity. However, the sovereign figure of the human in anticolonial thought was deconstructed by Spivak's and Bhabha's critique of the subject, a critique that Chakrabarty reads as a "deeper turning towards the human" (2012: 4). Chakrabarty argues now for us to deal with these "contradictory figures of the human" in an approach that alternates between conflations and separations of the person and the subject, and between ontological and non-ontological conceptualisations.

Differing somewhat from Chakrabarty's observations, but still with the intention to complicate flat conceptualisations of the human, what follows is intended to push posthumanism towards an engagement with a genealogy of the human traced from Fanon's sociogeny, through to Sylvia Wynter's sociogenic principle, before threading back to Fanon's depiction of the sociogenic materiality of the colonial city. A sociogenic reading of Rangoon is here advanced, wherein the cultivation of matter, Man, and Other-than-Man is explored from the city's colonial past, through to the coloniality of the present, in which claims to be already-human are also made material in urban life. My overarching appeal is for an understanding of how unequal (colonial) social relations become substantiated in material form over time and, conjointly, how the human is shaped within urban life by this specifically historical materiality which, as a product of historico-racial schemas, is fundamentally sociogenic.

Fanon and Wynter on meaning, matter, and Man

Fanon's distillation of nineteenth century thought on the human emphasises two genealogical traditions: first, axiomatic phylogeny (or species evolution), which was later followed by Freud's ontogeny (the development of the psychic and individual organism). Against Freud's isolation of the individual, and nineteenth century biology's determinism, Fanon sought to centre society in his analysis: "Beside phylogeny and ontogeny stands sociogeny" (Fanon, 1967a: 11). In *Black Skin, White Masks* he went beyond the accepted schema of the human's physical experience of the world – what Wynter refers to as its phylogenetic/ontogenetic dimensions (Wynter, 2001: 40; see also, Wynter, 1984) – and towards an understanding of how the human is socialised into a particular mode of sociogeny in which a person comes to experience herself as the defect of the human produced as 'normal'. Fanon's understanding of the human was developed through consideration of both conscious experience and physical processes, where consciousness in Fanon's conception is socially situated, and so differs according to social context. Following this formulation, to biological schema must be added an historico-racial schema if we are to understand the experience of the material self (Wynter, 2001: 41).

In *Black Skin, White Masks*, Fanon's own sense of self is cultivated within the colony in relation to the French sense of self, where the objective is to become more 'human', more like the colonisers. However, Fanon is profoundly altered by his journey to the metropole where the deep subjugation he experiences as part of the colonial social relation is more acute and wherein he becomes fixed in a subhuman status which affects him physiologically. He repeatedly describes his physical nausea upon falling under the white gaze in France. When a little boy repeats the words: "Look, a Negro!" his reaction is altogether bodily: "My body was given back to me sprawled out, distorted, recolored, clad in mourning in that white winter day" (Fanon, 1967a: 113). As Wynter remarks: "Meaning, negatively marked, has likely affected matter (that is physiology) negatively" (Wynter, 2001: 38). A change in meaning across sociocultural situations provokes a biochemical change in matter itself, and the self is experienced in an altered way by the colonised subject as a result. In the words of Scott, "displacement has a hermeneutical function" (2000: 132) due to the altered interpretation of the self upon a change in social and physical contexts.

Wynter takes the sociogenic baton from Fanon by examining the "socio-situationally determined" subjective experience (Wynter, 2001: 36) in which some people come to experience themselves as defects. Her central thesis asserts that each social order has a "culturally prescribed sense of the self or sociogenic principle" which is brought into being over "processes of social conditioning." These senses of self, for Wynter, are situated and "culturally relative modes of being" (Ibid.: 48), but here the material is always already enfolded within her understanding of the cultural. Wynter's concern is with how the human is "made to experience objects in the world" within the terms of that culture's frames of categorisation and perception (Ibid.: 49).

Wynter traces the European construction of the figure of Man and its globalisation as a "supracultural universal" (2001: 43) which comes to function as a disavowal of all other "modes of being human" (Silva, 2015: 91). Following Foucault's timeline of Man, and at the same time complicating it with the naturalisation of the racial colonial order of humanity, Wynter tracks the Renaissance secularisation and Enlightenment transformation of the human through homo politicus to homo oeconomicus to arrive at the present day figure of Man, the investor/overconsumer which "overrepresents itself as if it were the human itself" (Wynter, 2003: 260) and presupposes a necessarily 'dysselected'[4] and inferior human Other, or Other-than-Man.

Although Wynter broadly adopts his understanding of epistemes, she wholly rejects Foucault's antihumanism. Raised in the midst of the Jamaican anticolonial struggle, Wynter aligns instead with other anticolonial thinkers including Fanon and Césaire by maintaining a commitment to a reconstructed humanism: "a dissonant, a non-identitarian, but nonetheless a comprehensive and planetary humanism" (Scott, 2000: 121). This planetary vision of the human is the seam running through her broad canon of work, expressed in its multiple registers. Over and over again she poses the question of why some humans experience negation and as a result are persistently degraded as less-than-human. This, for Wynter, is the failure of humanism, a failure rooted precisely in the collective reproduction and overrepresentation of the ethnoclassed description of Man and the sociogenic principle which governs him and his Others. Outside of Wynter's figure of Man are not only the racialised but also a transracial collection of "the poor, the jobless, the homeless, the underdeveloped" who are persistently reproduced as Other-than-human (Wynter in conversation with Scott, 2000: 201).

If a planetary humanism is inhibited by the overrepresentation of Man, Wynter's work is politically productive because it leaves space for the retrieval of modes of being human beyond this prescribed figure. In this sense, Wynter's overall intellectual struggle comes to be one against the overrepresentation of Man and for a restatement of what it means to be human. This article's assertions on the material overrepresentation of Man and modes of materialising the already-human in urban Rangoon are intended to add to such a retrieval.

With reference to colonial and contemporary Rangoon, in what follows, I take up Wynter's 'socio-situational determination of subjective experience' as definitive in the struggle of modern urban life, wherein Man has become overrepresented in the material form of the mall, the executive office block, sanitised interior spaces, the rational and regulatory grid of reductive urban planning, and so on. Likewise, those represented as Other-than-Man, or as defects against a normalised trope, reclaim and repurpose overrepresentational urban material in restatements of their already-human modes of being. Beginning with this site and its tensions as productive of political potential, Wynter's work enables an understanding of the unfolding of meaning and matter in postcolonial Rangoon, and of how meaning and matter are co-cultivated in urban life in general. Ultimately, Wynter's thinking is an opening for analysis of the shifting epistemic separations in the human which are substantiated materially and in turn reproduced through material agency, but also subverted by already-human actions. One might also read from Fanon's passages on urban materiality and the sociogenic experience just how far he understood the entwining of matter and Man/Other-than-Man in colonial urban becoming.

Urban cultivation and the two towns of Fanon

A return to Fanon's famous passages from *The Wretched of the Earth* shows how he was attuned to the cultivation of urban life and to how the colonial social relation was both materially produced at the same time as being productive of material forms (see also Kipfer, 2011). These passages deserve repeating in full.

> The settler's town is a strongly-built town, all made of stone and steel. It is a brightly-lit town; the streets are covered with asphalt, and the garbage-cans swallow all the leavings, unseen, unknown and hardly thought about. The settler's feet are never visible, except perhaps in the sea; but there you're never close enough to see them. His feet are protected by strong shoes although the streets of his town are clean and even, with no holes or stones. The settler's town is a well-fed town, an easy-going town; its belly is always full of good things. The settler's town is a town of white people, of foreigners.
>
> The town belonging to the colonized people, or at least the native town, the Negro village, the medina, the reservation, is a place of ill fame, peopled by men of ill repute. They are born there, it matters little where or how; they die there, it matters not where or how. It is a world without spaciousness; men live there on top of each other, and their huts are built one on top of the other. The native town is a hungry town, starved of bread, of meat, of shoes, of coal, of light. The native town is a crouching village, a town on its knees, a town wallowing in the mire.
>
> (Fanon, 1967b: 30)

In Fanon's two towns, the colonial social relation had become substantiated in material form: stone; steel; asphalt; refuse; shoe leather, or the absence of shoe leather; the absence of coal, and of light, are all materialities politically implicated in, and given form by, colonial human relations. Yet his two towns are not contained within Saint Pierre on Martinique, or within Algiers; this is Fanon's poetic inscription of the colonial anywhere, the brightly lit town and the world without spaciousness, where material works in co-cultivation with colonial subjugation. In this sense Fanon's two towns are shot through with the sociogeny developed in *Black Skin, White Masks*, and describe the social cartography of cities across the (post)colonial world.

In a more recent depiction of a materially divided city, Bench Ansfield (2015: 124) extends "Wynter's metacartography of Man and les damnés" across urban space, detailing what might be called the co-cultivation of (in)humanness and (un)inhabitability in New Orleans. In such depictions, the matter and meaning of space, bodies, and signs unfold together and the old colonial binaries of torrid and temperate, uninhabitable and habitable, Other-than-Man and Man become mapped onto urban space in a "poetics of purity and taint" (Ansfield, 2015: 131). Taking cues from Fanon, Wynter, and interlocutors including Ansfield, the remainder of this chapter explores the historical and contemporary becoming of Rangoon as a sociogenic co-cultivation of meaning and matter unfolding within the shifting frames of Man/Other-than-Man.

Historicising the spatial present in Rangoon

Moving between the foundational colonial planning of Rangoon and present urbanisation, the following paragraphs explore how meaning becomes sedimented as matter, and vice versa, in the co-cultivation of urban life. This reveals the ways in which the city's becoming has been guided by the (post)colonial social relation between Man and Other-than-Man, in part through the purification of spaces for white colonists and the creation of worlds without spaciousness for labouring Others.

For much of Burma's postcolonial past, until around 2010 when sanctions against successive isolationist regimes were relaxed, authoritarianism insulated the country from many of the transformative economic forces that have altered urban spaces elsewhere in Southeast Asia and beyond (Morley, 2013: 604). As such, Rangoon has been largely preserved as the same Rangoon that Dr William Montgomerie planned in 1852, a little before the Haussmannisation of Paris (on which see Gregory, 1994: 224), and in the same overall mood: wide, rationalised avenues are plotted on a rectilinear grid to resemble what Ansfield (2015: 126) refers to as "disciplined, gridded urbanity". The colonial city, like so many others, is the expression of Enlightenment rationality in built space. Burma's blanket of isolationism began to lift with economic liberalisation in the 1990s, and continued vigorously with the gradual easing of sanctions from 2010 onwards. The preserved administrative buildings of Rangoon's colonial core were also left largely empty following the relocation of the capital to Naypyidaw in 2006 (Morley, 2013: 604), leaving something of a spatial vacuum expected to be filled by construction enabled by Burma's new inflow of investment.

A description of the architectural detail of downtown Rangoon might, therefore, begin by describing this colonial core of densely packed heritage buildings. It might take you on a tour of Rangoon's administrative and merchant-trading architecture, of its octagonal domes and ornate porticos, of its stuccowork, sculptures and rooftop lions, its colonial gin rooms of cool teak and its sweeping double stairways. But this would tell you little about Rangoon, if Calvino was right about the topography and architecture of a place: "[t]he city does not consist of this, but of relationships between the measurements of its space and the events of its past" (Calvino, 1997: 9). If the city's becoming can be understood as the co-cultivation of meaning and material within the frames of its colonial historico-racial schema, then Rangoon consists of relationships between the spatial and the historical in a Calvino-esque way.

Rangoon began as a swamp and many of its post 1852 buildings, such as the High Court with its towering clock and the Central Telegraph Office, were only built after the marshy land below was steadied with thitya hardwood logs and with deep plots of sand and cement (Rooney, 2012: 52, 57). Its architects designed it as an integrated whole within a wider global picture: "Calcutta should be Classic, Bombay Gothic, Madras Saracenic, Rangoon Renaissance" (Ibid.: 57) said James Ransome, Consulting Architect to the Government of India. As such, Rangoon was intended to be a polished ceramic segment set within the mosaic of empire. But, against this imaginary unscripted forces have consistently remade it. Set within a predominantly Buddhist society and under British rule, the dominant narrative goes, the city was viewed as tolerant surroundings for its 2,000 or so Jewish residents who sought refuge from the Ottoman Empire around the late nineteenth century (Goldstein, 2007: 103). Baghdadi Jews rose to positions of power in law and government (Ibid.: 104) and the community left architectural marks, like that in the form of the Sofaer Building on the corner of Merchant and Pansodan.

Rangoon's numerous Armenian residents also left their mark on the 'measurements of its space': the Strand Hotel along the bund of Strand Road, for example, was constructed by the Sarkies Brothers, an Armenian family business. Their hotel was then extended in verse across the globe by Rangoon 1927, Pablo Neruda's poetic representation of a colonial city with deep racialised divisions, embodied and reflected in its architecture: "where a white hotel for whites/ and a gilded pagoda for golden people/ was as much as happened/ and did not happen"[5] (Neruda, 1927). The Strand Hotel is today refurbished and explicitly references a colonial past: its grand white exterior is cool Victorian, and its interior palms and flourishes of teak a reminder of the jungle tamed. It remains more than a little reminiscent of Derek Walcott's Caribbean "Hostería del Mar" in his poem "Tropic Zone" (Walcott, 1986: 497–498):

> If the white architectural mode is
> International Modern, the décor must be the Creole's,
> so, in a terra-cotta lobby with palms, a local jingle
> gurgles of a new cerveza, frost-crusted and golden,
> right next to a mural that has nationalised Eden
> in vehement acrylics.

That the atmosphere of a Rangoon hotel could be echoed so authentically in a poetic representation of a Caribbean establishment suggests that similar images were repeated in urban space across the empire. In this sense, Rangoon, with its 'white hotel for whites' and other material and architectural segregations, becomes not just Rangoon but an allegory of empire, a spatial reification of a racialised hierarchy of power that stretched across the globe.

Writing 15 years or so later than Neruda, urban geographers Spate and Trueblood (1942: 70) also describe a deeply divided city:

> Buildings in the town proper are nearly all "pukka" (made of brick or stone), but few, even in the center, are more than three stories high. Many greatly overcrowded coolie dwellings are found close to the imposing buildings of the commercial core.

The Rangoon of the 1940s is portrayed here as unequal spatially and materially, with the privileged in buildings of 'pukka' stone and casual immigrant workers in low quality accommodation. Rangoon's material form here has been moulded through the historico-racial schema of the colonial project that binds Man to inhabitability, purity, and material comfort and durability, while binding Other-than-Man to uninhabitability in a 'world without spaciousness'. This is all Fanonian in a clear and simple way. But the authors go on to describe the complex layers within the city's divisions at the time (Ibid.: 71):

> Residence in the inner suburbs falls into three differentiated sectors, separated from one another by open spaces. In the east (Theinbyu East circle) are perhaps the most wretched slums of Rangoon, inhabited by the dhobis, or Indian washermen. The circle is very mixed racially. It has the third highest

> death rate and the highest infantile-mortality rate in Rangoon. [. . .] The middle sector (in Theinbyu West) is an area of large but fairly close-built houses, a stronghold of Anglo-India. In the west, the exclusively European Pegu and Gymnkhana Clubs stand in an area of first-class houses in large compounds spaciously laid out along Rosebank Terrace, which also carries Government House. Where this terrace falls steeply to the Ahlone flood plain on the west, the change in economic status and ethnography is immediately apparent.

Spate and Trueblood identified three towns within central Rangoon: one a concentration of non-white poverty; another inhabited by the higher status Anglo-Indian community; and, the third, elevated, spacious and exclusively European. The historical gradation of Other-than-Man appears in material form as the literal hierarchical ordering of space. Overall, writing in the midst of World War II and during the dusk of British imperialism, the authors illustrate the cementing of the colonial order in physical space at the point of imperial maturity. Colonial meaning had become material and in turn, that materiality continued to give meaning to those bodies and selves who were crammed into 'coolie' quarters and other 'worlds without spaciousness' marked by filth and uninhabitability.[6] However, the following section explores how spatial segregation and material inequalities continue to be cultivated today long after colonial agents of urban planning were overthrown in Burma.

Investment capital and the urban habitat of Man

> wherever the Empire has raised the standard of living by blinding high rises, gestures are made to the culture of a remorseful past.
>
> (Walcott, 1986: 497–498)

At this moment in Rangoon's spatial present, the political economic opening of Burma and the relocation of the country's capital to Naypyidaw has allowed for an unprecedented inflow of investment capital. The following paragraphs detail the ways in which the historical colonial ordering of space along the split figure of the human is echoed in the present by the ordering function of inward capital. Investor agency is materialising the overrepresentation of Man in built form in Rangoon, thus provoking the displacement and further marginalisation of those produced as Other-than-Man.

Literature on rapid economic liberalisation elsewhere is instructive here insofar as it speaks of the purification and sanitisation of city streets, and the creation of clean interior spaces – malls, arcades, and the like – out of untamed exteriors (see Harvey, 1985; Pred, 1990). More recently, inward investment has inflated land values in postcolonial urban centres, created bourgeois commercial spaces, and displaced impoverished communities, as Simone (2008: 189) attests in the case of Phnom Penh. In other words, investment capital has long played a role in the cultivation of an urban habitat suitable for Man, or the human as investor/overconsumer. The transformative role of investors is in evidence today in Rangoon as much as it was in Second Empire Paris where speculative investment and competition for land contributed to the socioeconomic division of urban space and the displacement of impoverished communities to the outskirts (Harvey, 1985). Fin-de-siècle Stockholm too (Pred, 1990) saw a similar spatial reordering and, like Paris, its wild exterior streets were tamed into visions of domesticated interiors.[7] Such material alterations under the advance of a capitalist market economy echo the segregations and subjugations performed under colonial rule, as alluded to in the two towns of Fanon.

The patterns of urban change in today's Rangoon recall these grand transformations in many ways. To begin with, the agency of investors is notable through its spatial effects as it serves to segregate, reorder, and displace Rangoon's residents. Most recently, a rush of investment after the end of sanctions has increased the demand for office space by those seeking geographic proximity to Burma's promised market. By the summer of 2013, annual rents in the three high-rise buildings built during the 1990s were reported to be over US$100 per square foot, more than corporate rent in downtown Manhattan at the time (Bloomberg, 2013). Since then, rising selling prices have been tempered somewhat by the introduction of more stringent taxes on high-end property deals introduced in January of 2014, although rents have continued to rise (Irrawaddy, 2014). This sudden hyper-valuing of the ultimate space of Man-as-homo oeconomicus – the sanitised high-rise office – provoked the rapid and ongoing reconfiguring of material and meaning in the city.

A general shortage of space means that companies are renting villas and hotel space to use as offices, and therefore driving rental prices up across all types of property (Bloomberg, 2013). These rising rents result in the displacement of residents to peripheral townships. One family from near the Ahlone Road area, for instance, told the Burmese press that their rent had more than doubled in just one year and as such they were forced to move out (Irrawaddy, 2014). To deal with such cases, the Yangon City Development Committee (YCDC), Rangoon's urban administrative body, is planning further "low-cost housing" on the city's periphery (Ibid.). But through such peripheral developments, communities are effectively segregated as the sprawl of the city's 'new towns' and Rangoon's worsening traffic problems can mean a two-hour journey from the suburbs to the centre (Morley, 2013: 603).

Meanwhile, developers plan to more than triple the amount of available office space in Rangoon within a year (Bloomberg, 2013). In the rush to do so, the demolition of old colonial buildings is much cheaper and easier than renovating them. Large red signs all over downtown indicate plans to demolish buildings judged to be "deteriorating", the residents of which are expected to move on (Morley, 2013: 606). While, in a sense, today's Rangoon continues to tell a global urban story once recounted in Second Empire Paris and fin-de-siècle Stockholm of the divisive and destructive nature of investment capital, on closer inspection, it is evident that investor agency is not total. The hollowing out of Rangoon's colonial core is, at the same time, restrained by various agents, including urban conservationists and Rangoon's existing residents.

Repurposing the urban material of Man

Residents marginalised as Other-than-Man also repurpose urban space in Rangoon, and so subvert its meaning and material. The assertion here is that urban life is partly cultivated by claims to be already-human in the city, that these claims take place in the form of spatial détournements[8] (deturnings) or the repurposing of space in contravention of its intended meanings. Détournements originally involved, in the words of Kristin Ross (1998: 42),

> using the elements or terrain of the dominant social order to one's own ends, for a transformed purpose; integrating actual or past productions into a superior construction of milieu. Détournement has no other place but the place of the other; it plays on imposed terrain and its tactics are determined by the absence of a "proper place."

A decolonising form of repurposing uses the 'terrain' of the (post) colonial social order for a subverted purpose – whether this terrain is composed of broad colonial avenues, palatial sites of power, dominating administrative buildings, or imposing squares intended for nationalistic military displays. Rather than destroying or damaging such architectural symbols of the dominant order, those positioned outside of the frames of Man instead alter the meaning of these symbols through their repurposing, and in doing so alter their material agency.

Urban repurposing has been effected in three main ways in downtown Rangoon by those who are marginalised as Other-than-Man and who perform their daily lives on Rangoon's streets and inside its buildings. These three means may be viewed through the acts of habitation, street activities, and protest. To begin with habitation, the main obstruction to the agency of investors is the ongoing occupation of private colonial buildings by residents who, due to long-term tenancy agreements and established communal life, resist any efforts to move them on. Thant Myint-U, Rangoon resident and historian of Burma whose own genealogy is deeply interwoven with the history of the country[9] explained this in an interview in July of 2013:

> In downtown Yangon [. . .] there are lots of smaller dynamics that no one is really in control of. For instance, if you take an average [. . .] private building, you could have the government controlling part of it, you could have six or seven tenant families who've lived there for a long time and who cannot be displaced. People have legal rights to stay, and because of rent control you could have tenants who are only paying the equivalent of a dollar a month. You could have multiple titles to that property, you could have an Indian family from whom that building was nationalised in 1964 who are now claiming it back, you could have other members of the same family counter-suing, and so all of this has locked everyone together, and nobody is fully in control of that. And this is replicated across the downtown area.

Simply by dwelling in Rangoon's downtown, then, residents positioned outside of the frames of Man not only repurpose and decolonise its colonial buildings, they also resist the next wave of coloniality in the form of transformative investment capital. Similarly Simone's (2008) observations of "city making" in Phnom Penh show how impoverished residents actively configure urban space even in the face of transformative effects of international investment capital, rapid land price inflation, and gentrification.

In the Burmese context, the Ma Ba Tha Buddhist organisation has increasingly encouraged anti-Muslim sentiment as well as political discrimination against Muslims in the country, resulting in the discriminatory 'Race and Religion' laws (see HRW, 2015). Yet a stroll through downtown streets reveals the cheek-by-jowl existence of Rangoon's various devout communities, Buddhists, Muslims, Hindus, Christians, for whom existing and interacting together on a day-to-day basis is a defiance of narratives of communal antipathy along religious lines. However, as investment capital displaces residents from downtown Rangoon, my fear is that increasing moves towards suburbanisation will create the atomisation of religious groups in peripheral satellite communities. And beyond this, spatial divisions along religious or racialised lines[10] may become bases for deeper economic divides as the increasingly open market economy begins to create more pronounced inequalities.

Otherwise, as noted with reference to European cities, spatial accounts of rapid urban change under an accelerating capitalist market economy tend to emphasise the creation of sanitised and controlled interiors out of the undisciplined exterior streets of the city. Yet today in downtown Rangoon a variant picture is apparent. Exteriors are made into interiors but through their re-appropriation by marginalised residents rather than through elite design. Colonial avenues become sites of domestic or informal commercial activity: food preparation, cooking, washing, reading, congregating, and relaxing on low stools under the shade of awnings. Side streets become living rooms and road-side trees become sites for miniature shrines, places of worship. Everyday life is a decolonial repurposing through the activities of the street, performing not only a redefinition of sanitised colonial avenues but also discomfiting incoming investors, as one office-hunting manager of a U.S. engineering firm laments: "We can't move into a space where someone's cooking up nasi goreng [sic] on the sidewalk all night" (Bloomberg, 2013).

But the most significant deturning of the intended meaning of space has taken the form of public protest. Rangoon has been one of the most controlled urban environments in the world with surveillance, barriers, and other spatial techniques employed to discourage political activity (Morley, 2013: 611). These are mainly still in place despite political liberalisation. Yet, the more that successive administrations have attempted to control and neutralise space, the more politicised and symbolic downtown space has become. Sites of national or religious significance have been redefined by political manifestations, and through protest the areas around the main Buddhist pagodas, the Shwedagon and the Sule, have become sacred for a different reason.

The series of protests now known as the '8888 Uprising'[11] had perhaps the greatest redefining impact on Rangoon's public spaces. Demonstrations that took place around August 8, 1988, persisted into mid-September and were only crushed after hundreds of lives were lost. Such protests create sites of symbolism to be revisited: in August of 2013 for instance, the '8888 Uprising' was commemorated with rallies and speeches around the Sule Pagoda (DVB, 2013). Thousands of Burmese including prominent political figures, activists, and Buddhist monks joined the commemoration of the pro-democracy uprisings to remember those killed by the regime. As such, downtown Rangoon, despite its imperial design, decades of autocratic spatial control, and the new agential power of investors, is still defined by the active material and meaning-making agency of those marginalised as Other-than-Man.

Conserving urban material

Alongside the struggles to reconfigure material urban life in Rangoon – played out between investors and the already-humans who repurpose space through their everyday actions – urban conservationists have also begun to actively influence the changing form of the city. The prominent Burmese historian Thant Myint-U has recently become Rangoon's most notable urban conservationist, and during 2012 he officially established the Yangon Heritage Trust (YHT). The Trust is focused on urban planning and the conservation of the city's most iconic sites and buildings, while intending to "offer up our own vision of what the city could be" (Interview with Thant Myint-U, July 2013). If the role of the historian has always been to preserve the past or some version of it through its representation in text then the YHT is best understood as a historian-led endeavour, preserving some version of the city in material form. In this sense, if we consider the textuality of the city itself – its layers of architectural inscription, the language of its street names, past and present, the changing grammar of its built form – the preservation of Rangoon then becomes historical writing in a different mode. As Thant Myint-U recognises: "by protecting and preserving these buildings we could help to recover and recognise the history of this city because [. . .] this is the space in which so much of Burma's history was made" (Ibid.).

The YHT actively seeks to mediate between the transformative agency of investment capital and the repurposing agency of everyday actors in the reconfiguring of the materiality of Rangoon. Thant Myint-U alludes to this positioning.

> [I can now imagine putting] in place the incentives, the policies to attract investment money to renovate these buildings so they can become hotels or restaurants. [Now] I'd like us to come at it from the other direction and [make sure] that people have the choice to stay if they want to stay, and not gentrify the whole area, but that we have actual policies in place that will encourage people to stay, to keep intact a lot of these very old neighbourhoods and communities as well.
>
> (Ibid.)

These words demonstrate an apparent tension between the desire to attract investment capital to rejuvenate Rangoon's decaying buildings and a discomfort over the homogenising effects of investor agency. Spatial/material homogenisation is folded together here with social homogenisation around the overrepresentation of the particular way of being favoured by investors. This discomfort might feasibly be rooted in a sense of social justice, or equally it may be rooted in concerns over the dulling effects of gentrification and the destruction of a city's authenticity in its social core. The idea resonates that 'these very old neighbourhoods and communities' – Other-than-Man spaces and bodies – might be part of the very fabric of Rangoon, part of what makes the city what it is, and therefore imagined as part of Rangoon's future as spectacle.

The YHT mission to preserve Rangoon's colonial core provokes debates over whether the decolonisation of urban space would not simply be best performed through the demolition of old British buildings rather than through their preservation and repurposing. Undoubtedly Rangoon's architecture tells the city's colonial history in an imposing way, and each façade and interior tells a different story. The late colonial City Hall, for instance, constructed in phases between 1925 and 1940, is itself testament to the long struggle for independence. The architectural design was left open to public competition when the building was first conceived in 1913. A European initially won this competition, but once construction began after World War I Rangoon residents and politicians began to demand the stronger emphasis of Burmese architectural features. The celebrated Burmese architect Sithu U Tin drew the final designs for the building, a fusion of neoclassical European features, pyatthat roofing and ornate stone serpents and peacocks (Rooney, 2012: 40). City Hall demonstrates how meaning has become sedimented in material layers, and the frames of the human have been shifted, from the figure of the European coloniser to a more racially inclusive but still restricted figure of Man, through anticolonial struggle and the continual repurposing of urban materiality. The Secretariat too attests to what has been overcome, its southwest wing the site of the assassination of Aung San – the independence leader and father of Aung San Suu Kyi – and six of his colleagues in 1947. Public opinion favours the preservation of the building, but plans to renovate and reopen it as a commercial hotel were inhibited by local resistance (Ibid.: 25).

Another part of Thant Myint-U's endeavour with the YHT is to populate the city with stories, to add to its architectural face those 'measurements of its past'. This has been achieved through placing blue plaques on heritage buildings "so we can tell different stories" (interview, July 2013) and through an online community mapping project to tell the family and community histories of buildings. The project will involve an interactive map to which Rangoon's past and present residents (at least, those privileged enough to have internet access) can upload their own family histories and photographs: "it could be a Gujarati family living in Vancouver, who'd say, we used to live here, this is what we remember of it" (Ibid.). But shaping the textuality of Rangoon and the stories it tells in this way may be seen as an aesthetic endeavour steered by metropolitan elites. It seeks to add the measurements of Rangoon's past, but only those of a selected past which denies the shaping agency of those of Rangoon's residents constructed as Other-than-Man.

Up to the time of writing, the YHT's blue plaque project had commemorated eight buildings, including the Jewish-built Sofaer Building, a number of British-built colonial administration buildings, and the Armenian Church on the corner of Merchant and Bo Aung Kyaw (see YHT, 2015). The particular layer of inscription impressed by the blue plaque project helps to cultivate a narrative of the city as a place written by British colonial agents but informed by tolerant Buddhist cultural influence. The Armenian and Jewish buildings, built by people persecuted elsewhere, are celebrated as material remnants of an era narrated in a false and nostalgic way as characterised by benign coloniality.

But more can be read into the present historical moment in Rangoon by the absences in the blue plaque project. No Muslim-built edifice has, as yet, been commemorated by the YHT. The organisation could choose, for example, from the Surti (Sooratee) Sunni Jamah Masjid otherwise known as the Great Mogul Street Mosque built in 1826 by Indian Muslims, or otherwise the Mogul Shiah Masjid built by Indo-Persians in 1854 (on which see Howard-Moore and Osiri, 2014: 170). Instead, the city's Islamic material text remains entirely absent from the blue plaque trail. This selective commemoration takes place against the present context of acute repression of the Rohingya people[12] and the political marginalisation of the wider Muslim population in Myanmar (see Lee, 2014). Absent the recognition of Muslim agency, a new set of concerns is opened up related to the co-cultivation of urban meaning and material: who is now becoming confined within the shifting frames of Other-than-Man in Rangoon in the spatial present? If Burma continues to politically marginalise its Muslim population the sociogenic material of Rangoon will come to reflect and reproduce this marginalisation.

The designs of colonial planners and contemporary investment agencies have cultivated the overrepresentation of normalised tropes of the human in the figures of the white coloniser through to the investor/overconsumer in Rangoon. This is the sociogenic materiality of the city in which those who experience themselves as defects of the figure of Man are confined to Fanonian worlds without spaciousness or displaced outside of urban life. Materiality gives meaning to bodies in quotidian life such that the overrepresentation of Man also brings Man into being, just as in Maung Chaw Nwe's poem, Rangoon appears as the dressmaker who trims and stitches garments to a pattern.

Conclusion

I have called for posthuman engagement with the figure of the human as advanced in sociogenic work from Fanon to Wynter. If posthuman thought involves the folding of subject and object, human and material, material and text into intra-active combinations engaged in active mattering, then understanding the sedimented material substance of the historically split human is vital. Revisiting Wynter's sociogenic principle reveals the overrepresentation of Man and the marginalisation of Other-than-Man, and the implications of these for the reconfiguration of material and meaning. Fanon had already entwined the material and the sociogenic in his 'two towns' passages in *The Wretched of the Earth*, showing us how the colonial social relation was substantiated in material form, which would act upon, and give meaning to, colonised bodies in the sense of "landscape known by phenotype and phenotype known by landscape" as Bench Ansfield (2015: 127, drawing on McKittrick, 2006: 5) phrased it.

In Rangoon the material sediments of Man/Other-than-Man have been traced here from its foundational colonial planning as a place divided, ordered, and gridded, to the present moment in which investor agency is materialising the overrepresentation of Man in the form of the mall and the office block, and through the marginalisation of Others. Yet investor agency is not total or totalising. Instead, it is confined by the everyday repurposings of Rangoon's Others, who, through dwelling and performing daily life and protest activities, add subverted meanings to matter in the city. Their actions reveal how the material becomes implicated in claims to be already-human.

At the same time, there are those who would preserve Rangoon as an aesthetic equivalent to the historic romantic European capitals, and therefore place it back in colonial time as part of the mosaic of empire. These urban conservationists sometimes ease the endeavour of investors. At other times they align with Rangoon's Others, but in remaking Rangoon out of colonial nostalgia they are also beginning to erase selected narratives from material space.

On the whole, meaning and matter co-cultivate urban life in Rangoon along the shifting boundary lines of Man/Other-than-Man, and in this sense, the City Boy Rangoon of Maung Chaw Nwe's poetic depiction is unsettled and contingent. The already-human constantly make their claims on the meaning and material of the city, always subverting its cultivation. And in many ways, Rangoon tells a wider story about the sociogenic materiality of the global city in which the overrepresentation of the normalised trope of Man is substantiated. This draws attention to the coloniality of city space elsewhere in which the very material of the urban acts to encourage the self to be experienced as defect in the case of those who do not fit the description of Man. The global negation and degradation of Man's Others – Wynter's prime obstruction to the realisation of a planetary vision of the human – is a material problem more broadly experienced in global urban life. Yet, as Rangoon residents show, materialising claims to be already-human offers a means of redescribing the human beyond the figure of Man and disturbs the sociogenic materiality of the city. Global urban life therefore becomes a vital site for performing the retrieval of the planetary human and bringing into being the kind of heretical new humanisms Wynter imagines.

Acknowledgements

I am grateful to Mark Jackson for greatly improving this chapter by way of his thoughtful guidance, and also to Yann Bigant for our many long conversations on urban life in Rangoon.

Notes

1 Although today the city is named Yangon, I mainly refer to 'Rangoon' in this chapter for consistency with the historical and poetic texts included.

2 Here I draw on Sylvia Wynter's (2001) extension of Fanon's sociogenic mode of seeing lived subjective experience as determined by the (post) colonial social relation, in which the sense of self is produced in relation to the dominant/colonial sense of self. Wynter traces the evolution of the Western ideal figure of "Man" through its

secularising incarnations from *homo politicus* to *homo oeconomicus*. Wynter's capital 'M' Man "overrepresents itself as if it were the human itself" (Wynter, 2003: 260) such that those who do not fit its ethnoclassed frames are pressured to experience themselves as less-than-human or Other-than-Man. In its most recent description, 'Man2', the human is confined to the homo oeconomicus figure of the "jobholding Breadwinner, [. . .] masterer of Natural Scarcity (Investor, or capital accumulator)" (Wynter, 2003: 321).

3 With the exception of some, including Todd (2014) and Sundberg (2014).

4 Wynter uses the term 'dysselected' in a Darwinian sense to refer to those who are seen as not "selected by Evolution" and thus outside of the dominant ethnoclass figure of the human (2003: 315).

5 My translation from the original: "en donde un hotel blanco para blancos/ y una pagoda de oro para gente dorada/ era cuanto/ pasaba/ y no pasaba".

6 With some similarities and in the same mosaic of empire, Chattopadhyay (2000: 157) describes nineteenth-century Calcutta as patchworks of spacious white towns alongside more confined black towns. As spatial divisions were fractured and unsettled, boundaries were created to protect British spaces in the form of "compound walls and railings that spoke a calculated language of exclusion".

7 See Gregory, *Geographical Imaginations*, 214–256, for a deeper discussion of the above two works.

8 On which see Debord and Wolman (2007).

9 See his work on Burmese history: Thant Myint-U, *The Making of Modern Burma* (Cambridge: University Press, 2001). And a personal history of the country: Thant Myint-U, *The River of Lost Footsteps* (London: Faber and Faber, 2008).

10 See also Walton (2013) on entrenched Buddhist Burman privilege over other ethnic groups.

11 Amidst protests against military rule and general economic turmoil Burma's military General Ne Win resigned in 1988. After student-led protests were brutally repressed by the military on 8th August of that same year in what became known as the '8888 Uprising', the Burmese people took to the streets en masse to protest against the regime (DVB, 2013). Many of the leaders of the uprising, known collectively as the '88 Generation' went on to form the political party led today by Aung San Suu Kyi, the National League for Democracy (NLD) (Lee, 2014: 330).

12 There are an estimated one million Muslim Rohingya living in Burma's Rakhine State, as they have done for many generations. Although in the immediate post-independence era they had full Burmese citizenship, under successive military juntas from 1962 the Rohingya were reconstructed as illegal Bengali migrants and excluded from citizenship. In 1982 the Rohingya were erased from the government's recognised classifications of ethnic groups (Lee, 2014: 325). During the 2015 elections, the Rohingya were prevented from having voting rights or parliamentary candidates, while the broader trend across Burma is towards discriminatory laws and practices against all Muslim groups (HRW, 2015).

References

Ansfield, B. 2015. 'Still Submerged: The Uninhabitability of Urban Redevelopment', In: K. McKittrick, ed., *Sylvia Wynter: On Being Human as Praxis*. Durham: Duke University Press, pp. 124–141.

Barad, K. 2003. 'Posthumanist Performativity: Toward an Understanding of How Matter Comes to Matter', *Signs*, vol. 28, no. 3, pp. 802–831.

Bloomberg. 2013. 'Yangon More Expensive Than NYC Sparking Boom: Real Estate'. www.bloomberg.com/news/2013-07-29/yangon-more-expensive-than-nyc-sparking-boom-real-estate.html [Accessed 02.08.2013].

Calvino, I. 1997 [1972]. *Invisible Cities*. London: Vintage.

Chakrabarty, D. 2012. 'Postcolonial Studies and the Challenge of Climate Change', *New Literary History*, vol. 43, no. 1, pp. 1–18.

Chattopadhyay, S. 2000. 'The Limits of "White Town" in Colonial Calcutta', *Journal of Architectural Historians*, vol. 59, no. 2, pp. 154–179.

Debord, G. and Wolman, G.J. 2007. 'Methods of Detournement', In: K. Knabb, ed., *Situationist International Anthology*. Berkeley: Bureau of Public Secrets, pp. 8–14.

DVB. 2013. 'Thousands Gather in Rangoon to Mark Anniversary of 8888 Uprising'. www.dvb.no/news/thousands-gather-in-rangoon-to-mark-anniversary-of-8888-uprising/31420 [Accessed 10.08.2015].

Fanon, F. 1967a. *Black Skin, White Masks*. New York: Grove Press.

Fanon, F. 1967b. *The Wretched of the Earth*. Harmondsworth: Penguin.

Felski, R. 2011. 'Context Stinks!' *New Literary History*, vol. 42, no. 4, pp. 573–591.

Goldstein, J. 2007. 'Memory, Place and Displacement in the Formation of Jewish Identity in Rangoon and Surabaya', *Jewish Culture and History*, vol. 9, no. 2–3, pp. 101–113.

Gregory, D. 1994. *Geographical Imaginations*. Oxford: Blackwell.

Harvey, D. 1985. *Consciousness and the Urban Experience*. Oxford: Blackwell.

Howard-Moore, E. and Osuri, N. 2014. 'Urban Forms and Civic Space in Nineteenth- to Early Twentieth-Century Bangkok and Rangoon', *Journal of Urban History*, vol. 40, pp. 158–177.

HRW. 2015. 'What Burma's Elections Mean for the Rohingya'. www.hrw.org/news/2015/10/08/what-burmas-elections-mean-rohingya

Irrawaddy. 2014. 'Rangoon Starts to Control Property Prices, But Rent Is Another Story'. www.irrawaddy.org/z_rangoon/rangoon-starts-control-property-prices-rent-another-story.html [Accessed 09.01.2015].

Kipfer, S. 2011. 'The Times and Spaces of (De-)Colonization: Fanon's Countercolonialism, Then and Now', In: N.C. Gibson, ed., *Living Fanon: Global Perspectives*. New York: Palgrave Macmillan, pp. 93–104.

Lee, R. 2014. 'A Politician, Not an Icon: Aung San Suu Kyi's Silence on Myanmar's Muslim Rohingya', *Islam and Christian – Muslim Relations*, vol. 25, no. 3, pp. 321–333.

Maung Chaw Nwe. 2013. 'Dressmaker Rangoon'. http://kennethwongsf.blogspot.co.uk/2013/10/dressmaker-rangoon-by-maung-chaw-nwe.html [Accessed 30.08.2015].

McKittrick, K. 2006. *Demonic Grounds: Black Women and the Cartographies of Struggle*. Minneapolis: University of Minnesota Press.

Morley, I. 2013. 'Rangoon', *Cities*, vol. 31, pp. 601–614.

Neruda, P. 1927. 'Rangoon 1927'. *Material de Lectura*. www.materialdelectura.unam.mx/index.php?option=com_content&task=view&id=274&Itemid=31&limit=1&limitstart=7 [Accessed 08.01.2014].

Pred, A. 1990. *Making Histories and Constructing Human Geographies*. Colorado: Westview Press.

Rooney, S. 2012. *30 Heritage Buildings of Yangon: Inside the City That Captured Time*. Chicago: Serindia Publications Inc.

Ross, K. 1998. *The Emergence of Social Space: Rimbaud and the Paris Commune*. Minnesota: University of Minnesota Press.

Scott, D. 2000. 'The Re-Enchantment of Humanism: An Interview With Sylvia Wynter', *Small Axe*, vol. 8, pp. 119–207.

Silva, D.F. da. 2015. 'Before Man: Sylvia Wynter's Rewriting of the Modern Episteme', In: K. McKittrick, ed., *Sylvia Wynter: On Being Human as Praxis*. Durham: Duke University Press, pp. 90–105.

Simone, A. 2008. 'The Politics of the Possible: Making Urban Life in Phnom Penh', *Singapore Journal of Tropical Geography*, vol. 29, no. 2, pp. 186–204.

Spate, O.H.K. and Trueblood, L.W. 1942. 'Rangoon: A Study in Urban Geography', *Geographical Review*, vol. 32, no. 1, pp. 56–73.

Sundberg, J. 2014. 'Decolonizing Posthumanist Geographies', *Cultural Geographies*, vol. 21, no. 1, pp. 33–47.

Todd, Z. 2014. 'An Indigenous Feminist's Take on the Ontological Turn: "Ontology" Is Just Another Word for Colonialism'. https://zoeandthecity.wordpress.com/2014/10/24/an-indigenous-feminists-take-on-the-ontological-turn-ontology-is-just-another-word-for-colonialism/ [Accessed 04.11.2015].

Walcott, D. 1986. *Collected Poems 1948–1984*. London: Faber and Faber.

Walton, M.J. 2013. 'The "Wages of Burman-Ness": Ethnicity and Burman Privilege in Contemporary Myanmar', *Journal of Contemporary Asia*, vol. 43, no. 1, pp. 1–27.

Wynter, S. 1984. 'The Ceremony Must Be Found: After Humanism', *Boundary 2*, vol. 12/13, no. 3, 1, pp. 19–70.

Wynter, S. 2001. 'Towards the Sociogenic Principle: Fanon, Identity and the Puzzle of Conscious Experience', In: M.F. Durán-Cogan and A. Gómez-Moriana, eds., *National Identities and Socio-Political Changes in Latin America*. New York: Routledge, pp. 30–66.

Wynter, S. 2003. 'Unsettling the Coloniality of Being/Power/Truth/Freedom: Towards the Human, After Man, Its Overrepresentation – An Argument', *The New Centennial Review*, vol. 3, no. 3, pp. 257–337.

YHT. 2015. 'Blue Plaques Project'. www.yangonheritagetrust.org/blue-plaque-project [Accessed 04.11.2015].

8 Ethno-linguistic cartographies as colonial embodiment in postcolonial Sri Lanka

Chitra Jayathilake

> [W]e have then two great classes: living beings (or substances) and apparatuses.
> And, between these two, as a third class, subjects.
> I call a subject that which results from the relation and,
> so to speak, from the relentless fight between living beings and apparatuses.
>
> —Agamben, 2009:14

In *Wasted Lives: Modernity and Its Outcasts*, Zygmunt Bauman describes the 'population of those who either could not or were not wished to be recognized or allowed to stay' as 'excessive' or 'redundant' – 'wasted humans' (2004: 5). He summarises the 'production of human waste' (Ibid.) as follows:

> To be 'redundant' means to be supernumerary, unneeded, of no use – whatever the needs and uses are that set the standard of usefulness and indispensability. The others do not need you; they can do as well, and better, without you. There is no self-evident reason for your being around and no obvious justification for your claim to the right to stay around. To be declared redundant means to have been disposed of because of being disposable – just like the empty and non-refundable plastic bottle or once-used syringe, an unattractive commodity with no buyers, or a substandard or stained product without use thrown off the assembly line by the quality inspectors.
>
> (2004: 12)

Bauman's metaphors of objectification refer to unwanted populations who show degrees of social uselessness. Non-utility is decided by the non-wasted. This chapter presents a close reading of the Sri Lankan playwright of the English language Ernest Thalayasingam MacIntyre's play-text *Rasanayagam's Last Riot: A Political Fiction for the Theatre* (1990) focussing on how populations are both 'declared redundant' and painted as 'wasted humans' in the ethno-linguistic cartographies of postcolonial Sri Lanka. It shows how ethno-linguistic embodiment cartographically produces biopolitical effects materialised through the diverse apparatuses that make up human agency and subjection.

An apparatus is 'literally anything that has in some way the capacity to capture, orient, determine, intercept, model, control, or secure the gestures, behaviors, opinions or discourses of living beings' (Agamben, 2009: 14). It is not only Michel Foucault's (1995) contrivances such as prisons, panopticons and schools which regulate human beings, but also,

> the pen, writing, literature, philosophy, agriculture, cigarettes, navigation, computers, cellular telephones and – why not – language itself, which is perhaps the most ancient of apparatuses.
>
> (Agamben, 2009: 14)

Agamben explains that apparatuses are 'rooted in the very process of "humanization" that made "humans" out of the animals' (2009: 16). The apparatus, for Agamben, as for Foucault, is not confined to linguistic forms; it refers to non-linguistic means as well, but he emphasises the political power of language in contemporary societies. He posits a comparative and productive process between the pejorative treatment of certain populations and languages. Those produced as waste (i.e. the migrant, the landless, the poor, etc.) are stripped of dignity and claims to human association because of 'the vicious entwining of language, people, and the state' (2000: 67); they are made 'peoples without a state', for instance, 'Palestinians and Jews of the Diaspora' by the language and exercise of the state (Agamben, 2000: 67). They are 'oppressed and exterminated with impunity, so as to make clear that the destiny of a people can only be a state identity, [. . .] the concept of people makes sense only if recodified within the concept of citizenship' (2000: 67–68). What I emphasise here is how language in the Sri Lankan postcolonial contexts mobilised by MacIntyre's play both enforces and resists enforcement through textual and para-textual assemblages.

As implied in the epigraph, Agamben (2009) problematises the way in which the binary between human beings and apparatuses is produced. The status of the human qua human resides in how the interaction of the apparatuses with what Agamben calls zoe or life manifests as legitimate, or as waste. To put it differently, human beings become 'human' as defined by colonial and postcolonial apparatuses through the praxis and actions of the apparatuses themselves, whether these be language apparatuses, or material, or, most likely, imbrications of language and materialities in and through practices.

This chapter explores the ways in which MacIntyre's play portrays the biopolitics of embodiment in Sri Lanka's recent history. It probes the involvement of transnational media, especially the British Broadcasting Corporation (BBC), as dislocated biopolitical apparatuses bound up with Sri Lanka's colonial legacy, and questions a minority English-speaking elite's perspectives on – and detachment from – ethnic tensions with which, largely, they do not dialogue. My analysis problematises received views of biopolitics as typically limited to state-population management, and gestures towards a reading of the play-as-text as a manifestly postcolonial experience and reproduction of the debris of colonialism.

Biopolitical apparatuses exercised through ethno-linguistic cartographies make populations irrelevant; specific people, individuals, emotional anchors are deprived of their political existence or life proper, culminating in undue death. I argue that it is as a result of colonial biopolitics that populations, or more precisely people, in postcolonial territories exacerbate ethno-linguistic differences, which end in turmoil and destruction. Debris legacies of colonial embodiment – the legacies of colonialism in the postcolonial present – imbricate distributed language devices such as aural materialities and tele-media to instantiate the agency of these violent apparatuses. The discussion problematises how biopolitics affects postcolonial human beings, makes them 'subjects' in contemporary socio-political milieux, and how resistance is manifest through creative interventions between human beings and apparatuses. I interrogate the legality of such biopolitical apparatuses, reflecting on the process of colonial embodiment and its diverse forms and aspects exercised through biopolitical apparatuses for future rectification. I aim to open a space to reflect meaningfully on the praxis of biopolitical apparatuses that remain operative in postcolonial nations. With this aim, the discussion ensues with a brief historical background to postcolonial Sri Lankan language policies and a particular event of violence that emerged in 1983; these are necessary to understand better the play.

Historical contexts

Playwright Macintyre's representation of the impact of state and language apparatuses on people foregrounds the relentless conflict that has been specific to the modern politics of postcolonial Sri Lanka. His 1990 English language play *Rasanayagam's Last Riot*, the focus of this chapter, is set in 1983, while also referencing the wider postcolonial era from the early 1950s to 1983. Despite the resolution passed on gaining independence from British colonisation in 1948 that English be replaced by both Sinhala and Tamil as official languages of the nation, what materialised through this Swabhasha (self-language) movement in postcolonial Sri Lanka is what has become a 'Sinhala-only language policy' enacted in the Official Language Act No. 33 of 1956 (see Neil De Votta, 2007, and Imtiyaz and Stavis, 2008). It must be noted here that this Act is commonly referred to as Swabhasha policy. Accordingly, only Sinhala (used by the majority Sinhalese ethnic population) was considered as the official language from the 1950s, until it was reformed later, according to the 1978 Constitution of Sri Lanka and its amendments in 1987.[1]

Although this policy was a political decision implemented in postcolonial Sri Lanka, its history is strictly bound up with colonisation. As Bill Ashcroft argues, '[the] post-colonized nation, that wonderful utopian idea, proved to be a focus of exclusion and division rather than unity; perpetuating the class divisions of the colonial state rather than liberating national subjects' (2009: 12). One effect of the language policy was that it created the structural, state-bound means to exacerbate fissures between the two dominant ethnic groups, Sinhalese and Tamil. As Imtiyaz and Stavis show (2008: 4), 'the contemporary pattern of ethnic relations in Sri Lanka have been largely shaped by its colonial history. Extant problems arose when colonial rulers favoured and allied with a particular group, often a minority, to help in colonial administration'; these created 'fissures' between ethnic groups (see also De Votta, 2007: 77). In the 1840s, British evangelicals compelled the colonial rulers to make Sri Lanka a 'bastion of Christianity'; the proselytisers 'vilified Buddhism and Hinduism' but were 'especially critical of Buddhist monks' (De Votta, 2007: 13). Moreover, aligned with the administration policy of 'divide and rule', British rulers 'disproportionately' provided Government employment to Tamils and supported Christians while leading the Sinhalese Buddhists to mobilise (De Votta, 2007: 14). Evidently, the British administration's policies favoured the Tamil minority and subsequently had a negative impact on the unity between Tamil and Sinhalese populations.

When independence was granted in 1948, Sinhalese political leaders attempted to recoup political and economic influence and extended their power to marginalise Tamils through linguistic means. De Votta summarises the reasons for the ostracism of Tamils and the implementation of 'Sinhala-only language policy' as follows.

> The marginalization of Buddhism during colonialism, Britain's divide-and-rule policies that favoured the Tamil minority, [and] the subsequent over-representation in the civil service, armed forces, universities and professional bodies [. . .] had contributed to the clamour by Sinhalese Buddhist forces for a Sinhala-only policy.
>
> (De Votta, 2007: 17)

The Sinhala-only or Swabhasha policy (1956) can be read as a counter-response to the colonial rulers' administrative policies. The political conditions of ethnic tension are inextricably bound up with the colonial era; in the contemporary Sri Lankan context, Foucault is correct when he writes that '[r]acism first develops with colonization' (2003: 257). For Foucault, racism emerges through explicitly biopolitical strategies of population representation, measurement, and control. As the discursive practices of European colonialism in Sri Lanka were predicated on an ethnicised politics of population management, linguistic cartographies were – and remain – key biological stratagems in terms of which populations were made 'subjects', during – and after – colonisation. As the chapter goes on to show, however, these linguistic cartographies and their resistances are materialised in the physical capacities of specific imagined and real bodies, as well as in how the distributed ecologies of real and imagined embodiment include transnational media, telecommunications, and, indeed, the text, paratext, and staging of MacIntyre's play itself.

Aligned with the Swabhasha policy since 1956, the medium of instruction in almost all educational institutions remained either Sinhala or Tamil.[2] Populations of the two major ethnic groups studied in their own first language and had scarce access to the language of the other, which led to the creation and extension of language barriers. Together with territorial demarcations and concentrations, with the Tamil minority pre-dominant in the country's north and eastern regions, divisive linguistic cartographies were further emphasised by the fact that English, previously the lingua franca of colonial government, was partly considered treasonous and un-patriotic within an explicitly postcolonial context. However, a small minority of upper middle-class elite from both Sinhalese and Tamils groups continued to have access to English education. Language was employed as a tool in violent ethnic conflicts that emerged from colonial legacy. An apposite case in point of overt ethnic violence is 'Black July'.

Black July

Black July was a Sri Lankan anti-Tamil pogrom exercised by Sinhalese mobs during July 1983 in reaction to Liberation Tigers of Tamil Eelam (LTTE) attacks. The LTTE was a separatist militant organisation that claimed an independent state for Tamils in the North East of the country. In Gamini Samaranayake's words, the LTTE was 'the most ferocious guerrilla organization in South Asia' (1997: 109). The impetus for the mob riots was the death of thirteen Sri Lankan Army soldiers killed by the LTTE in Jaffna, and the Government's subsequent plan to stage a state funeral for the deceased in Colombo (see Nira Wickramasinghe, 2006 and Basil Fernando, 2012). As Fernando asserts 'there were many culprits who caused the havoc that virtually destroyed the image of Sri Lanka and which gave justification for a prolonged period of violence' (2012 [n.p.], quoting Cooray, 2002: 60–63). Cooray (2002) has argued that, partly due to the powerlessness of the state and the questionable decisions undertaken by certain politicians, Black July intensified civil war tensions and the country experienced a significant period of chaos.

The state armed forces were outraged by the LTTE's actions and did not want the funeral to be held in Jaffna, the urban locus for LTTE political independence. The Sri Lankan President, influenced by the Army, decided to bury the dead in the main cemetery in Colombo, offering a state funeral while ignoring the objections of influential leaders including the Prime Minister. The standard procedure is to hand over the dead to family members of the deceased for burial or cremation in their home villages according to funeral customs.[3] The state funeral prevented the deceased's family members from deciding on burial locations and ceremonies. Moreover, this LTTE's attack and the opposing views for a state funeral created a space for socio-political opportunists to take advantage of the chaotic situation. As Cooray writes, 'there was an organized crowd [my emphasis] present [in the cemetery] making a huge show of grief' and prior to the arrival of the dead bodies from Jaffna, the crowd was 'ready'. He had 'warned' that the 'drama would end with a riot' (Ibid.). Cooray adds that

> [w]e could do nothing to stop the killing, the destruction. The President made a mistake in putting the Army in charge of restoring law and order. After the killing of the 13 soldiers the mood in the military was a very dangerous one and they were not really motivated in stopping the violence.
>
> (Ibid.)

If, as Imtiyaz and Stavis claim, 'elite political leaders . . . win support and strengthen their positions by mobilizing along ethnic cleavages' (2008: 7), then the actions taken by political elite were directly linked to the fuelling of ethnic tensions between some populations of Sinhalese and Tamil, and, in turn, the foment of divisions by colonial and bio-political legacies. Following Black July, the country experienced a 30-year civil war between the State and the LTTE that lasted from the late 1980s until 2009, with the death of the LTTE's leader. How the riot and eventual state of civil war become visible through a materialised biopolitics of language is explored in MacIntyre's play.

Rasanayagam's Last Riot

Rasanayagam's Last Riot premiered in 1990 in Australia. MacIntyre had migrated to Australia in the early 1970s, yet contributed to Sri Lanka's Anglophone theatre by writing plays about Sri Lankan socio-political issues.[4] Set in the midst of Black July, the play narrativises the brutal violence exercised by Sinhalese mobs against Tamil civilians living in Colombo, and problematises the state's involvement in this brutality. *Rasanayagam's Last Riot* is a thinly fictionalised account of the real events of Black July; MacIntyre asserts that his play is an allegory of the conflict and refers to actual events (1990). The play represents ethnic cleansing, specifically exercised through ethno-linguistic cartographies and biopolitical apparatuses that occur on a national level.

Rasanayagam's Last Riot had several performances and was well-received in Australia (Thangapandian, 2013), yet has been scarcely been staged in Sri Lanka for three reasons. First, performing ethnic violence in a country rife with ethnic violence is inadvisable. Second, as English is neither the first nor second language of the majority of residents, Anglophone plays are neither popular nor profitable. MacIntyre's play has been confined to English-speaking elite in Colombo and largely for educational purposes. Third, the political milieu of a country in negotiations of tense peace is largely unaccepting of critical portrayals of state and biopolitical complicities by exiled writers. Yet, in spite of its contested political resonances, the play remains understudied and begs further analysis.

My analysis shows that the play is a powerful metaphorical transposition of the biopolitical removal of populations, whose identity is tied to a country still dominated by a dislocated, though present, colonial legacy of material coercion. Although aspects of the media apparatus and its transnational legacy remain in postcolonial Sri Lanka, they function covertly; they are identified in this chapter as dislocated and distancing materialities.

Rasa: 'Human waste'

The status of 'human' ascribed, or not, to living beings is decided through the operation of discursive apparatuses. In Agamben's eyes, the biological existence of human beings becomes subject to political decisions and objectification, and is stripped from its political existence through biopolitical apparatuses. Its status is 'bare life' (Agamben, 1998). In this regard, *Rasanayagam's Last Riot* provides literary testimony of biopolitical apparatuses implemented in postcolonial Sri Lanka, but which function as colonial embodiments – apparatuses that deprive human beings of their political and biological existence. The discussion here expounds upon the ways in which Rasanayagam (Rasa), the protagonist of the play, 'a middle-aged Jaffna Tamil living and working in Colombo' (1990: 149), becomes a 'wasted human', a victim of biopolitical exercise and violence of language. I note here the symbolism of naming: in many South Asian cultural grammars, 'Rasa' refers to an aesthetic or 'agreeable quality'; hence, 'Rasa' the character embodies how an aesthetic value or favourable feeling is symbolically rendered waste and redundant via the intervention of biopolitics.

All three characters who appear on-stage in the play belong to the English-speaking, university-educated elite in Colombo. A middle-aged interracial-couple, Sinhalese Philip Fernando and Tamil Sita, attempt to protect their friend Rasa, a Tamil, from Black July (Sita is originally from Colombo, Rasa has moved from the Jaffna district in the North). Philip was Rasa's 'university room-mate in the mid-fifties and his good friend thereafter' (1990: 149). Rasa becomes a victim of a Sinhalese mob bent on retribution for LTTE actions because he refuses to camouflage his Tamil ethnicity. When the play opens, Philip and Sita are seen preparing to emigrate to Australia; they welcome Rasa and provide him with shelter from the violence. Outside their home, mobs are shouting as they, as stated in the play's stage directions, do their 'business of destruction and killing' (1990: 155). The first act contains a critical dialogue between the couple: Sita urges Philip to open up a conversation with Rasa about the ethnic tensions, which is reflected through an argument between the couple over the issue of the Tamil 'problem' in the country.

Philip: [. . .] Is your emphasis on the fact that we are married and have to live in the same home regardless of our positions on the Tamil question or is your emphasis on our rights to maintain our positions on the Tamil question regardless of its effect on our marriage? [original emphasis]

Sita: Putting it that way isn't helpful either, [. . .]

Macintyre writes, in a later reflection on the play, that,

> The allegory becomes very pronounced towards the closing moments of the first act. The audience is conscious, that it is the Sinhalese and Tamil races they see, struggling on stage, to survive in their marriage to each other, within the same island.
>
> (*The Muse*, Canberra as quoted in Macintyre, 1990: 151)

The second act begins with Rasa's arrival at Fernando's (Philip and Sita's) and his description of events outside. Despite Philip's unwillingness to send Rasa to a refugee camp, Rasa, escorted by two policemen, leaves of his own volition. He says:

> I think I go to the refugee camp. From 1956 I have been given safety of your home at every riot. We have enjoyed drinking and talking of old times, till the next riot, and the next riot, and so on and so forth . . . and it became a . . . (he gestures a sense of futility) . . . today I think I'll cross over, to see what it is like on the other side.
>
> (1990: 224)

Rasa's decision to 'cross over' is a denial of personal privileges owing to their friendship and granted by Philip and Sita to him. Rasa's departure is significant, for, as portended in his police escort, he is killed by the mob. His killing is narrated to Sita when Philip, who has gone to the refugee camp to return the briefcase Rasa left, arrives back. Fraught, Philip reveals Rasa's death to Sita,

> Sita listen,
> You listen to me carefully.
> Listen, how Rasa died.
> When I got there, he was already,
> A stiff burning log,
> On the ground.
> Just like a log, burning.
> And in the darkened street,
> There was no one else,
> Only I was there at his funeral pyre,
> No one else,
> His cremators had fled.
>
> And as I turned around to return,
> After the pyre was spent,
> I saw a policeman, weeping, coming towards me.
> He told me it was not their fault,
> He told me they were blameless.
> That as they entered Madangahawatte lane,
> A huge mob confronted them,
> With the BUCKET held in front.
> The policeman walked quickly to Rasa's side,
> And stood confidently,
> Waiting for the question and answer.
> And the mob pointed to the bucket,
> And asked, "What is this"?
>
> Rasa's chest heaved, a big heave,
> And the two policemen thought,
> He was preparing for the password, "BALDIYA"
> At the top of his heave,
> He slowly deflated,
> His head went limp,
> And bowed.
>
> He didn't,
> He failed,
> To use his knowledge.
> The huge crowd went berserk,
> The policemen lost control,
> They clubbed him on the head,
> He fell.
> They poured the petrol,
> They struck the match,

The policeman, weeping, fled,
Back to his station.

He couldn't understand,
Why Rasa had DISHONOURED,
The contract

(1990: 233–234)

I quote this passage at length in order to illustrate not only the poetic power of MacIntyre's language,[5] but also the coercion of one group of language speakers through linguistic discrimination, and how the materiality of the body is addressed as part of the apparatus of language.

Language matters

Language testing was used by mobs during Black July to identify Tamils. Sinhalese mobs tested Tamils' ability to speak in accurate Sinhalese, especially through the pronunciation of 'bucket'. The Sinhala word for 'bucket', 'baldiya' (/bɑːldɪjə/), beginning with the bilabial consonant /b/, was the 'password' for Tamil civilians. Unlike Sinhala, there is no /b/ in the Tamil alphabet. Tamils use the same letter to indicate the voiced bilabial stop and the voiceless stop sounds, and do not phonetically distinguish between the voiced and voiceless consonants, but depending on the position of the consonant, they vary the pronunciation. Although Rasa passed this test by accurately pronouncing it before coming to Philip's home, when Rasa goes to the refugee camp, he chose the Tamil pronunciation despite his ability to pronounce it in accurate Sinhala. In doing so, he deliberately identifies himself as Tamil and is killed. The apparatus is the division enacted around the performative utterance of an embodied linguistic ability and commitment – to choose or to refuse to speak a particular consonant. Tamils employ 'vaali' for the word 'bucket'. When Tamils use 'vaali' in Sinhalese, they pronounce it as 'vaaliya'(/vɑːlɪjə/). Philip later declares that this pronunciation is 'bastardis[ing] language' (1990: 194) and is a metaphor of illegitimacy. Although Tamils and Sinhalese are not confined to one geographical terrain, especially in Colombo, the refugee camp that Rasa intended to move into, indicates a geographical space that draws Rasa back in, linguistically.

One aim of racism is to separate people according to hierarchical population groups. As Foucault writes, hierarchies

> allow power to treat population[s] as a mixture of races, or to be more accurate, to treat the species, to subdivide the species it controls, into the subspecies known, precisely, as races. That is the first function of racism: to fragment, to create caesuras within the biological continuum addressed by biopower.
>
> (2003: 255)

Hierarchical divisions position and divest populations at lower class levels with little political dignity. But Rasa's active refusal to speak the bilabial stop, and his choice to die for it, indicates his political resistance and dignity. Rasa's death, however, provides testimony not only to a division among populations, but also to how biological fragmentation is materialised and resisted through the linguistic diversities present in Tamil and Sinhalese populations. Rasa lets himself be captured by refusing to camouflage his Tamil identity. His choice to identify himself with his language is not unintentional; he is well-aware of its consequences. His choice is made through a realisation that his sense of subjugation is proportionate to the powerlessness of the language which has been his mother tongue for years.

MacIntyre's emphasis on the use of the term 'bucket' is also poignant. A 'bucket' is a vessel, often connoting emptiness as indicated through the phrases such as 'a drop in the bucket'. Buckets are also commonly used to collect waste. This nature between signifier and signified – the bucket and its connotations – is employed in demonstrating the extermination of Rasa, and his ethnic cleansing as a form of 'human waste'. Rasa's redundancy is further materialised as dispersed through the burning of his body; the displacement of his corpse is figuratively presented within the spoken text of the play as a 'stiff burning log'. As his body is removed allegorically from the scene, he is absented in being re-materialised, although he is wrapped up in other forms of embodiment such as fire and the log. Rasa, the person, rather than simply the population as such, becomes the biopolitical victim.

Foucault's emphasis on how biopolitics functions as a form wherein life and death become circumscribed in a means of governance is pertinent. Biopolitics appears in two basic forms, as 'anatomo-politics of the human body' and as stratagems which 'foster[s] life or disallow it to the point of death' (1990: 138–139). The first is centred on the body as a machine and focuses on discipline, practised as apparatuses through confinement of bodies within institutions. The latter does not focus on individuals but populations as 'species' – 'control over relations between the human race, or human beings as species, and their environment, the milieu in which they live' (2003: 245). Foucault further calls this second form 'a "biopolitics" of the human race' (2003: 243), which seeks to regulate populations as a whole, while being controlled by political sovereignty. The state-deployed linguistic Swabhasha policy not only regulated the individuals and 'species', but culminated in torture and death: it becomes biopolitical as power over a group of populations, but whose effect is felt, in the context of the play, also in the specific aesthetic and person of Rasa.

It is necessary here to refer to the dialogue between Philip and Rasa (before Rasa's departure to the refugee camp), to discuss the biopolitical implications of Rasa's death. Philip recalls how Rasa has frequently sought refuge due to the continuous history of ethnic riots in postcolonial Sri Lanka since the 1950s; Philip recites the dates: ''56, '58, '61, '74, '77 '81' (1990: 162). Philip also recounts how Rasa was given protection from a 'gang – of seniors' at the university in the mid-1950s. Sita too recalls that in 1961 she 'had to tell the thug at the gate that there were no Tamils in the house, at the very moment that Rasanayagam was under the bed here' (1990: 157). Rasa rhetorically asks: 'who would have thought that you'll have to tell the same lie for me so many times in our history?' (1990: 157). Evidently, Rasa's body has long been subjected to acts of violence and subjugation. Beginning with an act of ragging (bullying) at the university where violence was inflicted as amusement, his body has intermittently been subjected to mobs' gaze and imposition. Rasa's tragedy – his death and prior subjection to fear since the 1950s – is an example both of perpetrated material and psychological violence on civilians manifested through an embodied, reinforced, and resistant language, and the way in which language functions as a legitimising apparatus to cleanse populations. Rasa's is also a specific, personalised, emotional loss.

Killing populations

Rasa's killing exemplifies the state's complicity in the extra-legal praxis; Rasa is killed by a mob armed with clubs, petrol, and matches in the presence of two policemen. The historian Cooray (2002) reminds us that the state police were powerless during the Black July riots. This is dramatically represented by MacIntyre when the policemen's expectations are shattered as Rasa heaves to utter the password and then fails. The mobs' 'business of destruction and killing' reworks the population into commodities and implies killing as a state condoned industry networked by hierarchical apparatuses and politically profitable.

Qadri Ismail (2005: 218–219) suggests in his analysis of MacIntyre's play that Rasa's choice to open himself to be killed – literally via his deliberate misvoicing – deserves further attention. Although the policeman could not understand '[w]hy Rasa has DISHONOURED' the contract, Rasa's gesture of 'futility' before his departure, suggests that he is ultimately refusing to live in disguise during the riots. In the opening dialogue between Rasa and Philip, Rasa says that 'Sri Lankan Tamil culture belongs in Sri Lanka', not to Tamil Nadu (the state in India, literally meaning 'the land of Tamils'), but 'pressure is applied to that culture by the Sinhalese' (1990: 219). This pressure ultimately induced Rasa to make decisions about being with other 'unprotected' Tamils. Before extending his wishes to the couple for their emigration, Rasa's last words on-stage allude to pressures on the Tamils to disguise themselves: "[w]hat about the people burning on the streets? They took no public political decisions. They didn't even deliberately choose to be Tamils. Many of them must have tried to pass off as Sinhalese" (1990: 225). Rasa chooses suicide in a dramatic gesture of solidarity both with Tamil citizens who have been subjugated and dehumanised, and with the Tamil language, which has been effectively exempted from state franchise.

The mob attack is not a crime committed directly by the state to bring law and order in the country. To identify Tamil residents, mobs found documentation containing the names and addresses of voters, information which is usually kept under the control of the state. Evidently, these mobs received access to such documentation through links with the state. As the play's narrative develops, we see political prisoners in jail attacked and killed, despite the assumption that a prison is typically a place where state security is high. Rasa is killed in the presence of two policemen. These instances illustrate how the mob violence is introduced as a 'business' of extermination, and suggest that mob violence is influenced and supported by powerful political forces. Rasa's death is a function of biopolitics, is enabled by state-sanctioned linguistic regulations, yet, almost ironically, is chosen in deliberate resistance as contingent embodiment.

Human Rights Watch named Black July as 'state sponsored' rioting (Imtiyaz and Stavis, 2008: 9), and the play reflects this judgment, for instance, in the killing of political prisoners. It is also reflected in Cooray's observation of the 'organized' crowd at the cemetery and his warning of a riot. In Rasa's initial description of the mobs before his departure to the refugee camp, he reveals that the crowd gathered at the cemetery, 'seem to be opposed to a state funeral' (1990: 190) because they want 'to hand the [soldiers'] bodies over to the next of kin' (1990: 190). Rasa observes that the 'crowds are violently filling up the graves with sand, to prevent the burials' (1990: 189). Although inspired by the denial of funeral customs, the play alludes to Cooray's reference to the organised crowd and political unscrupulousness. Ironically, Rasa's 'funeral pyre' is attended by Philip alone; even the 'cremators had fled'. All these instances point to a process of corruption whereby biopolitical imaginaries induce forms of legitimised violence implicitly or explicitly condoned by the state hegemonies, and become inserted into cycles of retribution and counter-retribution, yet are felt and narrated as intimate material imbrications: bilabial stops, buckets, burned logs, graves.

The play's epilogue mirrors this dance and counter-dance of death. In a similar fashion to Rasa's death, the epilogue depicts massacres committed by the LTTE against Sinhalese civilians. While Philip and Sita are seen waiting at Singapore airport on their passage to Australia, the news of Sinhalese civilians' deaths are read out across Philip's newspaper: 'Tamil Terrorists kill 150, wound 300, in ATTACK ON DEFENCELESS SINHALESE PEASANTS' (1990: 236). How Sinhalese civilians become targets of the LTTE attacks and oppression, and also susceptible to torture and violence, is evident. Despite Rasa's death, the epilogue is a pointed anti-climax. One may argue that what Macintyre attempts here is to follow the regulations of a well-made play. The epilogue is the playwright's attempt to 'present the "other side" of the conflict, the atrocities committed by the Tamil militants to counteract the play's depiction of the brutality of the Sinhalese during the 1983 riots' (Neluka Silva 2008: 10). But the massacre of Tamil civilians by Sinhalese mobs, and the execution of Sinhalese civilians by the LTTE demonstrate the intensity and extent of violence related to the conditions set in motion by biopolitical imaginaries.

Foucault theorises the killing of populations: he shows how war is waged not only on 'one's adversaries', but also on one's citizens. He argues that 'war is [. . .] not simply a matter of destroying a political adversary, but of destroying the enemy race' (2003: 257). We can argue that the LTTE aims to kill not only the soldiers who confront them in battle, but also to eliminate the enemy as racial category. A comparative link can be drawn between the LTTE and the Sinhalese mobs: each desires the other's elimination.

Distancing materialities

Although the play is about generalised extermination and intimate destruction, Macintyre does not situate dead bodies on-stage. By removing the body from the stage, the materiality of death and other experiences of violence are removed, even from off-stage narrativisation (as in the case of Rasa's body being represented and narrated by Philip as a log). This dematerialisation of the body is taken further in the narrative as death is announced and recounted through distancing material assemblages or apparatuses: the telephone, the BBC news, and the Singapore newspaper.

Cynthia – Sita's friend and wife of Anton, the DIG (Deputy Inspector General of Police) – telephones to give updates on the violence in Colombo, information she holds thanks to her husband's position. It is only through the ringing of the phone followed by Sita's exclamatory responses and revelation to Philip that the audience comes to know of the brutality and material destruction caused by the riot:

> (The phone rings, SITA picks it up)
>
> Yes. . . . . ah. . . . . hr. . . . hr. . . . hr. . . . hr. . . . finished . . . hr. . . . hr. . . . nopolice . . . hr . . . thanks. . . . keep us informed . . . ah. . . . . . ah. . . . thanks. . . . . (Now to Philip) MYSORE CAFÉ – GONE – CYNTHIA! (Like reading telegram).
>
> (1990: 164–165)

As Silva writes, '[t]he increasing intensity of the riots in *Rasanayagam's Last Riot* is mapped through a series of telephone calls, the Fernandos' link with the world outside' (2008: 3–4). MacIntyre relies on narrativisation to demonstrate embodied experiences because both an explicit and direct representation of embodied violence in a moment of war is problematic and difficult to stage, and because the play is a reflection on the apparatuses through which violence is perpetrated, legitimised, and itself staged. State violence is theatre made possible by numerous interdependent and sometimes conflicting discourses.

For instance, through the BBC, all three characters become preoccupied with ethnic violence. The BBC itself functions as a biopolitical trajectory in regulating the three characters emotionally. In the second act, before Rasa's departure to the camp, 'the internationally recognized signature tune of the [BBC] joins the action, sharp and loud, coming from the direction of PHILIP's room' (1990: 214). Philip 'emerges, carrying the radio', and explains how the BBC has explicitly revealed the violence including the killing and massacre of political prisoners in jail by mobs (1990: 214). Rasa's Tamil friend, Dr Rajasunderam, who is currently in jail having been 'arrested under the Prevention of Terrorism Act, but only for leading a demonstration' (1990: 215), is among the dead. This is confirmed through Cynthia's telephone calls. Rasa's gestures too – he 'remains seated, head down' (1990: 221), 'limp in his chair' (1990: 223) – demonstrate his emotional change, desolate and lifeless, presaging his death, and signifying the coercion of the BBC reporting and its power as a biopolitical apparatus to communicate and legitimise categories mobilised in and as violence. Rasa's despairing inquiry – 'how could this have happened inside a government prison' (1990: 222) – foregrounds the complicity of security forces and leads the audience to contemplate the precarity of, and trust placed in, biopolitical apparatuses like the state, law, and security.

The failure and precarity of law is conceptualised by Agamben. In contemporary politics, what he terms a state of exception is employed as the 'dominant paradigm' in maintaining political sovereignty (Agamben, 2005: 2). By referring to the politicisation of life and the logic of sovereignty, 'full powers' alludes to the

'the expansion of the powers of the government', especially the power 'to issue decrees having the force of law' and characterises the state of exception as acting outside of, or in spite of the law for a perceived public good (2005: 5). The mob's attack on the prison, where 'political activists' (1990: 222) like Dr Rajasunderam are incarcerated, provides an example of the 'expansion' and failures, in other words, the precarities, of the state's powers. Following Agamben, people, like Dr Rajasunderam and Rasa, indeed, everyone, is 'included in politics in the form of the exception, that is, as something that is included solely through an exclusion' (1998: 11). Rajasunderam, Rasa, Philip, Sita, the audience – all of us – are products within an apparatus, whether linguistic, economic, political, and aesthetic, that includes and renders potential through its exclusions.

MacIntyre shows the 'sharp' international intervention, as revealed through Philip's criticism of the BBC's explicit reporting: the BBC plays 'a jaunty' tune telling the 'whole world our agonies' by 'hover[ing] high above us, like carrion, [. . .] look[ing] down upon' Sri Lanka when the country is 'being torn apart'. The BBC reports the violence 'wholesale' (1990: 217): 'this kind of publicity is not reporting history, it adds to history by creating further division in the country' (1990: 216). Philip believes that the BBC is 'gloating' and thus making the situation 'worse' (1990: 215), as it is 'not woven into [Sri Lankans'] social fabric': '[o]ur own media is our society' (1990: 217). MacIntyre means this literally. The social is a product of how language is embodied as the distributed body made politic. The BBC's report is an apposite case in point: '[w]ithin one afternoon and part of a night, the physical division of Sri Lanka has occurred in the minds of the people' (1990: 216). The BBC plays a dual role of revealing a truth and making the situation worse through biopolitical constructs implicitly mobilised as information and truth-telling. Media publicise threats and propagate fear; while they focus on relatively rare violent acts, they also largely neglect structural, conditioning aspects of the first world. Violence and terror come to regulate human life in both the emotional and physical dimensions of the lived and represented body. It brutalises people while destabilising and restabilising social orders. Philip states:

> in twenty seven years this is the first time this has taken the lead over all other world news, just shows how the Tamil lobby has grown, it is they who are using the BBC [. . .] it's the international Tamil lobby that's doing all the damage!
>
> (1990: 215–216)

It is typical that narratives of war and genocide have representational limitations because they cannot be omniscient observers. Philip's statement prompts us to ask whether the BBC provides only a one-sided picture of the tension by favouring the 'International Tamil lobby', and complicates the image of a colonial legacy in the country and of its continuing legacies of global coercion. As Ann Laura Stoler (2008) argues,

> [t]o speak of colonial ruination is to trace the fragile and durable substance and signs, the visible and visceral senses in which the effects of empire are

> reactivated and remain. [. . .] It is also a political project that lays waste to certain peoples and places, relations, and things. To think with ruins of empire is to emphasise less the artifacts of empire as dead matter or remnants of a defunct regime than to attend to their reappropriations and strategic and active positioning within the politics of the present.
>
> (2008: 196)

Central to Stoler's (2008) argument is the distinction between ruin and ruination: ruins are the vestiges of imperial pasts; ruination entails 'what people are "left with"' (2008: 194). Ruination is alive and active in the present, and appears both in physical ruins and in mindscapes. Ruination does not end with independence granted by the colonised nations, but may continue to exist violently as implied through the 'vibrantly violent verb' – 'ruin' (2008: 194). Colonial processes of ruination leave their 'material and mental marks' and continue destroying post-colonised nations (2008: 204).

Stoler's reading can be linked to the representation of the BBC news. The play's use of strategies (narrated violence, off-stage sound, telephone conversation, newspaper headlines, etc.) to represent violence through the BBC is a portrayal of the invisible presence of the colonial legacy and the apparatus of its global transnationality today. Colonial processes are immanent within the social and political logics of postcolonial Sri Lanka, and they constitute the regulation of internal political legitimacies in postcolonial nations. The characterisation of the BBC in the play functions to highlight how a transnational biopolitical apparatus linguistically manipulates and affectually regulates human beings emotionally and corporeally. It represents both how political tensions between human life and apparatuses are produced and heightened, and how dislocated transnational grammars fuel pressures between ethnic groups initiated partly due to colonial legacies and linguistic cartographies. The way the three characters are shaken and their actions shaped by the BBC's 'triumphant' reporting of the violence reflects how pervasively mediatised apparatuses intervene as languages for living and dying, and so symbolise how colonial ruins return within postcolonial populations.

English-speaking elite: 'Irrelevant to this country?'

Tensions between human life and colonial apparatuses of language are further rendered visible through the play's representation of Philip, Rasa and Sita as English-speaking elites. English was introduced to Sri Lanka through British colonialism, and the play is, more broadly, a critique of Anglophone middle-class elite attitudes. But it also problematises how, as a colonial language, English marginalises elites from national and postcolonial concerns.

While describing the catastrophic situation of Black July, Rasa, paradoxically, also happily adds: 'I didn't know how long it would be this time, so I put in a few extra bottles [alcohol]. As soon as I got the first wind of this [riots], I rushed to Victoria Stores and got an ample quota' (1990: 189). Rasa's preparation implies that he has been trying to celebrate Black July: both Philip and Rasa have enjoyed the riots in the past, such events represent an opportunity to spend time discussing their university days over drinks during the relative peace of government-imposed curfews. Despite the cataclysm, the riots have been anticipated occasions for these two representatives of ethnic difference. Both Philip and Rasa believe that talk of current ethnic violence is not necessary; theirs is a strong friendship that 'transcends' Sinhala-Tamil politics (1990: 178). Yet their equanimity represents the English-speaking elites' poor grasp of the strength of the ethnic tensions surrounding them; happy meetings during the previous riots and their unwillingness to open a dialogue of ethnic pressures at personal level signifies political and cultural blindness, as well as their ambivalence towards, and detachment from, national issues. English, the preferred mode of class-bound communication, functions as an apparatus that marginalises them from national concerns regarding the ethnicised division of indigenous languages, and their associated linguistic cartographies in postcolonial Sri Lanka. Sita admits:

> I have never taken an interest in the language policy, the colonization schemes, the university admissions system, the employment ratios in the public service, Tamil kingdoms of the past, and the so-called traditional homelands and all that kind of thing. I have no feel for these things.
>
> (1990: 170)

Sita is politically blind to the language issue and the ensuing ethnic violences manifest through language divisions. Her almost wilful blindness is due to the coercive position of English-speakers in regulating governmental attitudes. Such regulation manifests a subtle and affective biopolitical register in shaping embodied and emotional responses to violence, and by those excluded by linguistic elitism and ambivalence. Sita confirms this reading of language as both regulatory and affective when she remarks: '[B]eing of the English speaking middle classes is like being irrelevant to this country . . . I hope it's a false perspective, but the heightening of things has certainly produced that feeling' (1990: 171).

A retired university lecturer of English literature, Sita pours out her inner feelings towards the political scenario: 'the state is in default of its duties to its Tamil citizens, when it expects its citizens to hide their identity inside a bucket, to be able to remain alive' (1990: 230). She opens up a relatively critical dialogue about ethnic tensions, winning over Philip and Rasa's reticence. She manifests disgust at belonging to the English-speaking middle-class (1990: 171) while being married to a Sinhalese, and not belonging to the majority Tamil population: 'I was only a nominal Tamil' (1990: 167). Her emotional and despairing voice resists the violence exerted on the bodies of Tamil populations. As Ismail asserts, Sita happens to be the 'most significant actant' in the play (2005: 212).

Sita becomes proactive about the racial and ethnic tensions embodied by the riot only after she is hurt personally in two particular instances. The first reckoning occurs to her at an Old Girls Association meeting during which she was excluded and ostracised from the discussion about the burning of the Jaffna library because she was married to a Sinhalese man.[6] The second instance occurred when her colleague induces her to question her own inability to use the Tamil language. Sita starts to self-question her linguistic place and marital status only after her exclusion and failure to conform to ethnic and social expectations: she had been politically blind to the ethnic riots and 'national level' (1990: 165) politics until she became personally affected by them. In both these instances, language intervenes as a qualifier of biological life: she crossed the Tamil-Sinhala boundary by marrying a Sinhalese and becomes a political victim of ostracism. Sita's distance from the language she is biopolitically and ethnically bound to renders her a social victim, or even waste. Her link with the colonial language leads her to become 'irrelevant' to the country's politics and national issues. Sita's life is subjugated within the colonial linguistic apparatuses, but in different ways than Rasa's.

Macintyre reflected that '[t]he middle classes is [sic] the class that I know. I could never write of the village people because I am not able to write about them' (as quoted in Athique 2006: 184). He refers specifically to the educated, English-speaking middle-class people living in cities, especially in Colombo. Hence, the play is a critique of the Anglophone elite's attitudes towards ethnic tensions. As Imtiyaz and Stavis write, '[n]either the Sinhala ruling elite nor state institutions openly condemned or took any meaningful immediate measures to prevent the violence against the Tamil civilians from spreading to the other parts of the island from Colombo' (2008: 10).

The ignorance of the Anglophone elite's involvement toward national concerns has led the country to a desperate situation, as indicated through the play's title 'last riot': The multiple connotations of 'last riot' complicate death, and encourage scrutiny of the processes by which living bodies become dead bodies due to the political conditions. This is Rasa's last riot with the Fernando family because they are due to emigrate to Australia; it is also, of course, Rasa's last riot as he is killed. The word 'last' is significant because after this crucial riot when many others died, the country underwent civil war for almost 30 years. This is the beginning of the civil war and the end of minor riots. 'Last' demonstrates that the elite's response to, and production within, linguistic and ethnic division is too late. As Rasa pronounces,

> ([u]ttering his thesis as obliquely as possible) From 1956 it has been slowly coming to this. [. . .] The next stage has to be a war between the Sri Lankan army and the armies of the Tamils. It is the last riot.
>
> (1990: 214)

The title signifies the peak of embodied experience: brutality, violence, massacres, and a future war which engulfed the country. Rasa is a soothsayer for the country. He resembles Cassandra in The Aeneid (Virgil, 1990) as she prophesies the downfall of Troy.

Conclusion

In the last scene, when Philip remarks to Sita that many Sinhalese expect Tamils to 'take an accommodating position', Sita, by '[s]miling warmly at him', responds not to expect it from her. She reiterates that 'there must be something called a Sri Lankan' (1990: 238); she intimates towards unity on a personal, but metaphorically, also a national level. Echoing Sita's optimism about Tamil and Sinhala speakers' unity, Sri Lanka amended its language policy providing the equitable use of Sinhala and Tamil languages in all areas in the country. For instance, as the thirteenth amendment of the Constitution (1987) mandates, the national languages (Sinhalese and Tamil) are used as the medium of instructions in the (state-run) school curriculum: English is taught as a Second Language while Sinhala and Tamil is taught as the 'Second National Language'.[7] Moreover, the sixteenth amendment to the Constitution certified in 1988 made Sinhala and Tamil as the languages of Courts and the languages of Administration, while and an officer is required to pass an examination in the second language before 'promotion' – 'the 2nd Language will be Tamil for a Sinhalese Officer and Sinhala for a Tamil Officer'[8] These provide evidence to the attempts taken to recognise the limitation in the language policies as biopolitical apparatuses in the country and rectify them.

Nonetheless, it is commonly observed, in the 2000s, that a high minority of Sri Lankans are capable of understanding and communicating in both national languages. Sasanka Perera's (2011) research conclusion made in 2010 – '[w]hile government circulars received by this office as a rule come in all three languages and sometimes in two, a great majority of routine communication from government agencies continues to be in Sinhala' [original emphasis] – also implies the extent to which the language policies have been ineffective and the strong negative repercussions of the Swabhasha policy as legacies of colonialism. What is required is a further critical reflection on colonial embodiment in postcolonial territories. This embodiment needs to realise its imbricated agency within material apparatuses that make possible biopolitical and intimate violences, and their resistance.

My critical analysis of the intervention of ethno-linguistic cartographies in regulating human life, as represented in *Rasanayagam's Last Riot*, highlights how, in postcolonial contexts, bodies, as distributed forms of embodiment and dissociation, become enrolled in, and produced by regulatory, colonial apparatuses. Resistances too, of course, also take the form of dispersed yet embodied agencies, however subtle and with whatever devastating consequences: the refusal to speak despite knowing and performing differences in pronunciation; dissociative alliances with and against objects like lips and buckets; colonial ruins and their return as ongoing ruination; and, transnational media forms, including aurally and visually embodied, yet distancing technologies like telephones and newspapers. These, also enhanced and amplified representationally as off-stage narration devices, performatively denote how embodied forms of coercion and colonial ruination signify the 'relentless fight' and constitutive tension between life and apparatus. Distinctions between biological existence (zoe) and the political life (bios) of human beings are produced in exposing bare life to the brutalities qualifies bios. Life becomes wedged between the state of exception and the target of violence; life as de-territorialised force is simultaneously set outside the domain of the political, but nevertheless implicated and subjected to politics as a force to be contained and managed. In Sri Lanka, this status is a result of a biopolitical continuum, particularly experienced by postcolonial subjects since colonisation through state-bound imbrications of linguistic and non-linguistic apparatuses. It is important to articulate new critical postcolonial perceptions of how biopolitical processes transform subjects. Reading the colonial past and representing its ruination in the postcolonial present calls for novel ways to understand the status of the postcolonial, the tensions between apparatuses as productive of human beings, and the resultant reflections on the effects and sites of biopolitics.

Notes

1 The 1978 Constitution of Sri Lanka states that '[t]he Official Language of Sri Lanka shall be Sinhala' and '[t]he National Languages of Sri Lanka shall be Sinhala and Tamil': it adds that '[a] person shall be entitled to be educated through the medium of either of the National Languages'. The thirteenth amendment of this constitution which was certified in 1987 states 'to make Tamil an official language and English the link language'. See 'The Constitution of the Democratic Socialist Republic of Sri Lanka' and 'Amendments to the 1978 Constitution'.

2 The Tamil Language (Special Provisions) Act, certified in 1958, states the following amendments: a Tamil pupil is 'entitled to be instructed through the medium of the Tamil language'; a person educated through Tamil medium is 'entitled to be examined through such medium at any examination for the admission of [. . .] the Public Service, and the Tamil Language shall be made a medium of instructions for University education for those who have been educated through Tamil Language prior to university admission. See 'Sri Lanka's Laws: Tamil Language (Special Provisions) Act', 1958.

3 According to Sri Lankan funeral customs, death ceremonies are highly elaborate and conducted by the family members; whether to bury or cremate the deceased is also usually decided according to traditional family customs.

4 Macintyre's nationality is stated as 'Sri Lankan': his middle name, 'Thalayasingam', signifies his Tamil ethnicity.

5 What is explicit here is MacIntyre's skill in expressing a tense movement of dialogue within the play poetically: this recalls Berthold Brecht's alienation effects (1964).

6 The Jaffna library was burned during communal violence in 1981. Its destruction was a brutal instance of ethnic-biblioclasm.

7 See 'Way Forward of Bilingual Education Programmes in Trilingual Sri Lanka'

8 See 'The Gazete [sic] of the Democratic Socialist Republic of Sri Lanka'

References

Agamben, G. 1998. *Homo Sacer: Sovereign Power and Bare Life*. Trans. D. Heller-Roazen. Stanford: Stanford University Press.

Agamben, G. 2000. *Means Without End: Notes on Politics*. Trans. V. Binetti and C. Casarino. Minneapolis and London: University of Minnesota Press.

Agamben, G. 2005. *State of Exception*. Trans. K. Attell. London and Chicago: University of Chicago Press.

Agamben, G. 2009. *What Is an Apparatus? and Other Essays*. Trans. D. Kishik and S. Pedetella. Stanford: Stanford University Press.

Ashcroft, B. 2009. 'Beyond the Nation: Post-Colonial Hope', *The Journal of the European Association of Studies on Australia*, vol. 1, pp. 12–22.

Athique, T.M. 2006. 'Textual Migrations, South Asian-Australian Fiction', A Thesis Submitted in partial fulfilment of the Requirements of University of Wollongong for the Degree of Doctor of Philosophy, New South Wales, University of Wollongong.

Bauman, Z. 2004. *Wasted Lives: Modernity and Its Outcasts*. Cambridge: Polity Press.

Brecht, B. 1964. *Brecht on Theatre: The Development of an Aesthetic*. ed. and Trans. J. Willett. London: Methuen.

'The Constitution: The Constitution of the Democratic Socialist Republic of Sri Lanka' [Online]. www.priu.gov.lk/Cons/ 1978 Constitution/ Chapter_04_ Amd.html [Accessed 13.08.2015].

'The Constitution: Amendments to the 1978 Constitution' [Online]. www.priu.gov.lk/Cons/ 1978 Constitution.html [Accessed 13.08.2015].

Cooray, B.S. 2002. *President Premadasa and I: Our Story*. Colombo: Dayawansa Jayakody and Company.

De Votta, N. 2007. *Sinhalese Buddhist Nationalist Ideology: Implications for Politics and Conflict Resolution in Sri Lanka*. Washington, DC: East-West Center.

Fernando, B. 2012. 'Sri Lanka: On the 23rd of Black July, One Man Destroyed His Nation', Asian Human Rights Commission [Online]. www.humanrights.asia/ [Accessed 25.07.2012].

Foucault, M. 1995. *Discipline and Punish: The Birth of the Prison*. 2nd ed. Trans. A. Sheridan. New York: Vintage Books.

Foucault, M. 2003. *Society Must Be Defended: Lectures at the College De France 1975–1976*. eds., M. Bertani and A. Fontana and Trans. D. Macey. New York: Picador.

'The Gazete of the Democratic Socialist Republic of Sri Lanka'. (1988) [Online]. www.pubad.gov.lk/web/images/stories/slas/Old_Service_Minite/Sri%20Lanka%20Administrative%20Service.pdf [Accessed 13.08.2015].

Imtiyaz, A.R.M. and Stavis, B. 2008. 'Ethno-Political Conflict in Sri Lanka', *The Journal of Third World Studies*, vol. 25, no. 2, pp. 135–152.

Ismail, Q. 2005. *Abiding by Sri Lanka: On Peace, Place, and Postcoloniality*. Minneapolis: Minnesota University Press.

MacIntyre, E. 1990. 'Rasanayagam's Last Riot: A Political Fiction for the Theatre', In: E.F.C. Ludowyk, ed., *Jaffna and Colombo: A Century of Relationships in Three Plays*. Colombo: Vijitha Yapa Publications, pp. 137–238.

Perera, S. 2011. 'Reflections on Issues of Language in Sri Lanka: Power, Exclusion and Inclusion' [Online]. http://groundviews.org/2011/10/24/reflections-on-issues-of-language-in-sri-lanka-power-exclusion-and-inclusion/ [Accessed 13.08.2015].

Samaranayake, G. 1997. 'Political Violence in Sri Lanka: A Diagnostic Approach', *Terrorism and Political Violence*, vol. 9, no. 2, pp. 99–119.

Silva, N. 2008. 'Ambiguities and Certainties: Ernest MacIntyre's Rasanayagam's Last Riot and He STILL Comes From Jaffna', *Colombo Review*, vol. 1, no. 2, pp. [n.p.].

Stoler, A.L. 2008. 'Imperial Debris: Reflections on Ruins and Ruination', *Cultural Anthropology*, vol. 23, no. 2, pp. 191–219.

Thangapandian, T. 2013. *Island to Island: The voice of Sri Lankan Australian Playwright: Ernest Thalayasingham Macintyre*. Chennai: Emerald Publications.

'Way Forward of Bilingual Education Programmes in Trilingual Sri Lanka' [Online]. www.moe.gov.lk/english/images/subject_related/Bilingual_Education/bilingual_edu_e.pdf [Accessed 13.08.2015].

Wickramasinghe, N. 2006. *Sri Lanka in the Modern Age: A History of Contested Identities*. London: C. Hurst & Co. Publishers Ltd.

9 Immanent comparisons and posthuman perception in the filmic sensorium of Apichatpong Weerasethakul

Carlo Bonura

Syndromes and a Century, the 2006 feature-length film by Thai film director Apichatpong Werasethakul, opens in the office of Dr Toey, a young doctor in a rural hospital. Toey is methodically administering a personality test to a nearly emotionless Dr Nohng, who has just joined the hospital and, it turns out, is averse to the sight of blood. After the test, Toey is called to another part of the hospital. As she leaves the office with an assistant, the camera does not follow them down the open hall, but pans out slowly over a balcony that is visible from Toey's doorway. The camera continues to pan out past the frame of the balcony until the entire screen is consumed by the green of the neighbouring rice field and a strip of jungle in the distance. The greenery serves as the background for the title credits positioned to the left of the screen so as to not entirely obstruct the audience's view of the field.

The titles scene lasts for over two and a half minutes. During this time the conversation between Dr Toey and the assistant remains audible. The scene is long enough, however, for the audience to lose track of what the characters are discussing. In fact, by the end of this long-shot on the field, when the credits are complete, and without any formal end to the scene, the actors begin chatting out of character. They begin discussing the scene itself, and the audible element to the scene concludes with one person complaining: 'it only takes five takes.'

These early moments in the film contain important elements of Apichatpong's cinematic style. The visual focus of the title sequence inaugurates an openness toward nature found across all of Apichatpong's films. In the first half of *Syndromes* the various representations of rural or semi-urban settings consistently incorporate images of the countryside or encounters with plants and flowers. Such images, however, do not present a kind of naturalism, founded on a moral respect for nature. Rather, Apichatpong's films frame the everyday co-inhabitance of both the human and the natural.

This co-inhabitance, however, extends beyond nature to other non-human living and spiritual forms that populate the three films examined in this essay: *Syndromes and a Century*, *Tropical Malady* (2004), and *Uncle Boonmee: Who Could Recall His Past Lives* (2010). The immanence of nature and the perception of the non-human in Apichatpong's films allow for a deeper understanding of one of the primary concerns of this volume, namely 'how human beings are entangled ontological aspects of wider relational and ecological processes'. Yet, in this case, such ontological aspects are not merely mundane. Apichatpong's cinematic vision insists that any understanding of posthuman ontologies must not foreclose the possibilities of the supernatural and the spiritual in 'wider relational and ecological processes.'

Beyond this challenge of how to think the posthuman in non-realist terms, Apichatpong's filmmaking employs complex cinematic structures such as non-linear narratives and slow long shots that confound any representational readings of his films. The title sequence of *Syndromes* with its interval between sound and image appears to 'educate' the senses of its viewers, to borrow an expression used by Henri Bergson (48–49). In the first chapter of *Matter and Memory*, Bergson discusses how human perception attempts to bring together a 'complete image of the object' despite the fact that there may be gaps across a human's senses in the perception of an object (49). Out of these divergent sensations comes the near complete image. As such, it is precisely by dividing sound from image in the title sequence that Apichatpong requires the audience to recognize, to rephrase Bergson, the 'whole of the cinematic object' assembled in *Syndromes*.[1] Thus, what appears in the gap between sound and image is not their 'disconnection' but their relation in a new arrangement. I will argue in this chapter that Apichatpong's cinematic arrangements present audiences with an image of the posthuman that challenges them to perceive the non-human in natural, spiritual, and magical forms.

Apichatpong's films stage, what Laurie Sears has referred to as 'the struggle between disenchantment and enchantment' (2013: 69). Both the characters within his films and their audience must grapple with a sensorium of various enchantments, whether spiritual, natural, fantastic or spectral. As Sears' analysis of the Indonesian literary archive demonstrates, it is wrong to presume, however, that disenchantment is the sole affective orientation available to modern subjects, as if modern disenchantment and the enchantment of the world were discrete opposites. In fact, Walter Mignolo rejects the widely accepted association of modernity with secular disenchantment. He argues that modernity is itself predicated on 'enduring enchantments' (2002: 927–928). By enchantment, Mignolo is referring to the forms of knowledge involved in the production of cultural and civilizational difference that served as one source of colonial power. This 'colonial difference,' which excludes non-Europeans from a claim to history and self-determination and at times ended in genocide, gradually becomes carried over, as an enchantment, into the core structures of modern philosophy, political economy and culture (947).

The form of the binary opposition, as the articulation of colonial difference, is an important element in the epistemic power of such difference. The force of binary oppositions is itself one enduring enchantment of the modern world (927). Although Mignolo is not focused on the excluded status of the non-human in the constitution of colonial difference (accepting that he does discuss the environment more generally), it is important to recognize the power of such oppositions in the divide between the human and the non-human. In examining the origins of the division of nature and culture, Felipe Descola calls into question the commonplace historical understanding of the creation of this division:

> nature has not revealed its essence thanks to the combined efforts of a cohort of great minds and ingenious craftsmen. Rather, it has been constructed little

> by little as an ontological tool of a particular kind, designed to serve as the foundation of the cosmogenesis of modernity.
>
> (63)

The creation of the exclusive categories of nature and the non-human, in part through the knowledge practices of colonial anthropologists, can be understood as an important element of colonial difference. Moreover, as Descola suggests, the nature/culture divide remains a powerfully enduring enchantment of modernity to this day.

The openness toward the non-human in Apichatpong's cinematic vision challenges the exclusiveness of these categories. The diverse enchantments found in his films refuse the conditions of modern knowledge that deny the reality of spirits, the immanence of nature, or communication between humans and non-humans. In this way Aphichatpong's filmmaking can be understood as an instance of what Mignolo describes as 'border thinking' (2000). The three films discussed in this chapter straddle the boundary of the human and the non-human, but they do so in a manner that does not privilege the knowledge and ontology of one over the other. In Mignolo's terms, Apichatpong's filmmaking contains a deep commitment to understanding the non-human as a 'locus of enunciation' of knowledge and ontology that is as legitimate as the enunciation of modern epistemology (2002: 939). This is a critical aspect of border thinking for Mignolo. As such, the films discussed here contain a creative vision of decoloniality, in which the epistemic privilege of modern knowledge and the force of colonial difference is interrupted.

In examining Apichatpong's posthuman images, I will employ Giles Deleuze's discussion of montage found in his two Cinema volumes (1986, 1989). Montage, as a form of a film's composition, is important for Deleuze because the process of composing through combining different elements provides a means to conceptualize the constantly 'divided' nature of film and filmmaking, whether it is related to the assembly of different shots and scenes, or by the very nature of the cinematic image itself – 'cinema' as a neologism from the Greek for 'movement'. The figure of a montage is the whole through which all of these differing parts are brought together in what Delueze refers to as 'a plane of immanence' (1986: 61, 71).

A key feature of the montage at work in Apichatpong's films is the use of juxtaposition in the assembly of cinematic elements. Apichatpong's films often employ bifurcated narrative flows (splitting films into two sections, for instance) resulting in a comparative aesthetic in which the narrative flow of the film leads viewers through a series of open comparisons consisting of imaginative oppositions. This aesthetic is another indication of Apichatpong's cinematic border thinking. Mignolo has identified a 'key configuration of border thinking' as 'thinking from dichotomous concepts rather than ordering the world in dichotomies' (2000: 85).

James Quandt describes the bifurcation at work in *Tropical Malady* and suggests that the two halves of the film:

> seem a study in contrasts, the first, untitled, frequently sunlit, social, talkative, urban; the second, entitled 'A Spirit's Path,' set in deepest, leafy darkness, a

> remote, wordless night isolation populated with bloodthirsty leeches, phosphorescent cattle phantoms, angelic fireflies, a telepathic baboon, trees that speak in electronic gargles, and a tiger that appears to the soldier's destiny, at once his lover and destroyer.
>
> (2009: 64)

These contrasts and comparisons are internal to the film. They are immanent comparisons that often stand in for conventional narrative arcs, as in the juxtaposition of the rural and the urban in Syndromes that forms one of the organizing orientations linking together its two halves. Comparison becomes part of the montage of cinema. It is one technique available when making and thinking about cinema to understand the arrangement of a wide array of parts of the whole of a single film. In this way, comparison is not strictly representational in the more commonplace sense of comparison as a positive register for the representation of identity and difference.

Rather, comparison as a form of montage is grounded in the differential relations at work in montage, in which elements are brought together in non-linear or heterogeneous ways yet remain subject to the aesthetic judgements and interpretations of an audience. The identity of individual elements is dependent upon their relation to other elements and their setting within the whole of the montage. Montage opens up possibilities for new kinds of commensuration of disparate matter as well as the possibility of perceiving difference anew, in this case the differences that constitute the non-human.

My aim is not to reduce our understandings of cinematic composition or montage to a logic of comparison (where all relations of difference are thought of comparatively). Rather, Apichatpong's dualisms and heterogeneous temporalities found throughout demand new styles of analysis of which an expanded and reoriented notion of comparison is just one.

To begin my analysis, I will focus on the immanent comparison of the rural and the urban, together with the implications of this apparently discrete spatial difference found in *Syndromes* and in Benedict Anderson's reading of *Tropical Malady* in his essay 'The Strange Story of a Strange Beast: Receptions in Thailand of Apichatpong Weerasethakul's Sat Pralat' (2009). As suggested previously, however, comparison as a structure internal to film is always an element of montage, which itself can undermine the representational certainty of comparative categories. As such, I will turn to Deleuze's discussion of montage in *Cinema 1* and Apichatpong's use of repetition in the divided structure of *Syndromes*. This will lead me to address the question of perception in Deleuze's analysis of cinema and perception-images in Apichatpong's films. I will conclude that Apichatpong's feature length films add to the 'inventory' of images that Deleuze assembles in his *Cinema* volumes by presenting a posthuman image, or an opening to a perception of the posthuman.

The importance of Apichatpong's cinematic vision, however, is not merely that it develops a novel image in Deleuze's 'image-ontology' not included in the *Cinema* volumes.[2] The posthuman image also raises a question of political ontology insofar as Apichatpong's films deny the presence of only a 'single ontological domain,' namely one grounded in European modernity (Blaser, 2009: 878, 881–882). They accomplish this by relying on a wide variety of sources and frames including vernacular Buddhist imaginaries, the jungle as the site of violence and social memory of the military campaign against the communist insurgency in northeast Thailand in the 1970–1980s, aesthetics of Thai myth, the vitality of nature, and the magical staging of communication between humans and non-humans. Each of these sources opens up a world to viewers that calls into question the 'political moment that created the ontological divide between humans and nature' (De la Cadena, 2010: 352). As such, Apichatpong's audiences are challenged on two levels. First, the use of what I have described as montage in Deleuze's terms and immanent comparisons enables viewers to perceive difference in new ways. Second, the perception of the non-human that as a core element invites audiences to decolonize their perception of the moving-image by allowing new forms of being and being in the world grounded in the countryside of North-east Thailand to enter their field of vision.

The immanent comparison of the rural and the urban

One of the most prominent juxtapositions found throughout Apichatpong's films is the comparison of the rural and the urban. Of the three films I address in this chapter, the difference between the rural and the urban is most pronounced in *Syndromes*, where the first half of the film is organized around life in a rural hospital and its countryside, while the second half focuses on the daily interactions at a Bangkok hospital surrounded by cityscape. The contrast Apichatpong depicts is stark. As described above in the introduction, in the first half the surrounding country plays an important role as a setting for the film. Not only is it nearly always visible from the interior of the hospital, but the main characters picnic and, subsequently, flirt in the countryside. The second half is set in the hyper-urban environment of Bangkok, where the only access to greenery is along the walkways outside the hospital. Here the ambient sound is not that of birds or a breeze as in the first half, but that of the ubiquitous sound of air conditioning.

Part of Apichatpong's explicit critique grounded in this comparison is that the cityscape dominates the imagination of its dwellers. In the second half of the film, a new Dr Nohng appears, this time no longer timid and weak-stomached in the face of blood, but confident and career focused. In a hospital office high above the city, Nohng's lover tries to convince him to move with her to a new city that is being built. To do this she has brought along pictures of the city to show Nohng. They consist of unattractive industrial landscapes, including a petrochemical processor and a coal power plant. Although Nohng's lover describes the city as new, the images are a random collection of aging industrial sites with no indication that they could be central to a new suburban development. Instead the images appear to caricature an urban imagination in which such industrial blight is acceptable and even aesthetically attractive. The images are supplemented at the end of their conversation when his girlfriend adds: 'there is a lake too.' This suggestion prompts Nohng to reply, 'don't try to tempt me with nature.'

A second example that clearly demonstrates the contrast between country and the city appears in one of *Syndromes*' repeated scenes.[3] In this scene, a monk, Sakda, visits Dr Ple for a dental check up. In the rural clinic, the relation between the dentist and the monk is friendly and could easily be read as flirtatious. The two socialize, and afterward Sakda discusses his secret desire to be a DJ and the 'mysterious force' that keeps him in his robes. Dr Ple, himself, admits to being a Thai country singer. He breaks out in song in an attempt to impress the monk. The large clinic room is colourful with brown panelling. Most importantly, the room's windows are open to the outside breeze and the swaying greenery of nearby trees is clearly visible. In contrast, the dentist office in Bangkok, when the scene repeats in the second half of the film, is absolutely clinical. With little engagement with the dentist the monk-patient is visibly anxious in his chair. The only conversation in the scene is between the dentist and his assistant, who did not appear originally. The massive examination room is almost entirely white with patches of grey and completely enclosed. The room is clearly a sterile, not a social, field. It is a setting organized to prevent the discussion of mysterious forces.

It is possible to read this characterization of difference as a crude critique of urban modernity, or an uncomplicated comparison that depends upon basic stereotypes. There appears to be a certain degree of ontological certainty in these characterizations of urban and rural life. Rachel Harrison has suggested that moralistic depictions of urban and rural life are a commonplace in Thai filmmaking after the 1997 financial crisis (2005: 334). In this regard Apichatpong's films explore a dominant motif in Thai film that has grown with the emergence of Thailand's 'new wave.'[4] In her discussion of the 2001 film, *Mon-Rak Transistor*, by Pen-ek Ratanaruang, May Ingawanij describes this motif in recent Thai independent film.

> This [i.e. Bangkok/rural] contrast has above all been articulated in ethical terms: though devices such as characterization and mise-en-scene, the ideal inherent rural goodness, morality, and beauty – the rural pastoral – is configured as a limited form of ethical critique of the metropolis.
>
> (2006: 81)

Both Harrison and Ignawanij identify this depiction of life in metropolitan urban environments as alienating, devoid of nature, dominated by concrete cityscapes, whereas the provincial countryside is more tolerant both of nature and of the spirituality found in daily life. Apichatpong's aim, therefore, is to explore the 'provincial object world, its noises, dialects, music, events, images and people' (Ingawanij and MacDonald, 2006: 122).

Yet, with this exploration comes, in part, a valorization of the provincial. The historical context for this is critical. Since the 1970s, both political elites in Thailand's north and northeast as well as the electorate in these regions have been gradually gaining power and, especially in rural constituencies, recognition in national politics (Pasuk and Baker, 1997; Keyes, 2014). The drastic political polarization that has shaped Thailand's national politics since the early 2000s has involved a reaction against, in particular, the political enfranchisement of rural constituencies in the north and northeast (Walker, 2008; Mills, 2012). The government of deposed Prime Minister Thaksin Shinawatra and subsequent Thaksin-aligned governments have all depended upon rural constituencies for their absolute majorities in parliament.

The most vocal reaction to the newly found parliamentary power of rural constituencies is typically characterized as a metropolitan, Bangkok middle class chauvinism toward a mass of rural poor understood to be undeserving of such power (Pavin, 2014). Hence, the emergence of an internationally acclaimed film director from the Northeast who attempts to 'humanize' rural social relations and openly challenges middle class values can be viewed as politically very meaningful. It is precisely this political role that Benedict Anderson has highlighted in his analysis of the tepid domestic reception of Tropical Malady in 2004.

In his essay 'A Strange Story of a Strange Beast' Anderson suggests that *Tropical Malady* is 'not 'about' the world of the *chao bann* [villagers] of Siam, but rather 'from inside' that world, from inside its culture and its consciousness of itself' (2009: 165). Hence the nonplussed reactions of metropolitan audiences are a result of the naïve, underwhelmed, and at times confused eye of the 'Bangkok person,' in contrast to the knowledgeable and accustomed eye of the chao ban, who has no trouble understanding Apichatpong's complex cinema. Anderson's description cites a political relationship in which aesthetic interpretation reveals broader political dynamics of the Bangkok-centred Thai polity. 'City people of today's air-conditioned Bangkok,' Anderson argues, find *Tropical Malady* ' "difficult" and "mysterious." They are accustomed to films about themselves and their social superiors, with chao baan included only for local colour or comical side-effects.' (166).

In this reading, the descriptions of the city and the countryside in *Syndromes* are not crude, but rather explicitly political characterizations in their attempt to challenge hegemonic conceptualizations of Thainess and political modernity. However, engaging Apichatpong's films at this level, as an inspired defence of rural life, becomes more difficult when the aesthetic and cinematic complexity of his work is considered. Apichatpong's treatment of cultural elements involves a substantial amount of creative assembly. Such cultural crafting taken together with radically non-linear plots makes it difficult, and perhaps inappropriate, to place representative demands on Apichatpong's work.

As such, the vernacular (whether found in his films in the form of dialect, locality, or religious practice) is both a source of the films' 'authenticity' and yet always subject to some kind of creative fabrication, at times involving a point of playfulness. The metaphor Ingawanij uses to describe Apichatpong's bringing together of disparate elements is one of a 'catalogue' of 'references, citations and allusions – an ensemble that presents within itself a diverse array of pre-existing media, texts, myths, stories, rituals, and other communicative practices' (Ingawanij, 2013: 93). The clearest example of this is found in the opening of the second half of *Tropical Malady*, the very source of mystery for Anderson's Bangkok middle class. Without any closure to the first half, Tropical Malady's second half begins abruptly and employs a traditional mural style to animate the textual introduction to the new segment of the film. As Arnika Fuhrmann describes: 'what Apichatpong does here is improvise on a 1950s adventure story an turn it into a

Buddhist – coded folk tale. What is more, his story refigures the conventionally female tiger spirit of Thai folklore as a queer male being' (2016: 150). The result is a mix of appropriated elements from earlier Thai film (with its own reworking of culture) and roughly constituted religious and cultural 'coding.'

Aesthetically, these codings appeal to an array of authenticities of 'rural life,' regional identity in northeast Thailand (Isarn region) and everyday religious belief and practice. Regarding religion, Fuhrmann has emphasized the quotidian Buddhism that appears across Aphichatpong's films encompassing 'a broad spectrum of beliefs and practices that include Theravadin Buddhist orthodox forms as well as magic practices and popular Buddhist conceptions' (35). On the one hand, his films serve to document a cultural complex that has often been excluded from nationalist definitions of Thainess or treated in Thai national popular culture as marginal. On the other hand, Apichatpong's rich use of elements from this complex presents his audience with a cultural montage that includes that fabrication of cultural forms and necessarily complicates any effort to read Aphichatpong's films as purely representational. Thus, the crude differences found in the immanent comparisons of Apichatpong's films cannot be as understood as standing apart from the cultural montage that actively challenges their ontological certainty. The point here is not to challenge the authenticity of Apichatpong's films, but to recognize how the aesthetic of authenticity (or the careful crafting of cultural forms to illicit a recognition of authenticity from an audience) is one element among many employed in Apichatpong's cinematic montage.

Montage and repetition

In his writings on cinema, Deleuze considers montage to be a fundamental element of all film in so far as the moving image itself is made up of a series of static images (in the technical means of creating the movement of images in film); parts of film are arranged to make a whole film. The effects of montage are most readably apparent in arrangements such as the use of segments to organize a film, the use of stories within stories, techniques of framing, and split shots (or parallel editing) in which narrative is accompanied by static images only loosely based on the narration. For Deleuze, however, montage as a concept involves the 'assemblage of movement-images, hence the inter-assemblage of perception-images, affection-images, and action-images' (1986: 72). All film employs a variety of kinds of images, described by Deleuze in the Cinema volumes. Within the scope of *Cinema 1*, most scenes fundamentally convey perception, affect and action. As such, the assemblage of these different images, each enabling the presentation of different visual effects, is itself a montage that makes the overall film possible.

Montage in this case, however, is not simply about the arrangement of elements, but also, to the contrary, it is also about the nature of cinema to be 'continuously dividing between things' (22). Deleuze refers to this aspect of cinema as its 'dividual' nature (16). In fact, upon reflecting on the relationship between the part and the whole in montage, Deleuze emphasizes that it is the arrangement of parts, as well as their dividing into other parts, that leads to a cinematic process that is continuously 'reuniting things into the whole, and the whole is continuously dividing between things (the Dividual)' (22). This 'dividual' aspect, therefore, can be understood as being divided while being part of a whole. The aesthetic effect of montage is based on the way that divided parts contribute to the perception of the whole (as in an example of montage-based art) precisely because they remain divided in the perception of the whole.

For Deleuze, an individual film serves as a whole because it is a 'plane of immanence' through which all kinds of divided elements (story parts, shots, frames, images) are brought together to become what they are to the whole of the film. These elements could not exist independently of the film, and if they did they would exist for a different film altogether. In this sense, the screen provides a form of commensuration:

> the cinematographic image is always dividual. This is because, in the final analysis, the screen as the frame of frames, gives a common standard of measurement to things which do not have one – long shots of countryside and close-ups of the face, an astronomical system and a single drop of water – parts which do not have the same denominator of distance, relief or light. (16)

The varying shots found in films, which capture differing scales and complexities, remain capable of conveying stories and expressing affects despite images that are continuously dividing. The screen enables differing shots and frames to be instantaneously perceptible precisely because they are part of a cinematic whole. The idea of the 'dividual' simultaneously incorporates the constant dividing of images and the fact that their division, even in the case of the most experimental cinema, will always to some degree be made commensurable.

An example of such commensuration appears in the cinematic structure employed in *Syndromes*. The film is divided in its middle into two seemingly discrete halves, one rural and one urban. However, this divide is not a boundary employed to represent the incommensurable difference of the rural and the urban (in a manner akin to Anderson's critique of the Bangkok middle class, where the motifs of the countryside are completely alien to a metropolitan audience). Instead, in a variety of ways the two halves are brought together by common cinematic standards of measurement.

Most importantly, the two halves are repetitive. The openings of the two halves are nearly identical in terms of their dialogue. Despite a different setting, the second half begins with mostly the exact same characters having exactly the same conversations as they had in the first half. This identical element in the two halves, however, soon recedes as the conversations of the second half slowly diverge from the first half. Apichatpong provides no explanation for this within the diagesis of the film. The opening of the second half of *Syndromes* occurs without any formal introduction, and viewers are expected to recognize the nearly identical nature of its opening conversations and scenes without any prompting.

Apichatong's critique of urban life, therefore, utilizes a complex comparative aesthetic in which the difference of the city becomes clear in the context of a cinematic repetition. Although both halves are discrete in their depiction of difference (of the urban and the rural), the halves are based upon a virtual repetition of scene and dialogue in contrast to the actual substance of the environment and the changes in personality found in each setting. This play of repetition and differentiation as the primary organization for the film demonstrates Deleuze's notion of commensuration as a principle of cinema. The repetition here is dividual as it reunites the different halves of the film.

The measured pace of the scenes also demonstrates how the two halves are not entirely discrete, but rather part of a single 'plane of immanence.' The pace of scenes, and Apichatpong's continuous use of slow tracking shots, remains constant across the two halves. Life does not speed up in the metropole, as could have been one option for representing a difference across these two settings. Rather, the camera moves at exactly the same pace in each half. The juxtaposition is maintained without a change in cinematic style. Moreover, the openness of Apichatpong's shots to ambient sounds and colours also remains in the second half as in the first. However, these are not the sounds and colours of a natural, living outside, but the sounds and colours of a sterile field, whether of the modern clinic or, more metaphorically, of the city itself with its soulless human interactions.

There is another aspect to this openness at the core of Apichatpong's cinematic style. Whereas the common measured pace (a common temporality) serves to bring the bifurcated film together as a whole (as a montage), it is also related to the form of cinematic perception that Apichatpong develops in his work. In his films the camera enables a vision of the movement of nature, the atmosphere and of the non-human itself. To clarify the nature of this perception and its importance, it is once again fruitful to return to Deleuze's treatment of montage and the fundamental importance of perception to cinema.

Perception images in Apichatpong's films

Deleuze's philosophical commentary on cinema recognizes the importance of the role perception plays in the structure of cinema. In his engagement with the early development of cinematic principles and theory found in *Cinema 1*, Deleuze identifies perception as one of three fundamental 'images,' in addition to affect and action, that enables the movement of a film (or 'movement-images'). In general, this movement consists of the assembly of different kinds of shots and editing that convey perception, affect and action. These three 'movement-images' not only structure particular scenes, but also their being brought together over the course of a particular film can shape the nature of its story and its narrative flow.

In *Cinema 1*, a perception-image is a scene or a shot in a film (its movement) that is defined by some aspect of perception, whether the perception of the camera, usually determined in Deleuze's typology of images by the long shot, or the perception of the characters in a film or even of the audience itself. In the former case, Deleuze's elaboration of the perception-image follows Bergson's approach to matter and perception in *Matter and Memory*. As Deleuze summarizes:

> the thing and the perception of the thing are one and the same thing . . . but the perception of the thing is the same image related to another special image

> which frames it, and which only retains a partial action from it, and only reacts to it mediately.
>
> (65)

This 'special image' is, in Bergson's own account, the 'interest' of an individual (1991: 49). When an image (meaning any kind of, for Bergson) is perceived by an individual, only the aspects of the image that are of interest to the individual (defined by its needs rather than by a rational calculation) will result in some kind of response or action. In Deleuze's cinematic recasting, the interest of the individual 'frames' the image of the thing, excluding some aspects and being attentive to others. This is the most elementary aspect of the perception-image in film, namely that the camera and the editing process enable a framing of the moving images before an audience shaping their perception of the matter presented to them.

It is the case of the perception of the characters and the audience, however, that dominates Deleuze' attention in *Cinema 1*. The complexity of cinema generally lies in its ability to present multiple sources of perception, namely those that include a character's own perception ('subjective' images) and those that have no relation to a character's perceptions (apparently 'objective' images). In addition, there are 'semi-subjective' scenes in which audiences watch a character being watched, or in which a scene is shot to frame the perspective of a given character, but not formally as a 'point of view' of that character. The use of these different perception-images can have a substantial effect on how characters relate, a narrative flows, or how certain elements are revealed to an audience (what is visible and invisible according to Deleuze), in other words, the very movement of a film.

Whereas Deleuze identifies the long shot as the primary form of the perception-image, in Apichatpong's film-making it is the slow pace of scenes in which the shot can be still or slowly tracking that allows for new kinds of perception-images. Apichatpong's long shot is not a shot measured by distance but by duration. The slowness of these shots does not merely account for the passage of time, but they present a living 'stillness' that contains the continuous presence (a perception-image) of the something-other-than-human.

The slowness of scenes continuously displays flow and movement, both human and non-human. In many long segments of his films, non-human nature is in movement, particularly in shots that dwell on the breeze of the wind or the restless movement of trees, the swaying of forest canopies or rippling water. A lengthy shot of a car ride in Uncle Boonmee shows Jen sitting in the back of the car by an open window. The window serves as a frame for the trees passing by, but it also is the source of intermittent sunshine that illuminates Jen's face and floods the back seat of the car. The scene is composed of this fleeting fluctuation of light and dark as the trees on the side of the road shade and expose the passing car. The sun blinds the viewer as it dominates the screen only to be rapidly followed by the muted light and shadows of the car's interior. The scene is disorienting because it quickly becomes difficult to discern whether the camera is fixed on Jen or on the sunlight itself. In this example, it is the duration of the shot that leads to the cinematic perception of movement of both the car ride and light, wherein the light itself becomes an agent in shaping the production of Jen, and hence the perception image – the matter – of the film itself.

The length, and at times stillness, of Apichatpong's scenes also dwells in the pauses in such movement. Pauses such as breathing, the long pause in answering a question, waiting, awkwardness (which provides further opportunities for silences), or the awkward pauses and stares required when flirting. Even if these pauses are the product of human activity, they usually allow for surrounding noises to be heard, particularly the wind, birds, insects, or the night. Even in the second half of *Syndromes,* where the depiction is of the lifelessness of urban existence, ambient noise still fills the silences, particularly in the sound of air conditioning and the industrial operations of a hospital.

The pace of Aphichatpong's long shots and the saturation of sounds and light (or darkness) provide the basis for a new kind of perception-image. It is a style of film making that presents a perception-image of the non-human (whether the non-human be spiritual, natural, or monstrous) and culminates in a cinema of the posthuman image. Such a cinema 'goes beyond perceptions, in the sense that it reaches to the genetic element of all possible perception, that is, the point which changes, and which makes perception change, the differential of perception itself' (Deleuze, 1986: 85). Aphichatpong's filmmaking is original precisely because it 'reaches "another" perception' (88) through its continuous pursuit of the perception of the non-human.

His films, with their ethereal movements and pauses or their audio and visual saturations, develop a different kind of 'camera-consciousness,' in Deleuze's terms, where the camera enables its viewers to see (and hear) anew. Ingawanij and MacDonald focus on the saturation of sound in their analysis of the end of Apichatpong's film *Blissfully Yours*, which, by employing 'trance-like natural sounds . . . pulls us into sharing the characters' state of bliss. Ultimately, the basis of this shared experience is sensorial rather than one secured through emotional identification' (2006: 131). The film conveys bliss through a dense perception-image, a 'sensorial' experience, in contrast to an identification with individual characters (subjects) and their identities or their relationship to a narrative arc.

Seung-hoon Jeong pursues a similar reading regarding the mythical jungle scenes in the second half of *Tropical Malady*, which again employ a series of 'rarefied' images, in Deleuze's terms, in which darkness and fluorescence dominate the visual frames, while being coupled with a saturation of nocturnal sounds and at times white noise. In Jeong's description 'the nature-sense circuit' in *Tropical Malady's* second half 'brings about a 'pure optical and sound' situation' (2012: 212). In this case, however, the style of perception made available to the audience by this 'circuit' is also enacted by Keng the soldier who is tracking a tiger spirit. 'A soldier groping in the dark jungle,' as Jeong suggests, 'extends all his sensorial antennae towards the opulent sound, smell, and touch of nature' (212). Once again, similar to Ingawanij and MacDonald's interpretation above, the tension that grips the second half of the film is not grounded in the identities of the soldier and the animal spirits that inhabit the forest, but in this sensorium transmuted to the audience: the perception of the non-human and seemingly the non-human perception that Keng must employ as he pursues the tiger spirit.

For Una Chung, it is Apichatpong's short-film project *Primitive*, some footage of which appears in the final minutes of *Uncle Boonmee*, that provides evidence of a new kind of perception-image:

> the rhythms of the bodies and voices of the young men playing soccer are inseparable from the rhythms of light passing across or jumping different media. Through this astonishing work with light, film apparatus, and screen, we could say that we are given back to the play of preontological forces.
>
> (2012: 220)

In *Primitive* and, more generally, across Apichatpong's work, the cinematic style makes possible a perception of the 'play of preontological forces'[5] that include not only the vibrations of sound and movement and saturations of light and dark, but more broadly of matter perceived as an 'aggregation of images' as Bergson suggests in *Matter and Memory* (1991: 22). In light of this kind cinematic perception, however, how can the non-human matter that captivates Apichatpong's artistic attention be characterized? What does it mean to talk about this work as fostering a posthuman perception?

A posthuman perception image

One of the clearest perception-images among Apichatpong's films can be found in the opening half of *Syndromes* when Dr Toey, the young doctor introduced in the opening of the film, pauses to look out the window of her office. Technically the scene is a close-up that is shot through the glass of an open window into the office. The window is open at roughly a 45-degree angle and both Toey and the reflection of the green field outside her office appear in the shot. Despite being a close-up, however, the shot can simultaneously be rethought as a 'long shot' in two ways. First, it is a long shot in terms of distance because the reflection of the field and jungle is as clear in the shot, as is Toey's stare. But, also, the scene utilizes a long shot in terms of duration. The shot is 18 seconds long, and consists solely of Toey staring out of the window. Whereas, Deleuze classifies the close-up as the basis for affect-images (in which the affect of the character dominates a frame), in fact, in this case Toey's stare is arguably affectless. It is not possible to tell what motivates her staring and nothing up to this point in the film has allowed for the development of any emotional depth in her character.

The sole movement of the scene comes not from Toey but from the field that appears in the reflection in the window framing the shot. From the reflection the bright field and the swaying trees can be seen clearly. Because of the length of the shot the reflection becomes disorienting making it difficult to determine the focus of the scene as both the doctor and the outside field are equally prominent. The window acts as a frame, however, the object in the frame is indeterminate. As in the opening of the film, when the audible and the visual diverge to produce a simultaneous perception of the movement of nature (outside) and mundane small talk (inside), this scene accomplishes a similar perception-image mostly visual means. The reflection enables the simultaneous perception of the motionless doctor (in her office) and the movement of nature (outside).

Just as Apichatpong's perception-images require a reworking of Deleuze's understanding of the long-shot, where it is now necessary to consider both the time and the scale of a particular shot, the kinds of images and shots found in the three films analyzed here also demand a reconsideration of Deleuze's association of the semi-subjective shot and the perception-image. One basis for the perception-image is a cinematic depiction of the perception of perception, or most often a scene where the audience watches another character being watched. However, this emphasis on the semi-subjective shot as the basis of the perception-image found in *Cinema 1* fails in some cases to provide a means of understanding the perception of the non-human in Apichatpong's films.

Often scenes incorporating the perception of non-human beings are shot using what could be referred to as an 'objective' shot, where the camera views a scene from a position that does not reflect the point of view of any characters in the scene. These shots do not fit Deleuze's own account of the perception-image. Rather than seeing a character being watched by (and respond to) another character, instead the camera records the response of a human character when he or she perceives the non-human. This is the case in the veranda scene in Uncle Boonmee. One evening on the veranda at Uncle Boonmee's house, Jen, Boonmee and Tong, their guest from Bangkok, are visited first by an apparition of Huay, Boonmee's dead sister, and then again by Boonsong, Jen's son now turned as a wereman. Boonsong's transformation took place after he mated with a 'monkey ghost' that he tracked and photographed. Huay's presence is reassuring to Boonmee, as he himself is close to dead. Boonsong's arrival solves the mystery of his disappearance some time ago, although it poses the new reality that while not dead, Boonsong is no longer human.

The response to Boonsong is not shock, horror (as this film has at times been considered a horror film), or some kind of surprise at the fantastic nature of the encounter, but specifically intimate familiarity with the non-human beings. As Ingawanij suggests, 'the diegses of Tropical Malady and Uncle Boonmee present worlds in which characters knowingly inhabit non-synchronous time and respond readily, through verbal and gestural acknowledgement or bodily immersion, to appearances, presence and events of a marvellous nature' (Ingawanij, 2013: 97). The audience sees characters seeing the non-human without being struck by its non-humanity, acting and reacting in ways in which the posthuman demands no second thought; it is engaged with little, if any, anxiety.

The audience watches the characters watch the supernatural emerge on the screen and begin to interact entirely as if an apparition or wereman were mundane. In the tense tracking of the tiger spirit at the end of Tropical Malady, the fact that animals communicate with the Keng, the soldier, does not appear to unnerve him. Similarly, in the most fantastical scene of the three films, when Apichatpong introduces the tale of the princess and the fish in Uncle Boonmee the camera peers into the water to see the fish talking and later the fish seduces the princess. None of these scenes relies on a semi-subjective shot or a traditional long shot, however. All of them present perception-images in which the perception of the non-human is foregrounded and conveyed to the audience.

One scene that does rely on the semi-subjective shot and provides a specifically posthuman image appears at the end of *Uncle Boonmee* after Boonmee's funeral. Tong, who is in monk's robes, comes to Jen's room at night knowingly violating the proscription that prevents monks from socializing. Jen and Roong, a younger woman, are sorting through funeral donations as they watch TV. Tong teases them playfully as they suggest that he should not be in their room. Tong insists that even though it is late they should find something to eat. As they get up to leave, instantaneously a second set of Jen and Tong appear laying on the bed as the three of them watch TV in silence. Tong does a 'double take' and looks twice at the copies of himself and Jen and Roong before he seems to accept that there is a new alternate reality in which he has seamlessly found himself. In this case the camera clearly catches him, effectively watching himself. However, this new virtual reality is taken as the new objective reality in which the characters and the film's audience inhabit. The scene continues with them getting food outside the hostel.

The indiscernibility of the actual and the virtual is described by Deleuze in *Cinema 2* as a 'crystal image' that presents both the reflection of the mirror (the indiscernibility of the actual and the virtual) and the potential of this reflection (that the actual and the virtual emerged as separate images to begin with) (1989: 79). The scene in *Uncle Boonmee* described above provides the most radical vision of the posthuman, where every new moment contains the potential for a new reality and change appears in every instant. Yet Apichatpong guards his secrets and provide no clues as to how or why a new virtual path has opened up.

Nonetheless, this example raises the question of how precisely the 'posthuman' is being conceptualized, if each example of a posthuman image has, in fact, been oriented to some extent around human characters. How does Apichatpong's film advance a case for a 'cinema beyond the human' when it appears to be so closely attached to the human (Pick and Narraway, 2013: 4). The posthuman is never a clearly demarcated category in Apichatpong's films. The ambient nature represented in the film is not reducible to a vision of nature that is autonomous or beyond human intervention. In fact, its perception is usually made possible in the midst of human activity. The spirits that haunt the three films are not 'grounded' in places or objects, but we are clearly told linked to their people. As Jen reminds Boonmee: 'ghosts are not attached to places they are attached to people, to the living.' Humans remain at the centre of these stories, however, there is always a beyond human that is co-present in a manner that introduces a measure of uncertainty in what it means to be human. As Teh has suggested, Apichatpong's 'uncanny realism' enables a 'scrambling the codes of humanist representation' (604).

A key to understanding the intimacy of the human and the non-human in Apichatpong's vision can be found in Bergson's suggestion of the relationship between perception and memory. The dividing line between the human and the various perceptions of the non-human is always burred in Apichatpong's films because, as Bergson suggests, 'perception and recollection always interpenetrate' (1991: 67) Yet this moment of political ontology poses precise challenges for many members of its audience. The recollection that makes the intimate acceptance of the non-human possible is grounded in cultural elements (some actual and some composed in a fantastic manner) that may not be readily understandable to viewers who do not recognize these elements. At an abstract level, therefore, to perceive the non-human is always simultaneously to recollect what it means to be human. However, at the level of political ontology, to perceive the non-human in Apichatpong's films requires a practice of decolonization. This practice acknowledges how the epistemic privilege of modernity forestalls the kinds of recollections necessary to intimately perceive the non-human. Modern knowledges, whether enunciated through the division of nature and culture or scientific understandings of the world, or even as the basis for 'reasonable politics,'[6] intercede in the capacity to recollect the enchantment of the world. They actively prevent the perception of the non-human.

In *Uncle Boonmee*, Apichatpong actively explores this relationship of perception and recollection in encounters with the non-human, particularly in the scene of the film's first apparition, namely the veranda scene described above. After the humans and non-humans become reacquainted, both Huay and Boonsong look at pictures brought to the table by Jen. Huay looks at pictures of her own funeral. The picture book is itself an archive containing pictures taken at a different time. Here the act of recollection is clearly dramatized. Moreover, Jen's very human impulse to bring the picture book demonstrates an intimacy in this instance of apparition and of perception.

This scene, however, presents more than a staging of posthuman perception and recollection. It also is an image of a complex temporality. The temporal heterogeneity in this scene is rich, as several different times are co-present. Among the humans, Jen and Tong are visiting Boonmee because of his worsening health. In fact, the anticipation of death for Boonmee is a continuous theme of the movie, but it is an experience of time that is obviously distinct from Jen and Tong's sense of time. The other figures also present different 'other-worldly temporalities,' a phrase Fuhrmann uses to describe the spectral time of ghost stories and horror found in contemporary Thai film (2016: 4). Huay is a spirit from the recent past. Boonmee takes her presence as a sign of his own near death. The temporality of the wereman, however, is far more complex. His life as a human is a part of the past, but his existence in the present is as a becoming-monkey ghost.

The co-existence of this array of figures and characters (all openly accepted as existing by the human characters in the film) becomes a living comparative frame for these contemporaneous relations. To borrow again from Fuhrmann in her analysis of temporality in *Tropical Malady*, the veranda scene takes the form of an anachronism, but not in the pejorative sense commonly associated with the term, as a kind of judgement that denotes something past that does not belong in the present. Instead, Fuhrmann has reconceptualized the term to become 'a comprehensive, nonevaluative term that designates the coexistence of two or more divergent temporal elements in a given moment or period of time' (7). This scene, as a posthuman perception image, contains an immanent comparison in this exact sense of a non-evaluative comparison designating coexistence. The perception of the non-human here takes the structure of a comparison that brings together differentiated times. This is comparison as a montage. Apichatpong's comparative aesthetic here brings together multiple temporalities and maintains their difference without insisting on any definition of the human that would exclude the wereman and spirit as non-human or as un-real.

Conclusion

The 'post' in posthuman refers to the various social and philosophical ways of imagining a beyond for or to humans. Rosi Braidotti has described her understanding of 'the posthuman predicament as an opportunity to empower the pursuit of alternative schemes of thought, knowledge and self-representation' (2013: 12). It is important to recognize that within the 'post' of the posthuman lies an inquiry into the historical moment at which the question can itself be posed. This question is posed across a variety of historical encounters and as such the posthuman can never understood as a universal condition.

Apichatpong's films present one vision of the posthuman, but it is a vision that comes with its own sensorium. His films educate its viewer to perceive the posthuman, not as a predicament, but as a possibility. Moreover, as just described, the posthuman is as much grounded in the recollection of the human as it is in the perception of the unquestionable presence of an array of other creatures, spirits, animals in the world. This vision of the posthuman as spiritual or 'marvellous,' or even 'mysterious' as for Anderson's Bangkok middle classes, is often absent in much of the literature in recent approaches to the posthuman.

For instance, in the accounts of new materialism within the field of political theory, there is often an implicit avoidance of the spiritual. In Jane Bennet's *Vibrant Matter* there is a deep-rooted secularism in the various lists of actants and outlines of assemblages that often include human and non-human elements. These non-human elements included in Bennett's lists demonstrate an effort to cast an imagination of assemblages as wide as possible, from microbes to tools. However, there is often an avoidance of any suggestion of other energy or spirituality constituting the ever reconfiguring relationships of human and non-human. As such, it is possible to identify a specifically modern epistemic privilege in this conceptualization of the non-human that demands further decolonization. When Bennett suggests that even her speech is dependent on 'millions of persons, dead and alive' there appears to be nothing more here to this afterlife other than a claim that discourse has emerged through the interaction of a population of millions of people, some of whom have now passed away (2010: 36). Apichatpong's films, by contrast, portray complex instances of intimate communication with spirits and non-humans from the past. Apichatpong provides an opportunity for decoloniality in so far as he refuses to acknowledge the boundary between the human and the spiritual that continues to some degree to animate the conceptualization of the human in new materialism.

Similarly, William Connolly in describing the 'creativity' at play in the world suggests that what is necessary is the creation of an interface between fields of neuroscience, critical philosophy, artistic work and the human sciences:

> Maybe it is timely to transfigure our activities, disciplines and selves so that we pursue more thoughtful responses to the risks and promises of moments

of real uncertainty as they unfold into the shifting degrees of mystery marking the future.

(406)

What is the nature of this mystery marking the future? Will it be enchanted with anything other than a secular creativity and 'real uncertainty?' The three films I have analysed here pose the mystery that is enchanting present, past and the future in its reimaging of matter and perception of a posthuman image. Apichatpong's filmmaking is an instance of 'artistic work' that thoroughly rejects a secular foundation of the definition of the non-human. Rather, it suggests a perception of the posthuman that cannot exclude the complex temporality of apparition, whether this involves the appearance of the afterlives of Bennet's speakers or the return of family members in spectral or non-human forms. Eva Aldea argues that a turn toward artistic work is necessary because 'what art does is to think the univocity of being through an immersion in the world' (28). This is precisely what occurs in Apichatpong's cinematic vision, with its immanent comparisons and posthuman images. Apichatpong's rich exploration of a Buddhist imaginary, Thai mythic forms, and the enchantment of nature provides an 'immersion in the world' that allows viewers to perceive the vibrancy and interconnectedness of being.

Notes

1 Bergson's original phrase is related to the outcome of this process of the gradual education of the senses: 'to harmonize my senses with each other . . . in short, to reconstruct as nearly as may be, the whole of the material object' (1991: 49).
2 For a discussion of the image-ontology found in Deleuze's philosophy, see Schwab, 2000.
3 The non-linear repetitive structure of the film will be discussed in the following section as an instance of Apichatpong's cinematic montage.
4 Apichatpong's critical success, including winning the Palme d'Or at the Cannes Film Festival in 2010, is the most widely recognizable among a group of independent filmmakers who emerged in the wake of the 1997 financial crisis. These filmmakers, including Wisit Sasanatieng (*Tears of a Black Tiger*, 2000) and Pen-Ek Ratanaruang (*Fun Bar Karaoke*, 1997), are often referred to as directors of Thailand's 'new wave.' For an account of recent transformations in Southeast Asian cinema, including this growing in the Thai film industry, see Baumgärtel, 2011.
5 Chung uses the term 'pre-ontological' in the sense of non-representational, that is cinematic elements that are not related to the identity of characters or capable of sustaining a linear plot or narrative.
6 In his study of environmental politics involving indigenous communities in South America, Blaser identifies a commonplace pattern of national political elites rejecting indigenous claims regarding the spirituality or agency of non-human entities. These rejections are often phrased in terms of the unreasonable nature of indigenous claims, and according to Blaser reflect the presumption of an 'epistemologically superior standing' by many non-indigenous political elite (2013: 17).

References

Anderson, B. 2009. 'The Strange Story of a Strange Beast: Receptions in Thailand of Apichatpong Weerasethakul's Sat Pralat', In: J. Quant, ed., *Apichatpong Weerasethakul*. Vienna: Synema Publikationen, pp. 158–177.

Baumgärtel, T. 2011. 'Imagined Communities, Imagined Worlds: Independent Film From South East Asia in the Global Mediascape', *Transnational Cinemas*, vol. 2, no. 1, pp. 57–71.

Bennett, J. 2010. *Vibrant Matter: A Political Ecology of Things*. Durham: Duke University Press.

Bergson, H. 1991. *Matter and Memory*. New York: Zone Books.

Blaser, M. 2009. 'Political Ontology', *Cultural Studies*, vol. 23, no. 5–6, pp. 873–896.

Blaser, M. 2013. 'Notes Toward a Political Ontology of "Environmental" Conflicts', In: L. Green, ed., *Contested Ecologies: Dialogues in the South on Nature and Knowledge*. Cape Town: HSRC Press, pp. 13–27.

Braidotti, R. 2013. *The Posthuman*. Oxford: Polity Press.

Chung, U. 2012. 'Crossing Over Horror: Reincarnation and Transformation in Apichatpong Weerasethakul's Primitive', *WSQ: Women's Studies Quarterly*, vol. 40, no. 1–2, pp. 211–222.

De la Cadena, M. 2010. 'Indigenous Cosmopolitics in the Andes: Conceptual Reflections Beyond "Politics"', *Cultural Anthropology*, vol. 25, no. 2, pp. 334–370.

Deleuze, G. 1986. *Cinema 1: The Movement-Image*. London: The Athlone Press.

Deleuze, G. 1989. *Cinema 2: The Time-Image*. London: The Athlone Press.

Fuhrmann, A. 2016. *Ghostly Desires: Queer Sexuality and Vernacular Buddhism in Contemporary Thai Cinema*. Durham: Duke University Press.

Harrison, R. 2005. 'Amazing Thai Film: The Rise and Rise of Contemporary Thai Cinema on the International Screen', *Asian Affairs*, vol. 36, no. 3, pp. 321–338.

Ingawanij, M.A. 2006. 'Transistor and Temporality: The Rural as Modern Thai Cinema's Pastoral', In: C. Fowler, ed., *Representing the Rural: Space, Place, and Identity in Films About the Land*. Detroit: Wayne State University Press, pp. 80–100.

Ingawanij, M.A. 2013. 'Animism and the Performative Realist Cinema of Apichatpong Weerasethakul', In: A. Pick and G. Narraway, eds., *Screening Nature: Cinema Beyond the Human*. Oxford: Berghahn Books, pp. 91–109.

Ingawanij, M.A. and MacDonald, R.L. 2006. 'Blissfully Whose? Jungle Pleasures, Ultra-Modernist Cinema and the Cosmopolitan Thai Auteur', *New Cinemas: Journal of Contemporary Film*, vol. 4, no. 1, pp. 37–54.

Jeong, S-H. 2012. 'The Surface of the Object: Quasi-Interfaces and Immanent Virtuality', In: D. Martin-Jones and W. Brown, eds., *Deleuze and Film*. Edinburgh: Edinburgh University Press, pp. 210–226.

Keyes, C. 2014. *Finding Their Voice: Northeastern Villagers and the Thai State*. Seattle: University of Washington Press.

Mignolo, W. 2000. *Local Histories/Global Designs: Coloniality, Subaltern Knowledges, and Border Thinking*. Princeton: Princeton University Press.

Mignolo, W. 2002. 'The Enduring Enchantment: (Or the Epistemic Privilege of Modernity and Where to Go From Here)', *The South Atlantic Quarterly*, vol. 101, no. 4, pp. 921–954.

Mills, M.B. 2012. 'Thai Mobilities and Cultural Citizenship', *Critical Asian Studies*, vol. 44, no. 1, pp. 85–112.

Pasuk, P. and Baker, C. 1997. 'Power in Transition: Thailand in the 1990s', In: K. Hewison, ed., *Political Change in Thailand: Democracy and Participation*. London: Routledge, pp. 21–41.

Pavin, P., ed. 2014. *'Good Coup' Gone Bad: Thailand's Political Development Since Thaksin's Downfall*. Singapore: Institute of Southeast Asian Studies.

Pick, A. and Narraway, G., eds. 2013. *Screening Nature: Cinema Beyond the Human*. Oxford: Berghahn Books.

Quandt, J. 2009. *Apichatpong Weerasethakul*. Vienna: Synema Publikationen.

Schwab, M. 2000. 'Escape From the Image: Deleuze's Image-Ontology', In: G. Flaxman, ed., *The Brain Is the Screen: Deleuze and the Philosophy of Cinema*. Minneapolis: University of Minnesota Press, pp. 109–139.
Sears, L.J. 2013. *Situated Testimonies*. Honolulu: University of Hawai'i Press.
Walker, A. 2008. 'The Rural Constitution and the Everyday Politics of Elections in Northern Thailand', *Journal of Contemporary Asia*, vol. 38, no. 1, pp. 84–105.

10 Political ontology and international relations

Politics, self-estrangement, and void universalism in a pluriverse

Hans-Martin Jaeger

Introduction

Once perhaps considered an obscure domain of metaphysicians, ontology has emerged in somewhat different guises as a significant theoretical concern in the social sciences (especially anthropology, sociology, and political geography) in recent years. Remarkably, it has even installed itself at the center of International Relations (IR), a discipline that, not so long ago, one would not have suspected of such philosophical liaisons dangereuses, unsettling as these could be for the alternately positivist-scientistic or "hard-headed" practical preoccupations that have dominated (and shackled) it for much of its existence. At least since the emergence of Wendt's constructivist quest for a social ontology of international relations (Wendt, 1987, 1999), ontology has become an accepted subject for theoretical conversation in IR, whether in his own scientific-realist (Wendt, 1999), Wight's (2006) critical-realist, Jackson's (2011: 26–32) pragmatist, or other guises. However, what is instantly striking about these different ways in which ontology has entered the agenda in IR is how they submerge or defer politics (let alone a more demanding notion of the political); practices, processes and events; and the larger questions of whether ontology should be understood as singular and universal or as multiple and pluriversal. To the extent that these concerns surface, it is further striking that they continue to be narrowly circumscribed in terms of states and their "international" relations.[1]

Without denying the continuing significance of states and interstate relations, an articulation of ontology in these terms is little more than a nostalgic sublimation in a world-political context in which many of the traditional state-centric (and often colonial) categories of IR, from anarchy to interdependence, appear unsuited to coming to terms with phenomena such as transnational economic and ecological crises, migration, and "terrorist" political violence. To be sure, beyond the sometimes narrow confines of ontological debate in IR, numerous attempts have been made to come to terms with these (and other) developments through new theories of cosmopolitan democracy (e.g. Held), empire (Hardt and Negri), or world risk society (Beck), to name only three well known examples. Contrasting with such theories of a global (albeit differentiated or layered) social whole, Rosenberg's recent Trotsky-inspired theory of "uneven and combined development" argues that "the fact of societal multiplicity" (Rosenberg, 2016: 129) is the distinctive trait of the international (broadly understood). For Rosenberg, then, societal multiplicity is, arguably, the necessary ontological touchstone for IR to leave the "prison-house of political science" and "export" a broader contribution to the social sciences rather than merely importing from them. Along with the discussions of ontology in IR, the theory of uneven and combined development is of particular interest here too, because unevenness, development, and the problem of commensurability shares the attention to multiplicity with wider discussion of ontology in anthropology, human geography, and other social sciences to which it would likely wish to contribute (e.g. Blaser, 2009; de la Cadena, 2010; Millner this volume).

This chapter juxtaposes recent discussions of ontology and (Rosenbergian) multiplicity in IR with what scholars elsewhere have called "the ontological turn" or "political ontology."[2] The ontological turn seeks to take seriously the multiplicity of being (alternatively designated as radical alterity or pluriversality) by overturning the Western philosophical presupposition of one ontology ("reality" or "nature") and multiple epistemologies ("representations" or "cultures"). Rather than leading to a mere plurality (and relativism) of worldviews, the ontological turn asserts that practical (and simultaneously conceptual) connections between cultural representations (including human and non-human thought and sensation) and nature (or materiality in general) entail an actual, that is, ontological multiplication of worlds.[3] In a more explicitly and more politically decolonial vein, political ontology advocates a transition from the putative universe of modernity (or even multiple modernities) to a pluriverse of different ontologies, understood as performances of worlds and stories, including non-modern ones (see Blaser, 2009, 2014). Drawing on ethnographic studies of indigeneity, actor-network theory (ANT) and other posthumanist social theories and philosophies, the ontological turn and political ontology highlight the simultaneously conceptual, performative, political, and plural character of ontologies often elided in scholarship on ontology in IR. Vis-à-vis Rosenberg's insistence on societal multiplicity and inter-societal connectivity (Rosenberg's broad and non-Eurocentric understanding of the concept of society notwithstanding) they propose a deeper, not merely societal or cultural, but ontological multiplicity taking the form of "multinaturalism" (Viveiros de Castro, 2015a: 59, 196) rather than multi-societalism or multiculturalism. In other words, as is now extensively argued, sociality is not limited to human beings, for humans are relationally always already processes of more-than-human multiplicities. By entertaining the possibility of varying human-natural-material assemblages, multinaturalism also disengages the recognition of difference and co-existence from the developmental categories and trajectories implied by the "uneven and combined development" of societies.

The present intervention initiates a conversation between political ontology and IR along three lines (with a particular emphasis on the first of these). Following some preliminary remarks, it first clarifies what kind of politics is at stake if IR were to engage with political ontology by showing how Rancière's politics of disagreement, equality and alterity implicitly informs the latter (without however exhausting it). Second, it suggests that the move from a universe of modernity (or multiple modernities or societies) to a pluriverse of modern and non-modern ontologies also implies a "deconstructive self-estrangement" of Eurocentric ontologies/modernities, for instance, by attending to their stories or by understanding these through political theology. Third, it in turn provokes political ontology's (partly Schmittian-derived) notion of the pluriverse by pointing to Sergei Prozorov's (2014a, 2014b) "void universalism" as a possible postfoundational grounding of an alternative cosmopolitics for a pluriverse. By foregrounding the performance of worlds through non-anthropocentric ontologies (via ANT and ethnography), politics (via Rancière), self-estrangement (via stories and political theology), and an alternative, postfoundational (or "void-universalist") cosmopolitics, political ontology moves us beyond the (merely) international politics of states and the (mere) multiplicity of societies, and can potentially contribute to provincializing, politicizing and decolonizing IR (if I may continue to speak to and from this parish within the social sciences).

Preliminaries: Politics and ontology and political ontology

While the political implications of ontology have been surreptitiously noted in the anthropological debate around the ontological turn, the most explicit statement on the matter has been offered by Holbraad et al. (2014) who differentiate three possible "correlations" between ontology and politics. We might dub these a "conservative-affirmative" correlation (whereby politics is about discovering and reproducing what is), a "critical-prescriptive" one (whereby politics debunks the ideological pretense of ontology while simultaneously affirming what should be), and a "creative-transformational" one (whereby politics multiplies forms of existence that herald what could be). Aligned with the creative-transformational mode, the ontological turn is concerned with making visible alterity. Rather than proceeding in a normative vein (as would the "critical-prescriptive" mode), it is geared towards giving the Other(wise) "full ontological weight so as to render it viable as a real alternative" (Holbraad et al., 2014). For Viveiros de Castro anthropology accordingly becomes "the science of the ontological self-determination of the world's peoples" (2015a: 54) or "the ontological self-determination of the other" (2015b: 4, 11), "and . . . thus a political science in the fullest sense" (2015a: 54), which is simultaneously a "politico-cultural struggle" of "multiplicity-building" indispensable for decolonization and political self-determination (2015a: 41, 75). Holbraad et al. (2014) clarify that this perspective would be misunderstood if equated to a (mere) struggle for indigenous rights, to ethnic essentialization (since it means "giving the ontological back to 'the people', not the people back to 'the ontological'"), or to a pedagogical role of anthropologists. The creative-transformational politics of ontological self-determination thus disavows or exceeds a liberal politics of rights, identity, and education. This is very suggestive, but still seems to beg the question of how a politics conducive to ontological self-determination can be conceived in positive, "affirmative" terms; how, indeed, we can conceive of a political ontology.

Mario Blaser has perhaps made the most systematic attempt to define political ontology. Beginning with the 'ontology' in 'political ontology', Blaser notes that apart from its dictionary definition as a doctrine of what kinds of entities exist, and how they relate to one another, ontology has two further derivations and meanings. First, political ontology extends ANT's performative understanding of ontology to the notion that "ontologies perform themselves into worlds". Second, it builds on ethnographies which conceive ontologies as "stories" (Blaser, 2009: 877; see also 2014: 51, 53–54). Based on this conception of ontologies as enactments of discursive and non-discursive elements, Blaser (2009: 877) then specifies political ontology in terms of: (a) "the politics involved in . . . a particular world or ontology," and (b) the conflicts and power dynamics arising in the interaction among different ontologies or worlds. Substantively combining (a) and (b), political ontology is particularly concerned with a critique and revision of the particular conception, and simultaneous repression, of politics in Eurocentric ontologies and their colonial encounters with Other worlds, and a corresponding affirmation of a more demanding conception of "the political" in encounters within and between different worlds (e.g. ontologies of modernity and indigeneity).

Drawing on ANT and other posthumanist social theories, political ontology highlights that Western politics is founded on a fundamental, supposedly universal, distinction between humanity (culture) and nature, and a corresponding distinction between the modern and the non-modern. By setting off the realm of politics against the state of nature, the Western political imaginary (from Hobbes to Locke to Kant and others) excluded nature from modern politics and consigned those supposedly living in the state of nature to non-modern status, thereby dehumanizing them. Both of these moves underwrote a hierarchical relationship between culture/humanity and nature, the modern and the non-modern, the human and the "barbarous" which made properly political relations between European colonizers and indigenous peoples, and their respective worlds impossible. Leaving behind the provincial conceptions of Eurocentric ontologies, the project of political ontology then is concerned with writing a reformed conception of more-than-human life into politics, and thereby creating the possibility for a properly political relationship between modern and non-modern ontologies (Blaser, 2009: 887–890; de la Cadena, 2010: 341–347) based on equality and alterity, and one neither denying existing inequalities nor essentializing radical differences.

As intimated by Marisol de la Cadena (2010: 346, 350–352), Jacques Rancière's conception of politics as disagreement between worlds is particularly suited to pinpointing and reversing the denial of properly political relations between modern and indigenous ontologies intended by political ontology. Without superimposing yet another Western thinker on a decolonial project, the following section proposes to use the terms of Rancière's conception of politics as interpretative devices to further explicate the politics of the ontological turn and political ontology. However, we will see that political ontology also expands the terms of politics beyond Rancière in certain instances. As illustrated below, mostly by bringing Rancière into conversation with Viveiros de Castro, the politics in question here is both a matter of decolonial practices and discourses of indigenous activists and movements, and of the kind of contestation inspiring the ontological turn within academic discourses. Or, as Viveiros de Castro (2015a: 12) suggests, politics may lie precisely at the intersection of indigenous discourses (of "the native") and academic discourses (of "the anthropologist"): the anthropological affirmation of ontological alterity is not a cognitive or psychological problem; '[t]he problem is rather epistemological, which is to say, political' pertaining to 'the manner in which relations of order are established between these discourses.'

Politics in a dissensual and egalitarian key

Like other postfoundational thinkers, Rancière distinguishes genuinely political practices and relationships from the (often depoliticized) "official" processes and institutions usually called politics (governmental institutions, elections, international organizations, etc.). He borrows and broadens Foucault's (2007: chs. 12–13) notion of "police" to refer to the latter set of activities, while reserving the term "politics" for those actions which disrupt police. As with Foucault, "police" exceeds the operations of state apparatuses and includes broader social relations. However, for Rancière, the role of police is both more general and more specific than for Foucault. For Rancière police more generally defines the very "configuration of the perceptible" or "the allocation of ways of doing, ways of being, and ways of saying;" more specifically, it determines and legitimizes the "shares" (places, roles, positions, identities) of different parties in the social order (Rancière, 1999: 28–29; cf. 2010: 36).

Rancière, then, enables us to see the colonial order of European modernity as a police order defined by particular arrangements of the perceptible, including, for instance, the distinction between nature and culture, and associated ones such as those between body and soul, matter and spirit, world and mind. Such orders, in turn, made possible the assignation of different, often hierarchical, social and political positions and identities (e.g. as "civilized," "barbarous," and "savage" peoples). Additionally, he may also allow us to see the asymmetric epistemic relation conventionally presupposed by social scientific discourses as a kind of police order in which the "the anthropologist" or subject is always already placed in a position of epistemic authority vis-à-vis the object or "the native" (cf. Viveiros de Castro, 2015a: 3–8).

If Rancière's notion of police helps to frame the anti-politics of the "universe" of European modernity, coloniality, and its legitimizing scientific discourses, his conception of politics proper may provide an outline of the possible politics in a postcolonial pluriverse. Rancière understands politics as involving an agonistic relation termed dissensus or disagreement, as well as the emergence of political subjects, termed subjectivation. He further defines politics in terms of three related elements, namely as a staging of equality that exposes wrong.[4] Let us briefly consider these three elements in relation to political ontology.

First, diverging from thinkers like Aristotle, Arendt, or Habermas, who ground politics in speech, for Rancière, politics has less to do with speech as such than with what counts as speech. Even before any communication of interests, politics "is primarily conflict over the existence of a common stage and over the existence and status of those present on it" (Rancière, 1999: 26–27; cf. Hallward, 2009). Beyond representing mere demands for representation or inclusion, the emergence of indigenous peoples and "earth-beings" (such as particular mountains or rivers) in public demonstrations, displays, and confrontations has to be understood as a form of theatricality allowing for a genuinely political relationship between indigenous and modern ontologies (cf. De la Cadena, 2010: 343). Rancière locates the theatrical element of politics within "dissensus," that is, a dispute not simply over interests, opinions or values, but over the very common sense or "partition of the sensible" of: what is and what is not sayable (or visible); who can (and who cannot) say it; and what counts (and what does not count) as a political statement or political space. Dissensus simultaneously creates division and community by "placing . . . two worlds in a single world" (Rancière, 1999: 27; see also 22–25, 49–56; 2010: 37–38, 69). It thus involves the attempt to reconcile radical alterity without any common foundations with a pluriversal, rather than universal, refounding of cosmopolitical community.

Second, and perhaps most centrally, Rancière argues that politics involves a "presupposition of equality" by those not supposed to "have a part" in the community and yet identifying with it (Rancière, 1999: 61–63). Equality here does not refer to a mathematical proportion (as for the ancients) or a legal-political construct or project of economic levelling (as for the moderns), but rather to an assumption that "the people" can "verify" through action (Rancière, 1995: 45–48, 85; cf. 1999: 33). Who constitutes the people (demos) is not predefined but rather emerges in the practical validation of the assumption of equality itself (Rancière, 1999: 83, 99–100, 2010: 33).

Beyond Rancière, we may suggest that a presupposition of equality also animates the anthropological discourse of the ontological turn (which, of course, does not imply that academic discourse can play a directly political, let alone democratic, role). Namely, Viveiros de Castro (2015a: 6) proposes to suspend the epistemic asymmetry of conventional anthropological discourse by rhetorically asking "[w]hat happens if . . . we claim an active . . . equality between [the] respective discourses" of "the native" and "the anthropologist," if we assume that "we are all 'anthropologists'" (or in Rancière's terms, if we assume an "equality of intelligences;" Rancière, 1995: 51). What may happen in more concrete terms (and terms that take us beyond Rancière's anthropocentric politics) is, for instance, that we take seriously the personhood or subjectivity of non-human animals in Amerindian perspectivist cosmology (see Viveiros de Castro, 2015a: 200–206). The latter's always contextual presupposition of equality means investigating not "how monkeys see the world," but rather "what world is expressed through monkeys, of what world they are the point of view," and crucially, acknowledging the equal ontological status of that world (Viveiros de Castro, 2015a: 64).

Whether activated by "the people" or in academic discourse or even in interspecies relations, the validation of equality in practice occurs in a context of "disagreement," that is, a speech situation in which the parties "both understand and do not understand the same thing by the same words" (Rancière, 1999: x–xi). Without diminishing the importance of the content of speech, disagreement primarily consists in breaching assumptions about who is entitled to speak a certain language. For instance, who or what may appropriate the name of "the people," "the nation," "democracy," or "authority." Such an appropriation constitutes a "transgressive," genuinely political, and self-emancipatory action whose objective is

> [n]ot to found a counterpower susceptible of governing a future society, but simply to effect a demonstration of capacity which is also a demonstration of community. . . . not secession, but self-affirmation as a joint-sharer in a common world, with the assumption, appearances to the contrary notwithstanding, that one can play the same game as the adversary.
>
> (Rancière, 1995: 49)[5]

Similar to – but perhaps once again also exceeding – Rancière's notion of "disagreement," Amerindian perspectivist cosmology demands what Viveiros de Castro calls a "method of controlled equivocation". Equivocation here "is not just a 'failure to understand' (Oxford English Dictionary, 1989), but a failure to understand that understandings are not necessarily the same, and that they are not related to imaginary ways of 'seeing the world' but to the real worlds that are being seen" (Viveiros de Castro, 2015a: 64). Thus, where Rancière's "disagreement" is still partly about "the same thing" being expressed in "the same words" though with a different positionality towards "the thing," the "method of equivocation" implies that the same words may indeed refer to different things/worlds: "[s]ame representations, different objects; same meaning, different reference" (Viveiros de Castro, 2015a: 256). Viveiros de Castro's multinaturalism thereby radicalizes Rancière's notion of disagreement.

The presupposition of equality then receives its validation in disagreement. "Politics occurs," says Rancière (1999: 16–17; cf. 25), when supposedly natural orders of inequality, such as the colonial order of European modernity or the epistemic hierarchy of conventional anthropology, are "interrupted" by a presupposition of equality. "The essence of equality is . . . to declassify, to undo the supposed naturalness of orders and replace it with controversial figures of division" (Rancière, 1995: 32–33).

Rancière's generic term for such figures of division, and the third element of politics, is "wrong." Politics "is a function of the fact that a wrong exists, an injustice that needs to be addressed" (Rancière, 1995: 97). While wrong may concern historical injustices or the distribution of positions (e.g. rights to land or resources), its significance does not derive from immediate "victimization", or a tort that could be redressed juridically, or by enhanced representation. Rather, wrong indicates an irreconcilability of different parties prior to any specific dispute – an irreconcilability of "radical otherness" exemplified in Aristotle's figure of the stranger to a city (Rancière, 1995: 97; cf. 1999: 35, 39) or, for that matter, between worlds (see also Millner, this volume).

Drawing on Deleuze, Viveiros de Castro refers to the figure of the Other or "Autrui" as an "a priori structure" (2015a: 9) – "less a concrete, already actualized other, as against a self, than the structure that makes a Self and an Other" (2015b: 11) or the "expression of a possible world" (2015a: 9, 84, 2015b: 11). He concedes that the insistence on ontological alterity may involve tactical essentialization whereby "sometimes it may be pragmatically, i.e. politically, vital to describe ontologies as" incommensurable "sets of presuppositions" (2015b: 10). Contrary to critics of political ontology (Bessire and Bond, 2014: 442–445; Vigh and Sausdal, 2014) this does not mean that political ontology reifies modern and nonmodern worlds in (typically modern) binary fashion. Rather than relapsing into essentialisms of "authentic" subjects or identities, it arguably conceives the alterity of indigenous ontologies and movements as one, whereby these (much like Aristotle's stranger in the city) are located in modernity (though not necessarily Euro-modernity) without being of modernity; they express a possible world outside or beyond modernity through their actions. In her account of indigenous cosmopolitics in the Andes, de la Cadena captures this with the concept of "partial connections" borrowed from Marilyn Strathern. Andean indigeneity, and indigeneity elsewhere in Latin and South America, neither dissolved into, nor remained impervious to, the influences of Christianity and modern citizenship. Rather, "[p]artially connected indigenous-mestizos are, like fractals, self-similar even though, depending how you look at them, they also appear to be different" (De la Cadena, 2010: 347–348). Much like dissensus and disagreement then, wrong and the alterity it embodies "unites interlocutors while simultaneously keeping them at a distance from each other," thus establishing the "contentious commonality" of a simultaneous division and sharing of the world (Rancière, 1995: 104, 1999: 9).

Despite its anchoring in alterity, the exposition of wrong does not represent identity politics. Rather than presupposing identities, it politicizes social status and constitutes parties to a conflict as political subjects. Indeed, Rancière sees politics as a constitutive process of subjectivation: politics creates political subjects rather than vice versa. Instead of enacting pre-existing identities, subjectivation transforms "identities defined in the natural order of . . . functions and places into instances of experience of a dispute" (Rancière, 1999: 36, see also 11, 27, 2010: 95–96). Subjectivation is a particularly useful lens to understand the political in political ontology. Rather than relying on pre-existing cultural-civilizational subjects or identities, the political emergence of indigenous ontologies marks an instance of political subjectivation through "disidentification" (Rancière, 1999: 36). Holbraad et al. (2014) note that the ontologically motivated study of alterity is "less interested in differences between things than within them: the politics of ontology is the question of how persons and things could alter from themselves." When staging equality to expose wrong then, indigenous movements problematize the "natural" identities assigned to them in the colonial order of modernity. It is their engagement in politics (which may be rare; cf. Rancière, 1999: 139) that effectively constitutes their political subjectivity.

While Rancière's political theory may provide a vocabulary well suited to the concerns of political ontology, (at least) two critical questions might be raised about it. First, it is unclear if Rancière's politics can ever be anything other, or anything more than, an event or a series of events signifying a rupture. Indeed, were it to aspire to institutionalization, it would risk establishing another police order, a perhaps inevitable development all too often seen in Latin America and elsewhere when erstwhile political radicals are elected into political office. For activists, the important lesson from this is not to abstain from politics, as though this were a rarefied realm of pure agonism, but rather to be aware of the always-already entangled relationship between politics and police which will sometimes involve tactical concessions and will probably be most successful when involving tactical alliances, for instance, of indigenous movements and "earth-beings" with "scientific" environmentalists (as illustrated by de la Cadena, 2010). Such alliances may bring together urban and rural struggles, hacktivism and ancestral causes (such as buen vivir/"good living"), technopolitics, and indigenous cosmopolitics in "trans-modern" constellations beyond classical Western notions of social organization revolving around the politics of class or social movements. A case in point is the emergence of a variety of transnational networked movements across Latin America in response to the disappearance of 43 students from the Ayotzinapa Rural Teacher Training School in Mexico in 2014. This arguably involved "the emergence of a new political subject" whose "outbursts owe[d] more to 'syncs' than to solid ideological alliances" (González, 2016).

Second, despite the resemblance between Rancière's "disagreement" and the methodological equivocation of political ontology, one might also ask to what extent the latter effectively involves agonism in a more practical-political sense to realize its creative-transformational potentials. On the one hand, Bessire and Bond (2014: 448, 450) charge ontological anthropology with an abdication of practical agonistic politics directed against the hegemonic power of states, markets, and empire in favor of a "speculative futurism" of ontological multiplicity. On the other hand, Vigh and Sausdal (2014: 63–66) appear to suspect its insistence on radical and seemingly incommensurable alterity of fueling hyper-ant-agonistic, clash-of-civilizations like scenarios in the practical-political sphere.

It seems, however, that political ontology, in fact, steers a middle-course between these two positions. According to Viveiros de Castro (2015b: 9),

> [t]he affirmation of alterity . . . and equivocation . . . signifies . . . that ontological questions are political questions insofar they come into existence only in the context of friction and divergence between concepts, practices and experiences within or without culturally individuated collectives. . . . Ontological differences . . . are political because they imply a situation of war – not a war of words, as per the linguistic turn, but an ongoing war of worlds . . . in a context in which the world ('as we know it') is imposed in myriad ways on other peoples' worlds (as they know them), even as this hegemonic world seems to be on the brink of a slow, painful and ugly ending.

This statement, and especially its martial metaphor, while certainly not foregoing agonistic politics, may appear to tilt political ontology toward hyper-agonism. However, this impression needs to be qualified in two ways. First, Viveiros de Castro clarifies that the "war of worlds" is not one pitting different cultures or civilizations or sovereignties against each other, but rather one that is fought with "guerilla tactics" against the (hegemonic) "powers that be (I mean BP, Shell, Monsanto or Nestle)." Second and related, ontology is a "philosophical war machine" (my emphasis) that is "counter-cultural" not only in the sense of being counter-hegemonic, but also insofar as it does not target civilizational worlds of supposedly essential human differences but rather deploys contending (albeit partially connected) worlds traversing the human and non-human, and the natural and the cultural (Viveiros de Castro, 2015b: 9). Indeed, the main target of this philosophical war machine may well become the dialectics of ecological and economic crisis of the Anthropocene (Ibid.: 7–8).

Self-estrangement

If Rancière can help to clarify the agonistic political relationality involved in political ontology, a second contribution political ontology may offer for IR (and other fields in the social sciences) relates to political reflexivity with regard to Euro-modernity. Apart from mobilizing the political potential of radical alterity – of indigenous and other (modern and nonmodern) ontologies – in theory and practice, political ontology arguably also involves a verification of equality in a rather different sense.

As Viveiros de Castro explains, attending to radical alterity is not a matter of interpreting the thought of Others "but rather one of carrying out an experiment with it, and thus also with our own." This kind of experimentation establishes a "relation of intelligibility between two cultures" which performs "a double dislocation" whereby anthropological concepts become "vectors that always point in the other direction". The task then is "not that of explaining the world of the other, but rather of multiplying our world" (Viveiros de Castro, 2015a: 18, 20, 27) which includes "a redescription of ourselves in the light of alterity" (P. Maniglier quoted by Viveiros de Castro, 2015b: 7), and which effectively substitutes a pluriverse for a universe. It is this kind of experimentation, multiplication, or redescription this section will selectively engage in by verifying the status of Euro-modernity as one particular ontology in Blaser's sense of a series of "stories." Without the provincialization or "deconstructive self-estrangement" of Western modernity, it may continue to appear as though other ontologies/worlds may have stories, whereas "we" have History (philosophy of history for the more universally inclined and histories for the more pluralistic among "us"). Rather than providing an abstract account of deconstructive self-estrangement,[6] I will develop the notion in terms of three highly stylized illustrations here, moving from an ethnographic example to a historical illustration to a broader (though by no means exhaustive) philosophical strategy.

First, if we can observe that on the occasion of Evo Morales' inauguration in 2006 "an indigenous Andean ceremony . . . is added to the protocol" not as a "cultural" symbolization of indigenous politics but rather to "follow[ing] a precise protocol to summon non-human forces to help him govern Bolivia" (Blaser, 2009: 878–879), it would seem that a similar ethnographic lens ought to be trained on Western practices on similar occasions. To give an example, one may think of the "so help me God" formula (sometimes accompanied by the laying of the speaker's hand on a Bible) used in many Western countries during the swearing-in of a newly and, by all appearances, popularly rather than divinely elected government. This also could be seen as a mere ritual, as a mere "cultural-historical hangover" from times either more pious or more given to the public display of religiosity. While I would not want to dispute this rendering of the practice in every particular instance, I think at least in some instances it ought to be taken more seriously as a premodern (Euro-Christian) intrusion into Western modernity. At least it would not appear to be completely absurd to maintain that when George W. Bush or Jimmy Carter (to use "tabloid" examples) invoked the formula, they too were following a certain protocol to summon non-human forces to help them govern the U.S.A.[7] Presidential or (prime-)ministerial inaugurations more generally would appear to be a fruitful terrain for investigating further the intrusion of non-human, natural (as well as mundanely material), or supernatural forces into human affairs which the dominant readings of the dominant narratives of Western modernity so carefully separate from them. To be sure, there is a fair amount of pageantry, ritual, and folklore here, but perhaps these are also attempts not merely to govern things but to govern with things. Indeed, they could also be instances wherein a particular we, expressed in sign systems, is constituted by the things themselves.

A second more historically than ethnographically "deconstructive" observation of Euro-modernity pertains to the quite obvious fact that its ontological distinction between culture and nature – which delegates the government of the former to politics while relegating the latter to the domain of science – has been continuously violated in practice ever since the beginning of the early modern degovernmentalization of the cosmos and corresponding governmentalization of the state, to put this in the idiom of Foucault's (2007: 236–238) governmentality lectures. Despite Bruno Latour's astute genealogical tracing of the "great divide" between nature and humanity to the Hobbes-Boyle debate, it would overstate the case to say that "[a]ccording to the modern order of things science and politics . . . do not mix" (De la Cadena, 2010: 342).[8]

As Foucault (2007) has shown, the birth or rebirth of governmentality, insofar as it was assisted by physiocrats and liberals in particular, involved understandings of the social, civil society, public opinion, the market, the economy, and the population as quasi-natural domains with law-like regularities that could be ascertained by human, natural and social sciences, and tapped to govern politically in the name of the state or by its proxies in civil society. This form of governing in accordance with "nature" and its (scientific) "laws" broke with (though it did not entirely displace) earlier governmentalities which had indeed excluded "nature" in this scientific sense, or sought to subdue it with reference to moral and/or divine law (ironically dubbed "natural law"), positive laws of the state, or administrative regulations of the police. Perhaps the connection of phenomena such as the population or the economy to nature and their constitution as objects and media of government in these terms has been lost over time.[9] Their scientificity has become indexed to the social itself rather than anything "natural." However, by pointing to their governmental lineage as quasi-natural phenomena, we may recapture ways in which "nature" intrudes on politics in Euro-modernity as much as in other indigenous worlds. From a truly Euro-modern perspective this intrusion would have to be considered as impertinent and "irrational" in Euro-modernity as it often is in the case of non-modern indigenous worlds.

A third, more philosophically established strategy of the deconstructive self-estrangement of Western modernity, though not usually understood as such, is political theology or, since Agamben's (2011) The Kingdom and the Glory, political and economic theology. If the Foucauldian thesis of the governmentalization of the state challenges the ostensible Euro-modern separation of nature from politics, political and economic theology conversely challenge the Foucauldian assertion of the degovernmentalization of the cosmos with its separation of the super-natural from politics.[10] In its Schmittian articulation, political theology makes the assertion that all important modern political (and with Agamben, at least some economic) concepts are secularized religious concepts (Schmitt, 2005: 36). Schmitt found an easy (but by no means exclusive) evidentiary basis for this claim in the sovereign state of Hobbes' Leviathan, which despite its contractual derivation retains all the attributes of the sovereign God of Christianity and is owed the same obedience expressed in ten corresponding commandments (in Chapter 30). More recently, Agamben (2011) ingeniously traced modern notions of government and economy to trinitarianism and divine praxis. In terms of the self-estrangement of Euro-modernity in the service of political ontology, these examples show two things. First, much like Latour does but in a very different way they demonstrate that "we have never been modern" or at least, that we have never been quite as modern as we think because the non-human, here in the form of the supernatural and sacred rather than the natural, perpetually lurks behind our supposedly carefully sequestrated human political and economic affairs. Second, and here one would have to go into much more detail than I can do here, political theology clearly exposes Western ontology (say, of the state and the economy) as "stories," or perhaps more precisely, as deeply embedded in particular stories of creative powers, divine omnipotence, a celestial division of labor, bureaucratic organization, and many others based on the Western story (history, parable, poetry, prophecy, philosophy, law, prayer, hymn, eschatology, etc.) book par excellence: the Bible.

"Void universalism" and cosmopolitics

As previously noted, many attempts to bring ontology into IR have been plagued precisely by the absence of those issues which political ontology seeks to highlight, namely politics and practices that exceed the state and the international. This constitutionally prevents IR from articulating a true world or cosmopolitan politics whose affirmation would not unwittingly confirm what it seeks to transcend. This paradox can be observed from Kant to Held and Habermas, all of whom rely on the assumption of bounded communities, usually states, the very aspirational transcendence of which reaffirms these as the ontological starting-point (see Bartelson, 2009: ch. 2). Conversely, however, critiques or negations of world or cosmopolitan politics inevitably seem to relapse into a form of statism as well, as can be observed from Schmitt (2003) to Walker (2010). If we are envisioning the transition to a pluriverse, this cosmopolitan self-incapacitation in IR is perhaps all for the better, and we might be best advised to simply abandon any notion of world politics beyond ontological multiplicity.

However, this is not the conclusion drawn by Sergei Prozorov who perhaps offers the most interesting, and from the perspective of political ontology, most compelling, attempt to introduce ontology to IR in his recent dilogy of *Ontology and World Politics* and *Theory of the Political Subject* (Prozorov, 2014a, b). Prima facie, Prozorov appears to be making matters worse from a pluriversal perspective by attempting to do so in the name of a new universalism. As will become clear, however, his universalism is one that is entirely compatible with (and perhaps even ontologically necessary for) a genuine ontological pluralism that neither replicates the "unhappy" Schmittian pluriverse of friends and enemies (or geopolitical Großräume) (cf. Schmitt, 2003: 351–354), nor does it "happily" endorse a postmodern pluralism of proliferating identity politics. Drawing on Heidegger and Badiou in particular, Prozorov makes the case for a "void universalism" which can be seen as a condition of possibility for conceiving a pluriverse.

A good place to start might be the observation that Euro-modern accounts tend to define the non-modern in terms of a lack (say, of rationality, civilization, history, statehood, territorial closure, a separation of the immanent and the transcendent, etc.) (cf. Blaser, 2009: 885). Prozorov's attempt to defend the possibility of a non-particularistic and non-hegemonic universalism is based on the assumption that this is "only possible as an affirmation that is not founded on any positive principle, but takes this void itself as its point of departure – a void universalism" (Prozorov, 2014a: xvii): "there is a world in which an infinity of infinite worlds appears, and this World . . . is nothing but the void" (Prozorov, 2014a: 22).[11] Rather than being everything or something, the World as such is effectively a nothing. Void universalism is based on "the inexistence of the universe" (Prozorov, 2014a: 12). In what one could once again see as a certain philosophical verification of equality, Prozorov points out (with reference to post-Heideggerian and post-Lacanian political theories; cf. Prozorov, 2014a: 45) that rather than characterizing the category of the non-modern, a void or lack or nothing is the condition of appearance of any positive world – non-modern, Euro-modern, alter-modern, or whatever. To illustrate the point with a classical Euro-modern account of "world creation" (an illustration which Prozorov borrows from Roberto Esposito) Hobbes' founding and legitimation of sovereign power is entirely reliant on the assumption of an "originary nothingness of community" which is then converted into the "secondary, artificial nothingness" of the state of nature (conceived as a state of war) (Prozorov, 2014a: 143). Modern sovereignty can only appear on the basis of this dual nothingness.

While "hegemonic forms of universalism suppress difference" and anti-universalist "forms of identity politics valorize difference," Prozorov's void universalism "both radicalizes difference by insisting on the singularity of each being [including non-human being/s] and deactivates its exclusionary potential by insisting on the irreducibly common exposure of these singularities that remains in excess of any particularistic figure of community" (Prozorov, 2014a: xxx). The World as void makes possible "the proliferation of being as inconsistent multiplicity" by bringing forth "the ontological material for the construction of worlds that ensures that whatever appears is real without prescribing how it should appear." "[W]hatever appears in the world [or as a world] is necessarily" real, while "the mode of its appearance is absolutely contingent" (Prozorov, 2014a: 30).

Lest we mistake void universalism and the genuinely pluriversal "world politics" that follows from it for a merely abstract and indeterminate form of "nihilist anti-politics" though, Prozorov insists that the void of the World can be "the starting point for the derivation of universal political principles" or "axioms" of politics (Prozorov, 2014a: 44). We can read this statement and Prozorov's conception of politics as "a practice of affirmation of universal principles" or axioms (Prozorov, 2014a: 50) as a generalization of sorts of Rancière's conception of politics as the staged) verification of equality in response to wrong. Much like Rancière, Prozorov does not give his universal principles/axioms any intra-worldly status as ontological principles. Indeed, his principles are axiomatic precisely "in the sense of presupposing no grounds for themselves" (Prozorov, 2014a: 51).

> From an ontological perspective . . . they are grounded precisely in the void of the World, which is the condition of possibility for every positive world, and hence are valid for any world whatsoever. Thus, what the ontic perspective of the intra-worldly observer dismisses as frivolous groundlessness is from an ontological perspective recognized as precisely the abyssal ground of universalist politics.
>
> (Prozorov, 2014a: 52)

Echoing Rancière's presupposition of equality (which is also one of Prozorov's axioms of "void-universalist" politics), Prozorov illustrates this point by arguing that "[r]acial, ethnic, sexual or any other form of discrimination is not 'bad' because it offends a particular race, ethnicity or sexual orientation ['ontically'], but because it negates the principle of equality that a political practice axiomatically holds to be valid for the entire world ['ontologically']" (Prozorov, 2014a: 53).[12] Going beyond Rancière however, Prozorov (2014a: ch. 3) posits two further axioms of politics besides equality, namely freedom and community. Despite the unholy resonance of Prozorov's axioms with the holy trinity of liberté, egalité, fraternité of the French Revolution, it bears repeating that nothing intra-worldly, ontic, historical, or even anthropocentric is implied by these axiomatic presuppositions. There is no freedom, equality, and community as such. Rather these axioms can only ever be verified – against considerable obstacles – contingently (and perhaps only selectively and temporarily) in pluriversal practice.

In allowing for a plurality of worlds inhabited by human and non-human beings and capable of political relations of mutual agonistic respect, Prozorov's void-universalist "world politics" arguably converges with the kind of cosmopolitics envisioned by Isabelle Stengers and adopted by political ontology, namely "a politics where 'cosmos refers to the unknown constituted by these multiple, divergent worlds and to the articulation of which they would eventually be capable'" (Stengers, 2005: 995 quoted by De la Cadena, 2010: 346). Much like Prozorov's composite "world politics" which pairs the ontological void with his triadic axiomatic conception of politics, Stengers (2005: 994) insists on the simultaneous "distinction and inseparable nature of [her] political and cosmopolitical proposals." Unlike conventional cosmopolitan (say, Kantian) proposals, Prozorov's "world politics" and Stengers' "cosmopolitics" reject the universalist assumption that a common world already exists by conceiving the World/the cosmos as a void/unknown and viewing any existing world/cosmos as a contingent (and assembled) and therefore multiply realizable political accomplishment. Explicitly (in Prozorov's case) or implicitly (in Stengers') both world politics and cosmopolitics also borrow (albeit only partially and in different ways) from Rancière's conception of politics which, as I have suggested above, helps illuminate political ontology.

Prozorov's borrowing is most explicit in his "subtractive" conception of world politics and specifically (world-)political subjectivation in terms of disidentification or "the deactivation of one's worldly identity" which "disrupts the [police] order of distributed and differentiated places in its affirmation of freedom (from places), equality (of places) and community (without regard to place)" (Prozorov, 2014b: 15, 17). Akin to Rancière, Prozorov's world-political subjectivation also occurs through the exposition of "'the wrong' of exclusion" in the name of (though not necessarily enacted by) "the part of the community that has no part in the 'natural order of the world'" while "identif[ying] itself with the [world/cosmopolitical] community as such" (Prozorov, 2014b: 31). World politics (and by analogy, cosmopolitics) is therefore simultaneously negative in its "intra-worldly political practice guided by the idea of the wrong" and affirmative in its "void-ontological" presupposition of community, equality, and freedom. In consequence, world politics "is not determined by any particular culture or tradition but is rather made possible by a subtraction from it" (Prozorov, 2014b: 38). This also recalls the political-ontological interest (noted above) in ontological self-determination as a question of how things or persons can alter from themselves. In unison with the politics of political ontology, world politics "affirms that whatever is denied existence in the world nonetheless possesses being and ventures to grant this being maximal existence in the world in question" (Prozorov, 2014b: 2). This (cosmo)political register of world politics though should not be thought of as an act of charity but rather as a principled, always already presupposed, ontological egalitarianism.

Stengers sees the cosmos itself as "an operator of . . . equalization."[13] By this she means that it infuses political voices with an affirmative concern for the fact that "the political arena is peopled with shadows of that which does not have a political voice" (Stengers, 2005: 995–996). The equality imbued by the cosmos "precludes . . . any differentiation a priori between that which counts and that which does not" (Stengers, 2005: 1003). For Stengers, it is the unknown world, "[t]he cosmos distinguished from any particular cosmos" (Stengers, 2005: 995) (or Prozorov's "void of the World"), which seems to instantiate politics in Rancière's sense; namely, insofar as politics concerns itself with the practical verification of the equality of those without a part in the community, or with counting those not counted as equals by the prevailing police order. She also associates this kind of cosmopolitics with an implicit notion of "wrong" or radical alterity: "politics . . . construct[s] its legitimate reasons 'in the presence of' that which remains deaf to this legitimacy;" it demands that one think " 'in the presence of' the victims" of decisions, or " 'in the presence of' those who would otherwise be likely to be disqualified as having idiotically nothing to propose, hindering the emergent 'common account' " (Stengers, 2005: 996, 997, 1002).

Accordingly, political subjectivity, which Stengers associates with the Deleuzian conceptual character of the "idiot," the one excluded from the community, emerges when one "resists the consensual way in which the situation is presented" and bears witness to wrong by making the victims " 'present', not arguing in their names but conveying what it may feel like to be threatened by an issue that one has nothing to contribute to" (Stengers, 2005: 994, 1003). However, as Prozorov (2014b: 33) explains, in agreement with Rancière, rather than bemoaning a particular victimization this kind of exposition of wrong "breaks away from the immediate context of the incident, universalizing the plight of the particular victim as a problem of . . . or, more precisely, the problem with the entire world."

Politicizing, provincializing, and pluralizing IR

Recent theorizations and debates of ontology in IR would be well advised to look to political ontology to rethink and sharpen their often limited – indeed, often colonial and depoliticized – understandings of world politics as international relations. The decolonial and post-anthropocentric sensibilities of political ontology and postfoundational politics as proposed by Rancière and Prozorov can be good starting-points in this regard. This would help open IR up to plural worlds of world politics in much richer ways than Wendtian "cultures of anarchy" (Wendt, 1999: ch. 6) or even Rosenbergian uneven and combined development of multiple societies, both of which presuppose the ontological universe of Euro-modernity. I have also suggested, however, that one aspect of such a pluriversal opening would be to interrogate the very notion of an ontology (of agencies and structures, material and ideational elements, etc.) as a particular, rather provincial, endeavor in need of a "deconstructive self-estrangement." Without explicit appeal to a decolonial approach, or a pluriverse, such interrogations have already begun in the guise of new-materialist and posthumanist approaches in international political sociology (see e.g. the contributions to the special issue of Millennium introduced by Srnicek et al., 2013). They could also be pursued through political/economic-theological investigations of the conceptual apparatus of IR including its Euro-modern/Eurocentric concepts of sovereignty, security, democratic peace, development, human rights, ethics, and community. Deconstructive self-estrangements could in turn open space for a non-hegemonic and non-particularistic as well as postfoundational, postanthropocentric and postsecular cosmopolitics of a pluriverse of modern, non-modern, alter-modern, and partially connected worlds.

Notes

1 A proper substantiation of these claims would require another paper. By way of anecdotal evidence, let me merely point out that despite the appearance of (international or world) politics in the title of Wendt's (1999) and Jackson's (2011) books neither one of them features "politics" in its Index. Wight (2006) appears to either reduce politics to (an) ontology (of agents and structures, in the title of his book) or speak of "political implications" of ontology (Wight 2006: 6). His Index entry for "politics" tellingly defers the latter to "[see] international relations [where it appears 'vs. domestic politics']; ontology ['centrality to politics']; [and] states ['domestic politics']" (Wight, 2006: 336). Jackson (2011: 28) treats "philosophical [as opposed to scientific] ontology" as a quasi-methodological practice ("our 'hook-up' to the world") without politics. Wight (2006: 7) ontologically subordinates "processes, practices and events"

to agents and structures (insofar as the former "only occur in structured contexts and through the practices of agents"). Wendt (1999: 6, 22–23, 370, 372) says his book "is about the *ontology* of the states system," "the *ontology* of international structure," or "the *ontology* of international life" (my emphases).

2 Thanks to Cristina Rojas for first bringing political ontology to my attention.

3 This synoptic characterization follows Viveiros de Castro whose work on "Amerindian perspectivism" has provided one of the main inspirations for the ontological turn. For his seminal contributions see Viveiros de Castro (2015a).

4 Rancière (1995: 96–97) himself gives a similar definition referring to "appearance," "imparity," and "grievance."

5 Central to this idea of self-emancipation is the aforementioned "equality of intelligences" (Rancière, 1995: 51).

6 One version of a more abstract account of the latter can be found in Horkheimer and Adorno's classic Dialectic of the Enlightenment.

7 In an ethnomethodological reversal, the significance of the practice may also be gauged from justifications of and reactions to instances of its deliberate breach or omission, for instance, by former German chancellor Gerhard Schröder and several ministers of his cabinet. In this context, one might also mention the omission of the common phrase "God bless [America/you]" (used in many addresses by U.S. presidents) in British Prime Minister Tony Blair's address to the nation on the occasion of Britain's joining of the 2003 invasion of Iraq. Blair had intended to conclude his address with the phrase, but dropped it on the urging of his advisers. See Jenkins (2007: 38).

8 Of course, Latour's point is ultimately also to demonstrate the falsity of the great divide between nature and humanity, and nature and politics in practice – we have never been modern. Incidentally, "political science," at least as practiced by its positivist mainstream, is a paragon of the modern alchemy of science and politics.

9 In Foucault's account, this becomes explicit in the case of neoliberal governmentality which no longer assumes the "naturalness" of the homo oeconomicus and the market but rather sees these as being in need of cultivation through the institutional facilitation of games of competition (see Foucault, 2008).

10 Viveiros de Castro (2015a: 233) points out that the "ontological dualism of nature/culture . . . is the direct descendant of the theological opposition between nature and super-nature" which relies on a secularization of "spirit" into "culture" and may add a "cultural" theology to Schmittian political and Agambenian economic theology. However, from the perspective presented here the domains of culture, politics, and economy in fact retain their theological traces rather than consummating secularization in an absolute sense.

11 The relationship between the World and world/s is conceived in analogy to the corresponding Heideggerian distinction between the ontological and the ontic.

12 Prozorov (2014a: 66) submits that "[i]f, as Rancière argues, equality cannot be the foundation of any social order, then it must transcend every particular world and in this sense find its sole possible ground in the void that we term World."

13 I.e. mise en égalité, which she distinguishes from equivalence because this would imply "a common measure."

References

Agamben, G. 2011. *The Kingdom and the Glory: For a Theological Genealogy of Economy and Government*. Trans. L. Chiesa With M. Mandarini. Stanford: Stanford University Press.

Bartelson, J. 2009. *Visions of World Community*. Cambridge: Cambridge University Press.

Bessire, L. and Bond, D. 2014. 'Ontological Anthropology and the Deferral of Critique', *American Ethnologist*, vol. 41, no. 3, pp. 440–456.

Blaser, M. 2009. 'Political Ontology: Cultural Studies Without "Cultures?" ' *Cultural Studies*, vol. 23, no. 5–6, pp. 873–896.

Blaser, M. 2014. 'Ontology and Indegeneity: On the Political Ontology of Hetereogeneous Assemblages', *Cultural Geographies*, vol. 21, no. 1, pp. 49–58.

De la Cadena, M. 2010. 'Indigenous Cosmopolitics in the Andes: Conceptual Reflections Beyond "Politics" ', *Cultural Anthropology*, vol. 25, no. 2, pp. 334–370.

Foucault, M. 2007. *Security, Territory, Population: Lectures at the Collège de France 1977–1978*. New York: Palgrave Macmillan.

Foucault, M. 2008. *The Birth of Biopolitics: Lectures at the Collège de France 1978–1979*. New York: Palgrave Macmillan.

González, B.G. 2016. 'Latin America, From Cosmopolitics to Technopolitics', *opendemocracy.net*, 30 March. www.opendemocracy.net/democraciaabierta/bernardo-guti-rrez-gonz-lez/latin-america-from-cosmopolitics-to-technopolitics [Accessed 03.12.2016].

Hallward, P. 2009. 'Staging Equality: Rancière's Theatocracy and the Limits of Anarchic Equality', In: G. Rockhill and P. Watts, eds., *Jacques Rancière: History, Politics, Aesthetics*. Durham: Duke University Press, pp. 140–157.

Holbraad, M., Pedersen, M.A. and Viveiros de Castro, E. 2014. 'The Politics of Ontology: Anthropological Positions', *Theorizing the Contemporary, Cultural Anthropology*, 13 January 2014. https://culanth.org/fieldsights/462-the-politics-of-ontology-anthropological-positions [Accessed 29.08.2016].

Jackson, P.T. 2011. *The Conduct of Inquiry in International Relations: Philosophy of Science and its Implications for the Study of World Politics*. New York: Routledge.

Jenkins, P. 2007. *God's Continent: Christianity, Islam and Europe's Religious Crisis*. Oxford: Oxford University Press.

Prozorov, S. 2014a. *Ontology and World Politics: Void Universalism I*. New York: Routledge.

Prozorov, S. 2014b. *Theory of the Political Subject: Void Universalism II*. New York: Routledge.

Rancière, J. 1995. *On the Shores of Politics*. London: Verso.

Rancière, J. 1999. *Disagreement: Politics and Philosophy*. Minneapolis: University of Minnesota Press.

Rancière, J. 2010. *Dissensus: On Politics and Aesthetics*. London: Continuum.

Rosenberg, J. 2016. 'International Relations in the Prison of Political Science', *International Relations*, vol. 30, no. 2, pp. 127–153.

Schmitt, C. 2003 [1950]. *The Nomos of the Earth in the International Law of the Jus Publicum Europaeum*. Trans. G.L. Ulmen. New York: Telos Press.

Schmitt, C. 2005 [1922]. *Political Theology: Four Chapters on the Concept of Sovereignty*. Trans. G. Schwab. Chicago: University of Chicago Press.

Srnicek, N., Fotou, M. and Arghand, E. 2013. 'Introduction: Materialism and World Politics', *Millennium: Journal of International Studies*, vol. 41, no. 3, p. 397.

Stengers, I. 2005. 'The Cosmopolitical Proposal', In: B. Latour and P. Weibel, eds., *Making Things Public*. Cambridge, MA: MIT Press, pp. 994–1003.

Vigh, H.E. and Sausdal, D.B. 2014. 'From Essence Back to Existence: Anthropology Beyond the Ontological Turn', *Anthropological Theory*, vol. 14, no. 1, pp. 49–73.

Viveiros de Castro, E. 2015a. *The Relative Native: Essays on Indigenous Conceptual Worlds*. Chicago: HAU Books.

Viveiros de Castro, E. 2015b. 'Who Is Afraid of the Ontological Wolf? Some Comments on an Ongoing Anthropological Debate', *Cambridge Journal of Anthropology*, vol. 33, no. 1, pp. 2–17.

Walker, R.B.J. 2010. *After the Globe, Before the World*. New York: Routledge.
Wendt, A. 1987. 'The Agent-Structure Problem in International Relations Theory', *International Organization*, vol. 41, no. 3, pp. 335–370.
Wendt, A. 1999. *Social Theory of International Politics*. Cambridge: Cambridge University Press.
Wight, C. 2006. *Agents, Structures and International Relations: Politics as Ontology*. Cambridge: Cambridge University Press.

Index

For Product Safety Concerns and Information please contact our EU
representative GPSR@taylorandfrancis.com
Taylor & Francis Verlag GmbH, Kaufingerstraße 24, 80331 München, Germany

www.ingramcontent.com/pod-product-compliance
Ingram Content Group UK Ltd.
Pitfield, Milton Keynes, MK11 3LW, UK
UKHW020011200926
475702UK00014B/1827

9780367872786